AF333467

CURRENT APPROACHES TO OCCUPATIONAL HEALTH   2

# Current Approaches to
# OCCUPATIONAL HEALTH  2

Edited by
## A. WARD GARDNER
MD FFOM DIH

WRIGHT·PSG
Bristol • London • Boston   1982

*Published by:*
John Wright & Sons Ltd, 42–44 Triangle West, Bristol BS8 1EX, England
John Wright PSG Inc., 545 Great Road, Littleton, Massachusetts 01460, U.S.A.

*British Library Cataloguing in Publication Data*

Current approaches to occupational health   2.
　1.　Industrial hygiene - Periodicals
　I.　Current approaches to occupational medicine
　363.1′1′05　　　　RC967

ISBN 0 7236 0618 8

Library of Congress Catalog Card Number: 81-71444

*Printed in Great Britain by*
John Wright & Sons (Printing) Ltd, at the Stonebridge Press, Bristol BS4 5NU

*To Dr J. F. Eustace*
first Dean of the first Faculty of Occupational Medicine
with warmest personal regards and all good
wishes for the continuing success of the Faculty

# Preface

*Current Approaches to Occupational Medicine* was well received and appeared to fill a need. In the preface to that book I said, 'The theme that runs through this book is a wish to illuminate some current problems, to give perspective and to review progress — mainly in the fields of occupational health, occupational medicine and occupational hygiene'. This book has similar aims but has been re-titled in order to stress the wider area encompassed by occupational health rather than occupational medicine.

Like its predecessor this volume 'is a *potpourri,* the flavour of which is determined by the choice both of authors and of subjects'. I am grateful to reviewers and to colleagues for suggesting topics which should be covered in this book and would welcome further ideas for the future.

It is therefore with great pleasure that I record my thanks both for helpful suggestions and for useful discussions to Graham Bell, Archie Downie, Malcolm Harrington, Dennis Malcolm, Doreen Miller, Ken Nickol and Jim Sanderson.

This book begins with four industry reviews. These are followed by seven chapters each of which examines an occupational health problem. The remaining seven chapters are devoted to toxicology, epidemiology, hazard management, work, education in occupational medicine and health education at work.

The authors have laboured to produce their chapters within a short time span so that their contributions are topical and informative. I am grateful to them for their devotion to the task and for their tolerance in matters editorial. It is my pleasure to thank them all most warmly.

A.W.G.

# Contributors

| | |
|---|---|
| W. Ian Acton | MSc BSc(Tech) FIOA |
| R. M. Archibald | MB ChB DIH FFOM |
| James W. Bridges | BSc PhD CChem FRSC FIBiol MI Env Sci |
| John E. Burgess | MSc MB ChB MFOM DObstRCOG DIH |
| I. I. Coutts | MB ChB MRCP |
| James Dick | MB ChB FFOM |
| W. M. Dixon | MB BS MRCP FFOM DIH |
| A. Downie | MB ChB MSc MFOM DIH |
| Liam Gorman | BA PhD |
| Susan A. Hubbard | BSc MSc PhD |
| S. Kanagasabay | MSc MB BS MFCM MFOM |
| Owen Ll. Lloyd | MA BSc PhD MD MFCM Dip Comm Med Dip Ed |
| James McEwen | MB ChB MFCM MFOM DIH |
| Dennis Malcolm | MB ChB DIH FFOM |
| A. J. Newman Taylor | MSc(Occ Med) MRCP |
| R. R. Pearson | MB BS |
| C. A. C. Pickering | MRCP DIH |
| Jonathan Plaut | BS JD LLM MA |
| A. N. B. Stott | MB FFOM |
| William Taylor | DSc MD FRCP(Edin) |
| Charles A. Veys | MD FFOM DPH DIH Cert Occ Hyg |
| J. G. P. Williams | MSc FRCS DPhys Med |

# Contents

# Contributors

**Ian Acton** began his preoccupation with industrial noise and occupational deafness when he was appointed as research assistant in the Department of Industrial Health, Newcastle upon Tyne in 1962. He moved to the Clinical Audiology Group in the Institute of Sound and Vibration Research, Southampton University in 1965, and transferred to the Wolfson Unit for Noise and Vibration Control when that was formed in 1969. He is also a Fellow of the Institute of Acoustics. He has published over thirty papers, and is a member of technical committees of the British Standards Institution and similar bodies.

**Roy Archibald** entered occupational medicine in 1947. He spent 5 years in the chemical industry and then joined the National Coal Board in 1953. He was successively Area and Divisional Medical Officer, Deputy Chief Medical Officer and now Director of the Medical Service.

He is a Member of the Society of Occupational Medicine (Secretary 1970–76, President 1980–81), Founder Fellow of the Faculty of Occupational Medicine in London, Fellow of the Faculty of Occupational Medicine in Ireland, Member of the Specialist Advisory Committee on Occupational Medicine 1976–80 (Chairman 1979–80). He is a Specialty Adviser in Occupational Medicine to the North West Thames Regional Health Authority, Member of the British Occupational Hygiene Society and of the Ergonomics Society, Chairman of the Working Party of Medical Officers of Coal and Steel, ECSC 1976–81. He was a BMA Council Member 1977–79.

**James Bridges** is the holder of the only chair in toxicology in the UK and the present Chairman of the British Toxicology Society. He is the Director of the Robens Institute, a new institute which is concerned with health and safety research and training, and offers an advisory and investigative service for industry, particularly in the fields of chemical hazards and ergonomics.

**John Burgess** has worked as an occupational physician for over 10 years and has acted as medical adviser to several companies who manufacture and formulate pesticides. He has investigated working conditions in the field as part of product stewardship programmes and his particular speciality has been the investigation of the effects of organophosphorus compounds. In 1975 he was awarded an MSc with distinction in occupational medicine at the University of London and also received the prize of the Society of Occupational Medicine in that year.

A Norfolk man by birth, he has had a long and intimate knowledge of farming techniques and his earlier work in general practice in rural areas has enabled him to become acquainted with the varied problems in the agricultural industry.

**Ian Coutts** is Clinical Lecturer in the Department of Medicine at the Cardiothoracic Institute in London where he has investigated asbestos-related diseases, occupational asthma and hard metal disease.

**James Dick** is a graduate of the University of Edinburgh and worked as a chest physician until joining the NCB Radiological Service in 1959. Since 1974 he has been responsible to the Director of Medical Services for all X-ray activities. He is Medical Secretary to the independent panel advising on epidemiology of chest X-ray surveys (Panel on Survey Radiology), a member of the committee of experts on the revision of the ILO Classification of the Pneumoconioses, and a member of an ECSC Working Party on radio-diagnosis.

**Bill Dixon** graduated from Guy's in 1947, passed his MRCP in 1952 and specialized in chest disease until 1957.

Entering occupational health at Esso, Fawley, he set up the occupational health service at Esso, Milford Haven, and established the first group occupational health service in Dundee in 1962. Since 1966 he has, except for a period as a management consultant with John Tyzack, been successively Chief Medical Officer of Fisons, Chrysler UK and John Lewis.

He is past President of the Society of Occupational Medicine, an examiner for the Diploma in Industrial Health and Occupational Health Adviser to the South East Thames Regional Health Authority.

**Archie Downie** graduated from Glasgow University in 1958, and the next 14 years were divided between surgery and general practice. In 1972 he entered the field of occupational medicine in Libya. During 5 years spent at the Esso Medical Centre at Fawley, he obtained a DIH and MSc and since 1979 has been on loan assignment with Aramco in Saudi Arabia. Special interests include sports medicine, alcoholism and diving medicine.

**Liam Gorman** is Senior Lecturer in Organizational Behaviour at the Irish Management Institute. He has written several books on behaviour in organizations, has been a consultant to the OECD in management education and over the past 20 years has lectured and consulted widely in Europe, North America and Africa.

**Susan Hubbard** is a BP Research Fellow in Experimental Toxicology in the Toxicology Unit of the Robens Institute. She was formerly in the MRC Cell Mutation Unit at the University of Sussex. Her main interests lie in the area of genetic toxicology with particular emphasis on short-term screening methods for carcinogenicity.

**Bob Kanagasabay** qualified MB BS from the University of Ceylon in 1953 and joined the Royal Air Force in 1959. He served as Senior Medical Officer in Christmas Island, Northern Ireland and Gutersloh. In 1970 he obtained the MSc in occupational medicine of the London University and was appointed Deputy Director of Community Medicine in the RAF. In 1973 he was awarded the Lady Cade Medal of the Royal College of Surgeons for outstanding service in the practice of occupational medicine in the RAF. The award was in recognition of his work in designing a system of ventilation to protect personnel who service integral wing fuel tanks of aircraft.

He is currently Senior Medical Officer, Research and Development in the Directorate of Civilian Medical Services, Ministry of Defence (PE) and is responsible for occupational health services in research and development establishments of the Ministry of Defence (PE) relating to the Admiralty and the former Ministry of Aviation Supply. The work includes the medical aspects of research and development programmes in the fields of lasers, new chemicals and materials, microwaves and other non-ionizing radiation and noise.

He is Secretary of the Ministry of Defence Occupational Health Committee, and Chairman of its noise working group. He represents MOD on the BSI Committees on lasers and acoustics and is visiting lecturer at the London School of Hygiene and Tropical Medicine for the MSc in occupational medicine.

**Owen Lloyd** is a Senior Lecturer in the Wolfson Institute of Occupational Health in the University of Dundee. He holds concurrent appointments as Medical Director of the Scottish Occupational and Environmental Health Service Ltd, Dundee University and as Medical Adviser to an industrial firm in Fife. Recently he received funds from the Scottish Home and Health Department to establish the Epidemiology Unit for Environmental Cancer in the Wolfson Institute. His previous experience includes epidemiological studies in community medicine with the Lothian Health Board, and also a Lectureship in Physiology at Edinburgh University, where his research was in the fields of toxicology and neurobiology.

**James McEwen** is Senior Lecturer in the Department of Community Health, University of Nottingham and Honorary Consultant in Community Medicine, Nottinghamshire Area Health Authority. He also acts as a Consultant to the World Health Organization. He was formerly lecturer in the Department of Community and Occupational

Medicine, University of Dundee. His principal academic interests have been concerned with postgraduate training in community medicine and occupational health, although he has been involved also with undergraduate medical education. His research studies have been mainly in the fields of community medicine, occupational health and primary care.

**Dennis Malcolm** graduated in 1941 from the University of Edinburgh. Following a house appointment in Southampton he joined the RAMC in August 1942, seeing service in North West Europe, India and eventually in the Middle East where he commanded the 8th Field Ambulance. After taking the DIH (Edin) in 1947 he was offered the position of medical officer to the Chloride Group in Manchester by Professor R. E. Lane. In 1953 he was appointed Honorary Lecturer in Occupational Health at the University of Manchester. He served on committees and the Council of the Society of Occupational Medicine, becoming President in 1970. He has published papers on metal toxicity and acid erosion of the teeth. He has also been involved in health legislation in the UK, EEC, USA and other countries in which the Chloride Group operates.

**Tony Newman Taylor** is consultant physician at the Brompton and London Chest Hospital and is Senior Lecturer in Clinical Immunology at the Cardiothoracic Institute, University of London. He has a particular interest in occupational asthma, in extrinsic allergic alveolitis and in understanding the immunological reactions underlying these diseases.

**Ramsay Pearson** is a Surgeon Captain in the Royal Navy currently serving at the Institute of Naval Medicine, Alverstoke, Hampshire, as Senior Medical Officer, Underwater Medicine. He has been closely associated with diving in the Royal Navy for the majority of his naval career and has wide experience in all the occupational health problems associated with diving. As an adviser to the Department of Energy and the Health and Safety Executive, he has made a major contribution to legislation concerning medical aspects of the safety of commercial diving operations in the United Kingdom.

**Anthony Pickering** is consultant physician at Wythenshawe and Withington Hospitals, Manchester, and is Honorary Lecturer at the University of Manchester. He developed a particular interest in occupational lung disease while working with Professor Jack Pepys at the Cardiothoracic Institute in London where he developed many of the occupational tyne inhalation tests now in current use. Recently he has investigated outbreaks of humidifier fever.

**Jonathan Plaut** is General Manager, Environmental Services with Allied Chemical Corporation in Morristown, New Jersey. His responsibilities include overall corporate responsibility for the areas of: pollution control, safety and loss prevention, occupational health, product safety and medical services. He received his BS in Engineering from Pennsylvania State University, his JD from Georgetown University, his MA at the NYU Graduate School of Arts and Science and his LLM in International Law from NYU School of Law. He has published in the *Journal of the Royal Society of Medicine* and *Chemical Times and Trends*. He is a frequent university and environmental seminar lecturer, and he teaches a course in environmental management at the Fairleigh Dickinson Graduate School of Business. During the period of his government service, he was the recipient of a Superior Performance Commendation from the US Department of Commerce.

**Norman Stott** served as a Medical Officer at the United Kingdom atom bomb trials during his service in the Royal Air Force. He subsequently joined the Atomic Energy Authority in 1959 and is now the Chief Medical Officer. He has published many papers on aspects of radiation protection and other aspects of occupational health. He is a member of the EEC Advisory Committee on radiation protection research, Editor of the *Journal of the Society of Occupational Medicine* and Regional Specialty Adviser in Occupational Medicine to the Oxford Regional Health Authority.

**William Taylor** was born in Shetland, received his early education in Orkney and Caithness, and proceeded south to Edinburgh University, graduating BSc and PhD in 1937. He then joined ICI's Nobel's Explosive Division at Ardeer, Ayrshire, first in their research department and, during World War II, in the Ministry of Supply factories at Irvine, Dumfries and Powfoot. In these war-time factories manufacturing acids and explosives he was introduced to, and responsible for dealing with, numerous occupational health hazards. This interest in occupational medicine led to a return to Edinburgh University in 1945 to read medicine. To complete his medical education he followed the usual hospital posts with 8 years in a single-handed rural general practice in Castletown, Caithness, Scotland.

An opportunity to combine industry, chemistry and medicine presented in 1960 when Professor Mair set up an Industrial Health Unit at the University of Dundee. Commencing as Lecturer in 1960 he was promoted to Reader in 1968 and awarded a Personal Chair in Occupational Medicine in 1973. He was responsible for the Dundee postgraduate course in Industrial Health (DIH) from 1963 to 1977.

Professor Taylor retired from Dundee University in 1978. He remains active in the research fields of noise, vibration and the heavy metals.

**Charles Veys** is currently Chief Medical Officer for Michelin Tyre Company in Stoke-on-Trent, where he has worked since 1963. He is also an active member of the Health Advisory Committee of the British Rubber Manufacturers' Association. Award of the Heinz Karger (1968) and René Barthe (1975) prizes encouraged a developing interest in research, especially in the field of occupational cancers. An honorary appointment in the Cancer Epidemiology Research Unit at Birmingham University, together with some postgraduate teaching commitments, provide academic lustre and added interest to the day-to-day practice of occupational medicine.

**John Williams** is a Consultant in Rehabilitation Medicine and Medical Director of Farnham Park Rehabilitation Centre. His particular interest is in injury in sport, on which he has published many books and papers. He was for 10 years Secretary General of the International Federation of Sport and Medicine and before that Honorary Secretary of the British Association of Sport Medicine.

**Ward Gardner** is a graduate of the University of Glasgow and has worked mainly in the oil industry in refining, petrochemical manufacturing, transportation and shipping and research. He has written, contributed to and edited a number of books whose subjects have been occupational health and occupational medicine, first aid, safety and accident prevention, health information and education, absence from work attributed to sickness and a medical guide for the use of seafarers. These subjects have also been the main topics of his other published work. He has a special interest in alcohol-related problems at work and has both taught occupational medicine and examined for the Diploma in Industrial Health.

# 1. THE RUBBER INDUSTRY:
## REFLECTIONS ON HEALTH RISKS
*Charles Veys*

## INTRODUCTION

This chapter collates some recent findings about the possible health risks to workers in an industry using both natural and synthetic rubber as its basic materials. Polyurethane, vinyl chloride monomer, polyvinyl chloride and plastics are not discussed.

Rubber was one of the first substances to impress the early European explorers of the New World. Columbus, during his second voyage of discovery in the Santa Maria (1493), noticed how the natives of Hispaniola (now called Haiti) played games with solid balls. These were astonishingly resilient and elastic, and bounced much higher than the inflated leather balls used commonly in Europe at that time. The solid rubber balls were made from a dried milky liquid which could be obtained by cutting into the bark of certain wild trees. The South Americans called these trees 'heve' or 'cauchuc', which signifies weeping wood.

It was, however, several centuries before this new material was brought into commercial usage in Europe. Interestingly, rubber was first marketed not for its elastic properties, but to rub out pencil marks — hence the English word 'rubber' coined by Joseph Priestley in 1770. The first 'Macintosh' waterproof cloth was fabricated in 1823, but it was the discovery of vulcanization by Charles Goodyear in 1839 that revolutionized rubber manufacture.

Shortly afterwards R. W. Thomson invented the pneumatic tyre in 1845, but it lay dormant until its subsequent development by J. B. Dunlop, a Belfast veterinary surgeon, in 1889. At about the same time (1891) the first detachable pneumatic cycle tyre was evolved by the Michelin brothers in France. Marketing of Henry Wickham's Far East plantation rubber around 1900 made possible the rapid development of an important new industry using rubber at the turn of the twentieth century. By the end of the nineteenth century there were 5000 acres of rubber plantations in Asia. In 1910 planters, spurred on by the advent of Henry Ford's famous motor car and the great demand for rubber to make tyres, achieved a million acres of rubber plantation. Asia had now become the main supplier of rubber.

The industry burgeoned in the 1920s. World War II brought with it a blockade of the Far East shipping ports and gave the necessary stimulus for the development of the 'synthetic rubber' industry. Now more synthetic than natural rubber is being manufactured.

**Table 1.1.** United Kingdom rubber consumption (tons)

|  | Tyre and tyre products | | General rubber goods | | |
|  | Natural | Synthetic | Natural | Synthetic | Total |
|---|---|---|---|---|---|
| 1948 | 116 830 | 213 | 76 901 | 2 368 | 196 312 |
| 1958 | 79 753 | 46 809 | 95 707 | 16 170 | 238 439 |
| 1968 | 92 400 | 128 200 | 98 600 | 102 100 | 421 300 |
| 1978 | 84 400 | 118 600 | 54 800 | 194 800 | 452 600 |

Source: Rubber Statistical Bulletin.

*Table 1.1* outlines the expansion of the rubber industry in the United Kingdom. It lists the tonnage of natural and synthetic rubber used since 1948. It also shows clearly the reversal of the trend in raw material production, away from natural to more synthetic rubber, in order to meet the requirements of a rapidly growing post-war industry.

In 1978, the latest year for which comparable statistics are available, the total world-wide consumption of natural rubber was about 3 715 000 tons and that of synthetic rubber about 8 760 000 tons — together making a total of 12 475 000 tons of rubber used.

Tyre and tyre products still take up the greatest amount of the raw material, but the general rubber goods section of the industry is now also a substantial user of rubbers. In December 1978 some 108 100 persons (84 100 men and 24 000 women) were employed in the UK rubber industry, which, however, now shows some signs of contracting.

In Europe during 1978, France (459 404 tons), Italy (378 000 tons) and the Federal Republic of Germany (614 349 tons) all consumed less rubber than Japan (1 096 000 tons). China (385 000 tons), with an emergent rubber industry, used much the same as Italy, but more than Brazil (294 496 tons) and Canada (293 018 tons). The whole of the Eastern European block combined (2 825 000 tons) used more than the EEC (2 120 000 tons). Still the largest individual consumer, however, is the United States (3 253 760 tons). Such statistics are important because they show how sizable and widespread the industry is throughout the world. Furthermore, it is an industry that has until recently been little explored medically.

Thus, although our understanding of the health problems to be discussed in this chapter has derived mostly from study and research in the United Kingdom and the United States of America, the implications are world wide.

## THE RAW MATERIALS

Natural rubber latex is derived from the mature 6-year-old rubber tree grown on plantations that are now mostly located in Malaysia, Indonesia, Thailand, Sri Lanka, India and West Africa. Latex is a milk-like liquid obtained from the bark by using a special knife for tapping. It is composed of a hydrocarbon of basic chemical formula $(C_5H_8)_n$. After dilution with water, it is coagulated with dilute formic or acetic acid. The latex coagulum is converted into crumb rubber by adding a small amount of castor oil and passing it through a series of rollers; or it is formed into sheets of crêpe. After drying, the crumb rubber or crêpe is compressed into conveniently sized bales and made ready for shipping. These bales, sometimes covered in talc or calcium carbonate, or wrapped with polythene sheeting, are the essential raw material for industry. Production of natural rubber is still rising (at approximately 2·9 per cent per year) but less so than its synthetic counterpart, for which the demand is greater (approximately 7·8 per cent per year).

Synthetic rubber production is firmly linked to the petrochemical and oil industries; but present concern about the future availability and cost of petrochemicals, and the increasing use of natural rubber in radial tyres, has again highlighted interest in natural rubber.

In 1955 there were only four main types of synthetic rubber: styrene butadiene (used in tyre treads); butyl rubbers (with their low permeability for air and gases); nitrile rubbers and polychloroprene rubbers (with their oil, heat and solvent resistant properties). During the early 1960s advances in the synthesis of elastomers produced polyisoprene (a synthetic rubber more akin to its natural counterpart) and polybutadiene (with its resilience, high hysteresis — energy lost as heat generation — and outstanding abrasion resistance). Finally, the ethylene propylene rubbers, with their improved resistance to sunlight, ozone, ageing and weathering, together with a capacity to accept large loadings of aromatic oils and fillers without serious loss of physical properties, provided impetus for the great increase in the use of rubbers with added oil (called oil-extended polymers).

Other types of speciality rubbers (silicone, polyethylene, fluoro and so on) comprise only about 2 per cent of synthetic rubber production. The thermoplastic rubbers, which melt when hot and solidify when cold but without significant loss of their elastic property, make up a rather complex but ever-changing and rapidly developing field.

Natural and synthetic rubber should really not be viewed as alternatives but rather as individual materials each one with unique properties often mutually beneficial in compounding.

## COMPOUNDING TECHNOLOGY

Rubber alone,[1] either natural or synthetic, is quite useless for tyres or other rubber articles, because its consistency varies with changes in temperature and its elasticity is lost by the continued reapplication of tension. It has, therefore, first to be mixed with compounding ingredients before being vulcanized or cured. This is achieved by the application of heat which imparts durability, strength and a permanent final shape. This hot process can release a complex fume, the chemistry of which is still largely undetermined. Some cold curing processes are also used.

The compounding ingredients used by the industry are chemical and mineral additives.[2] They not only modify or improve the properties of the finished article, but they also facilitate its handling during manufacture, and thus help to reduce costs. For example, in order to improve the life of rubber it is customary to add to it antidegradants (antioxidants and antiozonants) in amounts of up to 1 per cent. Deterioration in rubber is largely due to oxidation by ozone and oxygen. The antidegradants act chemically and help to reduce flex-cracking due, for example, to the repetitive compression forces acting on tyres as the wheel turns. It was, incidentally, from this important group of compounding ingredients that the aromatic amine bladder carcinogens derived.

Other important classes of compounding ingredients are: vulcanizing agents (for example sulphur); accelerators; softeners, plasticizers and extenders (for example mineral oils, tars and waxes); fillers which extend and reinforce (for example carbon black, soft clays and resins); blowing agents; pigments and dyes; promotors and retarders; bonding agents; dusting powders (for example, talc and zinc stearate); solvents; stiffeners; abrasives and so on. All these form a very diverse and complex list of chemicals used by the rubber industry.

In particular, during the past 20 years mineral oils have been extensively used. These are usually of the highly aromatic type especially compounded from refining residues, and are employed for their miscibility, plasticizing and extending actions. Lighter coloured paraffinic oils, vegetable oils, coal tar, pitches and naturally occurring or synthetic resins are also incorporated for special applications.

The growth in the use of such materials derived from oil and the petrochemical industries, and the increasing use of synthetic and oil-extended polymers since the mid-1950s, are significant compounding changes that need to be taken into account when considering health aspects of the industry. In particular, for current tyre production, oil extension up to 35 per cent by weight of polymer may be incorporated. Pine tar oil was universally used prior to World War II, but restriction in supplies, economics and the increasing availability of aromatic mineral oils changed all that.

Another important compounding change has been the introduction of furnace carbon blacks (oil-derived). Previously, channel blacks (mostly produced from natural gas) predominated. Furnace blacks are of small particle size, and also have more polycyclic aromatic hydrocarbons (PAH) firmly adsorbed on to their structure. They have now almost totally replaced the channel blacks.

Sulphenamide accelerators and the paraphenylene diamine antiozonants have also assumed more prominence since the 1950s. Of some special interest is the introduction of some nitrosamine compounds into processing during the late 1950s and early 1960s. The full implications of this are only just being recognized within the industry.


## RUBBER PROCESSING

The modern tyre may be composed of over 200 different products: for example, 1·5 l of solvent for every 100 kg of tyre weight; 60 per cent of elastomer with extending or processing oils; 35 per cent of carbon black of different sorts; and 5 per cent of various other chemical compounding ingredients.

In order to provide a better understanding of some of the health problems currently encountered by the industry, *Fig. 1.1* depicts diagrammatically the main steps in rubber processing. Although this outlines tyre manufacture, other rubber articles are manufactured in a broadly similar way. The industry is labour-intensive: jobs tend to be physically orientated and often involve fairly heavy manual labour. Although there have been great improvements in mechanization and in the efficiency of the machinery used during the past two decades, the essential basic stages of rubber processing have altered little over the years.

First, the raw material consisting of bales of natural or synthetic rubber has to be cut into much smaller pieces by guillotining. Secondly, the rubber is subjected to the physical treatment of premastication to soften it, and thus to render it more suitable for accepting the various chemical compounding ingredients. These compounding ingredients are incorporated either on an open mill or in an internal banbury mixer. This latter can be likened to a food mixer used in the kitchen (set on its side) incorporating overlapping paddles but revolving much more slowly. Additionally, reclaim rubber may also be mixed in at this stage. Weighing, blending and then admixing with the various chemicals is the basis of the compounder's art.

The plaques of rubber stock are then either extruded hot through an especially shaped nozzle, for example to produce tyre treads, or they are reheated and calendered into rubber sheets. Such sheets are then joined up with other rubberized fabrics incorporating wire, rayon or cotton on the tyremaker's drum. Each tyre is constructed by hand. It is

# PROCESSING OF RUBBER (TYRE MANUFACTURE)

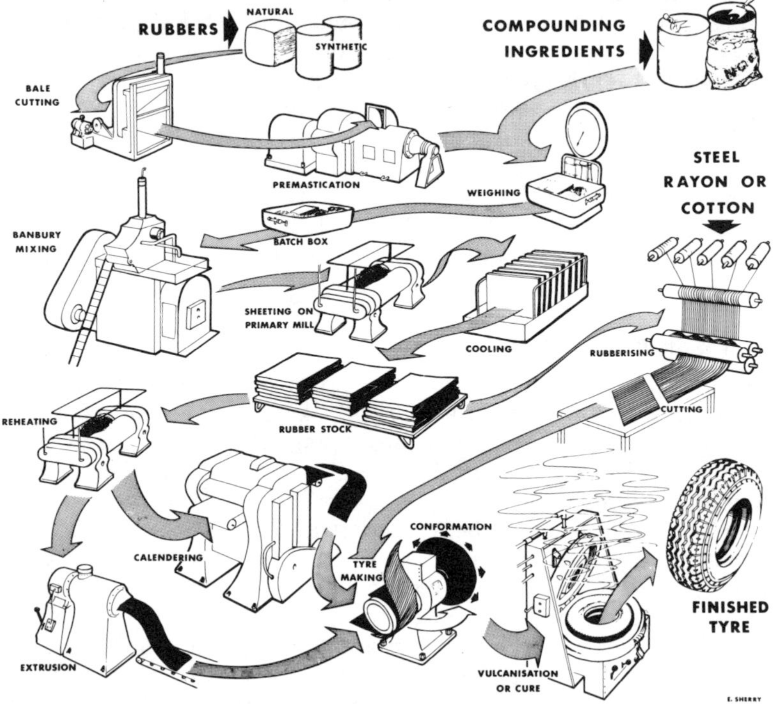


*Fig. 1.1.* Processing of rubber (tyre manufacture).

built on a horizontally positioned drum which revolves, while the operator superimposes the variously compounded fabrics in layers. Each fabric and rubber contributes its own particular property to the finished article. The tread rubber has to be hard-wearing and abrasion-resistant, whereas the rubber used for the sidewall has to be flexible and able to dissipate the heat build-up caused by the very frequent twisting and shearing forces arising within the casing of the modern radial tyre. Finally, the completed tyre carcass, which may be conformed into the more recognizable shape by inflation with air, or just left as a flat cylinder to receive its final shape during the cure, is then placed into an individual press to be subjected to heat and pressure in the curing cycle. The finished tyre is finally trimmed with a wet cutter, inspected and then stored before dispatch to depots, warehousing and sales outlets.

Inner-tubes are extruded as curved, seamless hollow tubes. These are cut to length, the valve inserted and the two ends finally joined by butt-welding before again being vulcanized by heat in an individual steam press. Tubes are now invariably made of butyl rubber because of its excellent air-retention properties.

Compounded rubber is elastic, flexible, airtight, watertight, long lasting and insulating, to mention just a few of its properties. There are thus thousands of products which can take advantage of these attributes in the modern technically advanced world; herein lies the reason for such a widespread and diverse industry. Virtually all wheeled vehicles run on rubber tyres. In the past the tyre industry predominated, but the general rubber goods section of the industry now merits equal consideration *(see Table 1.1)*.

## TRADITIONAL HEALTH CONSIDERATIONS

Health considerations in india rubber manufacture was alluded to in the Annual Report of HM Chief Inspector of Factories and Workshops in 1894. The predominant theme then was the use of naphtha and carbon bisulphide. In that report reference was made to a specific study by Dr J. T. Artlidge who, in 1892, had written one of the first detailed treatises on occupational health in England. His book, entitled *The Hygiene, Diseases and Mortality of Occupations,* was the first to mention health problems in the rubber industry. Apart from this trade being listed as one that set free offensive vapours, no specific occupational health risks were indicated however.

The India Rubber Regulations of 1922 laid down the conditions and duties of occupiers under which persons working with lead and in fume processes could be employed. The fume processes related to exposure to solvents, including carbon bisulphide, sulphur chloride, benzene, carbon tetrachloride and trichlorethylene. Later, the 1955 India Rubber Regulations prohibited the use of carbon bisulphide.

It was really not until 1950, however, that the industry was alerted to another rather more sinister health risk, that of bladder cancer, which is described in the next section.

The physical labour demand imposed by the nature of the work in the rubber industry commonly leads to muscular strains and sprains, tenosynovitis and, especially, to back injuries. Rubber plaques tend to stick together and rubber itself is a heavy material to carry. Occasionally, more serious injuries to the hands or upper limbs, due to their entrapment between the revolving cylinders on the mill, are encountered, but fortunately, with modern guarding and safe systems of work, these accidents are now uncommon. Skin burns from hot steam pipes, presses and friction are also encountered. The eyes occasionally suffer injury from particulate matter, from chemical splashes or from handling wire which flirts out. Modern eye protection and the advent of the Protection of Eyes Regulations in 1974 have resulted in a dramatic drop in these types of injuries, and the problem is generally now under control. Heavy machinery (especially mills and banburies) driven by large electric motors through gearboxes is very noisy, thus many areas within rubber workshops usually require a hearing conservation programme.

The skin is also subject to chemical irritation and insult because of the nature of the work. Some of the chemical compounding ingredients and solvents used can by themselves promote primary irritation. Less commonly, a direct sensitization leading to an allergic contact eczema occurs. The most common causes of skin trouble in the industry, however, are sweat, dirt, heat, friction, trauma and wetness. The International Contact Dermatitis Group (ICDRG)[3] lists some twenty chemicals which, in its experience, have a sensitization potential and have most commonly in the past caused an allergic contact eczema. There are now probably others, but rubber itself is not usually allergenic. The problem of skin disease in the rubber industry has formerly been little studied from within the industry itself, but an in-depth study is planned by the Institute of Dermatology in conjunction with the British Rubber Manufacturers' Association (BRMA).

The difficulty about ad hoc studies from hospital clinics is that they probably reflect a multiplicity of skin problems, including those from wearing apparel. They do not necessarily reflect the situation at the place of work. One investigation of rubber dermatitis[4] on over a hundred patients presenting at a London dermatology clinic indicated that mercaptobenzthiazole, thiuram sulphides and zinc diethyl dithiocarbamate were the predominant sensitizers. A study from France[5] implicated some of the paraphenylenediamines, whereas a report from Australia indicated that thiuram-mix, carba-mix, PPD-mix and mercaptomix most frequently gave positive results.[6] The whole subject of skin sensitivity to rubber compounding ingredients has recently been reviewed in an excellent monograph.[7]

At one large tyre factory with some 5500 hourly-paid employees, the incidence of newly acquired, occupationally related skin episodes dropped from 99 in 1973 to 40 in 1979. The majority of these cases were primary irritant rashes caused by one of the six common factors listed above. Other cases were caused by solvents, or by the direct action of the uncured compounded rubber on the skin. When rubber and solvents combine in their action on the skin, the irritation potential is often enhanced. This may occur commonly at the component building stage. At the factory studied, during the 7-year period, only three persons each year on average had to give up their usual job because of adverse clinical progress, or because of a proved sensitivity on patch-testing to one of the constituents of the ICDRG standard tray of test allergens. It seemed likely that a continued programme of education of the workforce, with early reporting of skin lesions, speedy and appropriate first aid treatment, a clean method of work incorporating a minimal or no-touch technique, good skin care and the use of after-creams, together with mechanization from improved processing changes, all played their part synergistically in bringing about the sustained fall noted.

Reference to *Fig. 1.1* will indicate that the other main health hazards likely to be encountered in rubber processing derive from exposure to dusts and fumes. Bales of natural rubber are covered in talc which is released when they are handled and cut. Talc is also a commonly used anti-stick agent, especially in tube extrusion. However, despite its extensive use, talc pneumoconiosis in the industry is an uncommon finding.

Weighing out the individual chemical compounding ingredients in the mixing room, and the sifting, blending and weighing of carbon black, can all create a dust problem. However, modern systems of tote-handling and well-designed local exhaust ventilation have greatly reduced exposure to the dusts of compounding.

The toxicity of the individual chemical compounding ingredients has received increasing attention during recent years,[8-11] especially from the Health Advisory Committee of the British Rubber Manufacturers' Association. This body, in conjunction with chemical suppliers, and more recently assisted by trade union representation, with encouragement from the Health and Safety Executive, has produced a code of practice and toxicity manual[8] aimed at informing both management and shopfloor operatives.

Both dust and fume exposure can occur around banbury mixers and their associated two-roll mills, unless efficient local exhaust ventilation is installed. Once the rubber is compounded, subsequent reheating, extruding and calendering are further sites for potential fume exposure, especially as the temperatures of extrusion rise.

In subsequent tyre building there is an exposure to rubber solvents and solutions. However, it is during tyre (or tube) curing itself that the greatest potential for exposure to fumes exists. Depending on work practices and plant layout, men employed as moulders, sprayers and inspectors may be similarly exposed to fumes at this final stage of processing. The control of curing fumes at the end of processing has not, until recently, enjoyed the same attention from ventilating engineers as the control of dust from early processing.[12-14] Likewise, the buffing and repairing of tyres, especially in re-treading operations, can cause considerable release of dust unless attention is paid to adequate local exhaust ventilation. This, however, can be particularly difficult to achieve, because the repair sites on the tyre carcass are not always compatible with the most efficient hood design.

The thermal environment of a rubber factory can at times be adverse, particularly during the summer months. The predominant cause is radiant heat from processing machinery, steam pipes and the ubiquitous hot curing presses. Steam leaks and condensates compound the problem, leading to an increase in both the wet- and dry-bulb temperatures.


## BLADDER CANCER

By mid-1949 it had become apparent that workers in the British rubber industry were contracting an excessive number of tumours of the bladder. The finding was almost fortuitous. Dr Robert Case, working as a research fellow conducting a large scale survey of occupational tumours of the urinary bladder in males employed in a section of the British chemical industry, found also an unusually high prevalence of these tumours in a certain county borough in central England chosen as a control situation. The county borough was coincidentally the centre of the British rubber industry.[15] A certain antioxidant (an aldehyde-amine condensate containing about 2·5 per cent of residual uncombined beta- and alpha-naphthylamines as contaminants) was suspected of causing bladder tumours among men in the chemical industry who were manufacturing it. Other antioxidants in use contained up to 13 per cent of these uncombined aromatic amines as residual contaminants. Generally, however, the residual free beta-isomer was in the order of 0·25 per cent, but this was sufficient to more than double the incidence of bladder tumours in the workforce of rubber operatives exposed.[16] The antioxidant concerned had been used in processing since the late 1920s. It was promptly withdrawn and stocks destroyed or returned to the manufacturer following receipt of a warning letter in 1949.

The BRMA set up a Health Research Unit with exfoliative urinary cytodiagnostic facilities in 1957, using a modified Papanicolaou technique. An increasing number of cases of occupational bladder cancer which had resulted from the pre-1949 exposures came to light. There were exchanges of correspondence in the medical literature in 1964, and especially in the lay press after the Lucy inquest in 1965 on a former cable worker who died of an occupational bladder cancer. Other similar inquests reminded everyone that an industry hazard had been heralded, but then had not been further studied or explored. As a direct result of these events the legislators published the Carcinogenic Substances Regulations in 1966.

The Medical Inspectorate of Factories, within the Ministry of Labour at that time, decided also to conduct their own detailed survey in order to study the prevalence of occupational cancers in the rubber and cablemaking industries, after issuing a warning card (F.2173) to former workers in these industries. Finally, there was medico-legal culmination at common law, when two rubber workers with bladder cancer sued the chemical supplier and the rubber manufacturer (their employer) for damages in a test case. Their claim was successful.[17] The industry itself then set up its own study in order to document more fully the resultant quantitative risk to the workforce exposed to the relevant carcinogenic antioxidants prior to 1949. Furthermore, there was an urgent need to ensure that any adverse effect had, in fact, disappeared as a consequence of the withdrawal of the putative carcinogens by 1950. A question remained — were there as yet other carcinogens to be rooted out?

## Factory studies

Three main groups set out to make their inquiries — individual companies, the British Rubber Manufacturers' Association and the Health and Safety Executive. *Table 1.2* shows the results of two longitudinal

**Table 1.2.** Bladder tumours in rubber workers

|  | By 1970 | |
| --- | --- | --- |
|  | Obs. | Expt.* |
| 3867 men employed in rubber factory A (1945–49) | 26 | 13·2 |
| 2081 men employed in rubber factory B (1946–49) | 23 | 10·3 |

Obs., Observed. Expt., expected.
*At local morbidity rates.

(also called cohort) studies carried out at separate factories. In each case the workforce employed, and also deemed to be exposed to the dangerous agents in use at the relevant time, experienced twice as many bladder tumours as expected during the 20-year period subsequent to their withdrawal in 1950. The difference between the observed and expected number of tumours was statistically significant in both instances ($P < 0.001$). The importance of both these studies is that they were done at factory level, using well-documented personnel records and a knowledge of the precise 'at risk' situation for operatives on the shopfloor. These studies also relied on morbidity data. Studies done at national level, and utilizing mortality data, would have given a less precise answer, because the mortality pattern of bladder cancer changed during the period of the study. The prognosis for this tumour was improving. Patients did not necessarily die of their tumour, but of some other intercurrent disease, so the bladder cancer was not necessarily recorded on the death certificate. Furthermore, the rules for coding the death to bladder cancer could exclude this diagnosis as the recorded underlying cause, if, for example, it was written under 'II' on the death certificate.

Much can be learnt from a study of the spatial distribution of the presumed occupational tumours that occur, using, for example, a

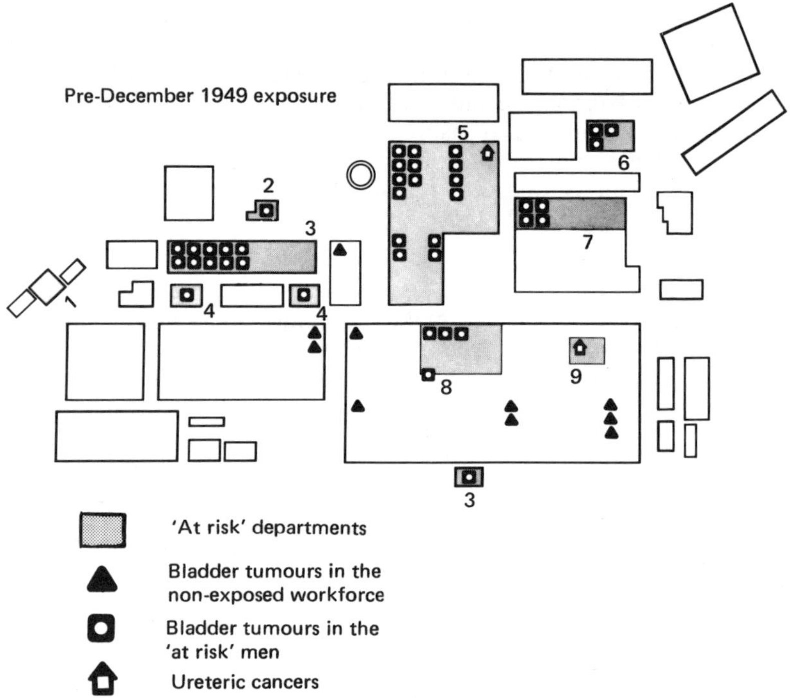



*Fig. 1.2.* A factory plan. 1, Scrap; 2, general maintenance; 3, engineering maintenance; 4, laboratory; 5, chemical stores, mixing and millroom; 6, despatch; 7, inner tubes; 8, calendering; 9, miscellaneous items.

simple factory plan. *Fig. 1.2* shows such a plan from one of the factories studied. It depicts the situation as at 1980. The shaded areas represent the 'at risk' zones for the exposed pre-1949 workforce who had worked for at least 1 year at the factory. The remaining white areas are the 'no risk' zones where exposure to the dangerous anti-oxidants could not have occurred. Each black square (with a white circular inset) represents an individual tumour in the 'at risk' population. The black triangles represent bladder tumours occurring in the concurrently employed but not exposed workforce. In the non-exposed group, 11 tumours have been registered. This number is still within expectation, but for the 'at risk' group, the 42 tumours depicted on the plan indicate a more than twofold excess of bladder tumours registered between 1950 and 1979. Reference to the plan indicates that some ancillary groups such as maintenance workers and fitters, laboratory technicians and dispatchers who periodically worked as spare labour in the 'at risk' zones, as well as the process workers themselves, contracted tumours. Such ancillary groups, even though only intermittently exposed, should not be excluded from screening in any evaluation of an 'at risk' situation.

Study of a similar map on to which all the known presumed occupational bladder tumours from the factory were plotted (some 51 in all) indicates that even men who worked for less than 1 year contracted tumours (5 cases). Tumours in women employees (4 cases) were found to be much less common, even after making due allowance for the usual sex difference in incidence and the smaller number of exposed workers. It is to be remembered that many of the women worked during the war years in place of the men, and in conditions of poor ventilation and black-out; yet few tumours seem to have resulted. This anomaly merits some further study.

Among the 51 presumed occupationally induced tumours there were two ureteric cancers, which are rather rare. These are also shown on *Fig. 1.2*. The occurrence of tumours at the more unusual sites (often called 'signal' tumours) should draw attention to the possibility of an industrial factor in causation. Occupational tumours of the urinary tract can arise at any anatomical site from the renal pelvis to the first part of the prostatic urethra, which areas are all lined by transitional urothelium.

An extension of the study at one of the factories has shown that for those who joined after 1950, when the purported aromatic amine carcinogens had been removed from processing, there was no longer any excess of bladder cancer evident up to the end of 1979. More studies are proceeding, but clearly there is already a reassuring contrast between the experience of those who worked in the industry before the end of 1949 and that of employees who joined for the first time only after 1950 when the dangerous chemicals had been withdrawn. The mean latent period of an occupational bladder cancer presenting in a rubber worker is about 22 years. There is thus now about 30 years of follow-up experience in the tyre sector, as yet without any real evidence of a continuing excess of bladder cancer. The subject has recently been reviewed.[18]


**Industry study**

In 1972 the British Rubber Manufacturers' Association (BRMA) working in conjunction with the Cancer Epidemiology Research Unit of the Department of Medicine at Birmingham University initiated their own study.[19] This involved thirteen large factories mostly engaged in tyre and tube production, but also some general rubber goods manufacture. Over 37 200 men were included. Three 5-year intakes of newly employed male operatives, joining between January 1946 and December 1960, were defined. The first intake represented men who might still have been exposed for 4 years to the now recognized aromatic amine bladder carcinogens. Despite this relatively short exposure, it was notable that among men who died at least 10 years after such exposure, 15 deaths from bladder cancer were observed against 9·9 expected. For the two post-1950 intakes combined (numbering 20 330 men) some 12 deaths due to bladder cancer were observed against 11·5 expected. These two intakes were not 'at risk' because the purported carcinogens had been discontinued before their engagement date. Although the follow-up period as reported up to 1970 was still rather short, the number of men was large. Furthermore, in contrast to the pre-December 1949 intake, no particular occupational groupings stood out.

## National studies

A census taken of the British rubber industry on 1 February 1967 by the then Medical Inspectorate of Factories within the Ministry of Labour enumerated 40 867 men aged 35 + years who had also worked for at least 1 year in the industry. The census covered 381 factories and 13 different sectors of the industry including tyres, remoulds, cables, adhesives, clothing, belting and hose, footwear and so on. For the analysis[20] the whole study population by sector was divided into three groups:

(*a*) those who started work before 1 January 1950 in a factory which had used the suspect antioxidants (12 779 men);

(*b*) those who started work on or after 1 January 1950, also in a factory which had used the suspect antioxidants (11 118 men);

(*c*) those who worked in factories which had *never* used the suspect antioxidants (9970 men).

The results indicate that bladder cancer is no longer an overall threat in the British rubber industry. Selective factors were operating to exclude men who may nevertheless have been exposed to the carcinogenic antioxidants, but who may not have reached the census date to be enumerated. Many men could have died, retired or left the industry between 1950 and 1967 when the census was taken. Nevertheless, for group (*a*), in non-tyre sectors there appears to be some cause for concern, because 23 deaths from bladder cancer were observed against 11·6 expected ($P < 0.001$). Excesses in cables and electrical goods, mouldings, motor accessories and mechanicals were evident and certain occupational categories were singled out.

Like the tyre sector of the rubber industry, the hazard of bladder cancer had been heralded for cable workers employed where rubber rather than plastic was used for coating the cables.[21] These results point firmly to the need for still more study and surveillance in the non-tyre sectors of the industry.

World wide there is a dearth of published work about bladder cancer in rubber workers. In Europe there has been one report from Germany and two reports from Switzerland suggesting a possible excess. Whilst most EEC countries list and define occupational bladder cancer in their industrial diseases schedules, very few statistics are available which define its extent. In the United States of America several studies[22-26] in rubber and tyre manufacturing companies concluded that there had not been the same risk of bladder cancer among American rubber workers as had been found in Britain. Although the risk appeared to relate to certain jobs, the suggestion was made that the excess of bladder cancer mortality occurred in men who had worked at least 35 years in the industry and who had died at the age of 75 years and over. From Australia it is reported that there is no evidence of a problem. Neither have researchers from Russia, China or Japan made mention of any similar bladder cancer hazard in their own rubber industries. Nevertheless, there have been sufficient studies and published data to indicate how best to identify and subsequently to control occupational bladder cancer in the rubber industry.[27]

## OTHER CANCER RISKS

In his report on occupational mortality based on the 1931 census, the Registrar-General reported that skilled workers in rubber had a probably significant excess mortality from all cancers. The cancer excess of 13 deaths was from cancers of the buccal cavity and pharynx, oesophagus and stomach, skin and other sites. A similar excess had also been recorded in 1921, but not in subsequent reports until the most recent one based on the 1971 census data.[28] An excess of registrations due to stomach and bladder cancer in rubber workers was especially noted, but not, as might have been expected, any excess of bladder cancer mortality. The techniques used to link mortality and occupation in such national inquiries are still too insensitive to pick out slight excesses in numerically small groups of workers.

In the course of the bladder cancer studies it became apparent that there was both a need as well as an opportunity to examine the total mortality pattern of workers in the industry. If one cancer risk was being scrutinized it was clearly important to exclude others at the same time. In view of the pluripotential nature of carcinogens, with the possibility of several target organs being affected, further reassurances were clearly required.

### United Kingdom studies

The British Rubber Manufacturers' Association set up its own industry longitudinal study in 1972, as outlined earlier. The preliminary results of their health research project were first reported in 1976,[19] and subsequently published in 1979.[29] Further analyses extending the follow-up to 1975 are now in hand. The study population included 37 221 male rubber workers who were employed in the industry for at least 12 months between 1 January 1946 and 31 December 1960. During the observation period to 31 December 1970 there were 4371 deaths; 23 548 men left the industry for other employment or by retirement, while 9302 were still employed at the study end-date. Overall more than 98 per cent of the population were finally traced. Deaths from all causes were less than expected because of the recognized healthy population effect.

Deaths from cancer were not elevated except in the latest intake group of 1956–60 (103 observed against 77·2 expected). Lung cancer mortality was increased (560 observed deaths, 520·7 expected, standardized mortality ratio (SMR) 107) as also was stomach cancer (165 observed deaths, 147·7 expected, SMR 112). There was quite marked variation between the three intake groups, between individual factories and according to whether a 5- or 10-year latency was taken into account, and whether regional or urban mortality weightings were allowed for. Of the remaining cancers, except for laryngeal cancer (11 observed against 9·4 expected) there were no excesses in either pancreatic, prostatic, brain, colonic or rectal tumours, and no excess of leukaemia. When broken down into entry intakes and into individual occupational groupings, however, some very small excesses became apparent in colonic, rectal, pancreatic and prostatic cancers. The numbers involved were too small for valid conclusions to be drawn about them.

A further analysis which took account of the detailed work history of men who died of certain cancers[30] does select out some possible occupational factors which might be enhancing mortality from both lung and stomach cancers in the industry. An exposure to fumes from curing could be relevant for lung cancer, whilst an exposure to the dust and fumes encountered in the earlier stages of processing (compounding and mixing) might be a relevant aetiological factor for stomach cancer. Further studies are proceeding, but no more definitive conclusions can be drawn from these data at this stage.

The Health and Safety Executive's study based on a more widespread census of the rubber industry (taken in February 1967) enumerated 50 867 men in 381 factories and in 13 different sectors of the industry. The mortality pattern of these men was finally evaluated[31] in 1980 after two interim reports.[32–33] The latest report revealed an excess of lung cancer deaths across the entire industry, with 822 deaths from this cause observed when 764 were expected. In particular several sectors highlighted the excesses: tyres, adhesives, rubber solutions and sealing compounds, belting, hose and rubber-with-asbestos flooring, ebonite and vulcanite.

An analysis by occupational category suggests that exposure to the fumes given off by the rubber mix when heated during processing may be implicated in the lung cancer excess, but it is also likely that some exposure to asbestos was relevant in the flooring section. In the tyre sector the lung cancer deaths numbered 326 against an expectation of 299·4.

An excess of deaths from stomach cancer was also shown for the tyre manufacturing sector of the industry. Among the 16 030 men in this sector who were included in the study, 91 stomach cancer deaths were observed against 73·9 expected. This excess was statistically significant ($P < 0.05$). The rather limited information available on job background did not permit any firm conclusions to be drawn about possible causative factors in the working environment.

Although the survey method had its limitations, due allowance was made for geographical variation in disease mortality, especially necessary as most of the study population worked in highly industrialized and urban areas. The large size of the study population, the 10-year follow-up period and the high proportion of successful tracing all give support to the conclusions drawn. Moreover, the findings of the HSE study are broadly consistent with those of the BRMA study carried out on selected individual member companies, despite the different methodological approach. Indeed, both studies also reflect the findings of concurrent studies carried out in the United States.

No outright excesses for other individual cancers stood out, nor were there any signal tumours suggesting areas for special scrutiny. A particular attribute of the HSE study was that it embraced a very wide span of the other (that is non-tyre) sectors of the industry and gave some insight into their mortality experience, which has been little explored elsewhere.


## Studies from the United States of America

A decade ago two main groups of researchers set about their own epidemiological studies in the North American rubber industry. Between 1970 and 1972 several major rubber companies agreed to support studies on the health effects of the work environment, after important trade union negotiations with the United Rubber, Cork, Linoleum and Plastic Workers of America.[34] These studies were finally undertaken by the department of epidemiology at the Harvard School of Public Health in Boston,[35-40] and by the occupational health studies group at the School of Public Health, University of North Carolina at Chapel Hill.[23-26,41-48] The mortality pattern of male and in some instances female workers in rubber manufacturing plants was explored.

The scientific literature now published on these US rubber industry studies is quite extensive, but the main conclusions to be derived from them are remarkably consistent with those from the UK — with some notable exceptions. In the US rubber industry an approximately twofold excess of deaths from stomach cancer, but not especially of lung cancer, was found in one large tyre manufacturing plant.[41] This excess was associated with jobs done early in the production process, where chemical compounding ingredients were mixed and milled.[26] Several of the other studies in the US tyre industry, however, have also confirmed an increased risk from stomach cancer mortality, possibly associated with an exposure to dusts.[45] Similarly, an excess of lung cancer,[24,26,37,39,45] possibly associated with exposure to fumes, was demonstrated.

The US researches have also highlighted an excess of prostatic cancer in the batch preparation areas,[47] an excess of colonic tumours,[26,37] skin cancers[39] and both brain[49] and biliary tract tumours.[50] These excesses, however, are not reflected in the UK studies.

In particular, an excess of lymphatic tumours and of leukaemia has been highlighted by the US researchers,[42,48] but again not reflected by their British counterparts. These differences are important and should lead to further exploration and study. The excesses of lymphomas and of leukaemia seemed to be associated with tyre building and tyre repair operations, in which an exposure to solvent is likely. Benzene itself has been incriminated, but the final position is still not entirely clear.

Only one report from Europe[51] has been published in which attention was drawn to a possible excess of brain and bladder cancer in the rubber industry in Switzerland, although a second unpublished study contains similar findings.


**Causative factors**

It is clear from all this research, with consistency of results from different studies in divergent geographical locations and employing varying epidemiological techniques, that a probable association/causation link between the cancer excesses and the working environment is implied. Not many months go by without the scientific literature announcing yet another positive Ames test, or according a label of carcinogenicity following animal testing to even well-established and traditional rubber chemicals. How then is all this information to be interpreted realistically? A proper cause and effect relationship, if there is one, needs to be further narrowed down if effective preventive action is to be taken. There have been balanced[38,52–56] but also some rather exaggerated and one-sided[57] reviews written. However, a recent purportedly educational document[58] from the Occupational Safety and Health Administration (OSHA) promotes the concept that the rubber worker is likely to get cancer because of his occupation, which in my view is unwarranted.

There are, indeed, some suspect compounds,[52] but the likelihood of pinning the problem down to one single agent — known or as yet to be determined — is remote, and unlikely in the short term. The searches should not demand unlimited time and financial resources for want of simply tackling the basic clean-up that much of the industry needs, and which has been commenced already by directing most attention to the control of dust and fume at its source.

Carbon black itself has been indicted,[59] but apart from one Russian study[60] the evidence against it is weak.[61] Moca (methylene-bis-ortho-chloroaniline) is used in the manufacture of certain polyurethane products, but also for solid rubber mouldings such as gear blanks and industrial tyres. It is a highly suspect bladder carcinogen[62] and the BRMA strongly recommends (in bulletins nos. 6, 8 and 16) that its use should be discontinued in manufacture, although the factory inspectorate still prefer to define precautionary measures.[63] The evidence against chloroprene monomer (2-chloro-1,3-butadiene) is unconvincing,[64] while that implicating styrene is under study,[65-66] but both form potential epoxides *in vivo*. Formaldehyde has recently come under scrutiny for possible carcinogenic potential in a study by the Chemical Industry Institute of Toxicology. Talc itself, still widely used by the industry, has also been suggested as a possible carcinogen.[67,68]

There are several potential carcinogenic polycyclic hydrocarbons and many related heterocyclic compounds present in compounded rubber, and they are released in the fumes emanating from processing, especially at cure.[69] Measurements taken in factories do not, however, suggest excessive levels, as compared to outside urban air. These polynuclear compounds derive from the furnace carbon black and the mineral oil adjuvancy. Commonly found are: fluoranthene, pyrene, perylene, benzofluorene, benzanthracene, chrysene, coronene, benzo(a)pyrene, phenanthrene and dibenzanthracene, as well as many heterocyclic compounds. The significance of these findings is still speculative — whether a particular combination of polycyclic and heterocyclic compounds is the pertinent factor, or whether there could be synergism between some of these weak carcinogens and the undoubted carcinogens present in the tobacco smoked by many rubber workers. At one factory where a smoking history for most of the lung cancer cases was available, there were only two non-smokers amongst 30 long service curers. Perhaps there is some enhancement of the neoplastic process initiated by the tobacco habit when rubber fumes are concurrently inhaled.

Attention has recently been directed to the presence of some nitroso compounds in rubber workshops.[70] Some of these are formed as unwanted by-products, or from the interaction between a constituent amine and a nitrosating agent. Although there is no direct evidence linking nitrosamine exposure to cancer in man, many of the nitrosamines are very potent animal carcinogens.[71,72]. The subject has been well reviewed recently.[73] Interest has centred particularly on the retarder N-nitrosodiphenylamine (NDPA). Initially, animal feeding studies had shown this to be non-carcinogenic, but more recent work from the National Cancer Institute in the United States has shown a positive carcinogenic effect in rats at a very high level of dosage. In processing, NDPA may react with any secondary amine evolved to give the corresponding nitrosamine. If used with tetramethylthiuram disulphide (TMTD) it may produce dimethylnitrosamine (NDMA) by nitrosation of dimethylamine. If used with 2(morpholino-thio) benzothiazole it will produce N-nitrosomorpholine (NMOR). Both these nitrosamine compounds have been shown to be highly carcinogenic in either long term animal or in short term *in vitro* testing. Manufacturers have recommended that NDPA should not be used in combination with any of these two compounding ingredients, or with zinc dimethyl dithiocarbamate (ZDMC or related dithiocarbamates).

The BRMA, however, strongly advises that the use of NDPA should now be discontinued[74] and an alternative material sought. Removal of NDPA does in practice sharply reduce the level of atmospheric nitrosamines in the workshop, but this will not eliminate their presence completely. Present evidence suggests that very small traces of nitrosamines, perhaps as initial constituents, or even produced from secondary amines by reaction with atmospheric nitrogen oxides, may still be present during rubber processing. Sophisticated modern analytic techniques (using the thermal energy analyser, TEA) can detect as little as 0·05 $\mu$g/m$^3$ of nitrosamine in the working environment of a rubber factory. The highest levels thus far recorded, based on personal sampling techniques, have been in the range of 100–400 $\mu$g/m$^3$ for NMOR and 10–100 $\mu$g/m$^3$ for NDMA. Here again, very cogently demonstrated, is the principle of the need to control the fumes of processing to guard against the as yet unrecognized but potentially harmful constituent. Past experience in the industry alone dictates this need to prevent unnecessary exposure, if the workforce are to be reassured confidently that there is no risk to their health at their place of work.


## OCCUPATIONAL HYGIENE

Concurrently with the rather intense epidemiological approach, it was necessary that industrial hygiene surveys of the working environment should also be carried out. The BRMA conducted an extensive survey in ten factories among its member companies during 1974 in order to measure the amount of total particulate matter present, to measure fume where this predominated and to assess the amount of benzo(a)pyrene as a marker. The results of the various analyses[75] for benzo(a)pyrene provided very little evidence of any significantly elevated environmental levels, and this has been confirmed yet again in research conducted in association with the Institute of Petroleum. In fact, the results were not found to differ markedly from those levels usually reported in urban atmospheres, and they were at least as much influenced by seasonal factors as by the particular factory processes. Despite these findings, however, it was often clear on simple visual inspection that there were many areas in a rubber factory that were excessively dusty and heavily laden with fume. The results of the BRMA studies and those subsequently carried out in 1979 by the occupational hygiene team of the Health and Safety Executive reflected this.

One special feature of the BRMA study was that three out of the five samples taken at each chosen site were stored under deep freeze conditions for possible analysis at a later date. One sample was further sealed under vacuum to provide an oxygen-free environment for long term storage. In order to collect adequate samples of fume particulates for analysis of its solvent soluble fraction, a special high volume static sampler was designed and commissioned.[69]

Likewise, renewed activity and research into the chemical analysis of the fumes emanating from rubber processing was initiated.[76-80] The exact mechanism of the chemical interactions is still largely unknown, and in view of the recent epidemiological findings they merit much detailed further study.

In the United States similar extensive occupational hygiene surveys[81,82] were conducted in order to quantify the working environment for particulate and solvent vapour exposures. There are two outright advantages to be gained from these surveys: first, as an accompaniment to future epidemiological studies they provide a necessary part of the data to be collected if the cause and effect relationship is to be properly elucidated; and secondly, they form a baseline against which to measure any improvements.[83]


## EFFECTS ON RESPIRATORY FUNCTION

Increasing interest is focusing on the possibility that exposure to dust and fume in the rubber industry may also affect health adversely by a direct action on the respiratory tract. The evidence for this mostly comes from the United States, but there is also some confirmatory evidence from Egypt.[84,85] These studies have yet to be repeated in the United Kingdom to delineate the operating conditions in the British industry, which may differ from those pertaining in the United States. Peters and Fine, in a series of papers[86,87] during 1976, outlined their results when conducting a case/control study on 121 curers. The curers had a higher prevalence of chronic bronchitis than the 189 control men in three plants. Some 25 per cent of curers with more than 10 years' exposure were classified as having chronic obstructive lung disease. The smoking habits of the men were considered insufficient to explain the discrepancies, which were attributed by the researchers to tyre curing fumes emanating from manually operated automobile tyre presses. As a group the curers were noted to have lost more measurable lung function over a year than their controls. They also had less satisfactory lung function overall, and the divergence was more marked for those exposed to the fume for more than 10 years. The conclusion drawn was that heavy exposure to curing fume adversely affects pulmonary function. Similar results were found in workers exposed to processing dust and fume originating in the banbury and chemical mixing areas, and also in workers exposed to talc during rubber tyre manufacture.

Another study[88] suggested that amongst 1820 production workers the four work areas most obviously associated with chronic respiratory symptoms were milling, calendering, tube curing and tube inspection. Further work demonstrated a possible potentiating effect between smoking and exposure to dust and fume within these same work areas. Possible late effects are reflected in retirements due to pulmonary disability in rubber workers.[89]

Other studies suggest an adverse effect on respiratory function in rubber workers exposed to phenol-formaldehyde[90] and to thermosetting resins.[91] Although there seems to be little evidence of an overt pneumoconiotic effect by carbon black on lung tissue, there are indications of some symptomatic and functional changes promoting airway disease in those exposed to high concentrations of carbon black dust.[92] The most marked changes occurred in those who smoked habitually,[93] indicating yet again a possible synergism between smoking and inhalation of dust and fume in the working environment.

## THE FUTURE

The search in chemical terms for the elusive carcinogenic influence will undoubtedly continue, necessarily accompanied by further epidemiological studies. Both approaches are required. Meanwhile a new development that may improve the monitoring of populations at risk has been reported. Recent research has demonstrated the feasibility of detecting mutagenic metabolites (thioethers) in the urine concentrates of exposed workers[94-96] using a bacterial fluctuation test. In a chemical plant the highest thioether excretion was found in rubber workers, and especially in radial tyre builders, when a comparison was made between these latter two groups and clerks, plastic monomer mixers and footwear preparers.[95] Smoking and medication both tended to increase thioether excretion, but it was shown that urinary thioether determination may prove to be a useful tool in assessing exposure to mixtures of chemicals regardless of their route of absorption.

Recent advances in the field of lymphocyte immunoreactivity are likely to improve the surveillance of workers with a past history of exposure to bladder carcinogens in the industry. A twofold or greater increase in reactivity to a bladder cancer target cell in a group of 68 workers[97] was strongly associated with the subsequent development of abnormal urinary cytology indicative of malignant change. The findings support a premise that changes in lymphocytic immunoreactivity might be useful in predicting the subsequent onset of bladder cancer in workers previously exposed to bladder carcinogens. Such screening is also likely to augment the existing techniques of surveillance using routine urinary exfoliative cytology.

A more radical and practical approach, by tackling the problem at source, is to consider both reformulation and process control as methods for the reduction of rubber curing fume. For example, a measurable reduction in the quantity of fume released at cure can be achieved by adjusting the curing temperature. Testing methods which define the chemical constituents in the fume of particular compounds under specifically controlled conditions of cure have recently been developed.[80]

An awareness of the problems being faced by the industry and of the essential need for a much stricter control of dust and fume emanating from processing is reflected in the formation of two new groups. First, a National Industry Group within the Factory Inspectorate Branch of the Health and Safety Executive is already active, and secondly the setting up of an Industry Advisory Committee is planned. In addition, the International Agency for Research on Cancer (IARC) plans to devote its second industry monograph to rubber manufacturing processes, in continuance of the series 'Evaluating the Carcinogenic Risk of Chemicals to Humans'.

## CONCLUSIONS

It is difficult to escape the conclusion that the rubber industry has environmental health problems, some of which have only recently come to light. These may nevertheless relate to past rather than to present conditions only. This is the broad picture presented by the industry as a whole. However, the situation varies considerably between individual factories, between countries and between different sectors of the industry — accepting that it is mainly the tyre sector which has been studied. It is perhaps these differences which now need to be studied more closely, in an attempt to seek out possible aetiological factors and to suggest appropriate preventive action.

The evidence points, in both the British and US rubber industries, to a small excess of both stomach and lung cancer mortality. In the British tyre industry at least, the once evident bladder cancer hazard has now largely disappeared, but that may not be the position throughout the rubber industry as a whole. Respiratory function appears to be adversely affected in some work situations within the US rubber industry, suggesting that the position in Britain also requires further study.

There is debate as to whether the environmental conditions are synergistically enhancing the normal incidence of lung and stomach cancers, which are relatively common, or whether weak carcinogens in their own right may be acting independently. These questions are unlikely to be answered satisfactorily in the short term.

The solution, however, is more easily defined and can be put into practice now. It is to apply much stricter controls both to the physical working environment with regard to dust and fumes, and to the chemistry of compounding. If the operative is not exposed to excess dust or to excess fume emanating from rubber processing operations, either because these are adequately controlled at source by efficient general and local exhaust ventilation or by designing them out with modern processing techniques, then whatever might otherwise have been potentially harmful to the workforce is no longer likely to be a hazard.

The success of preventive measures taken today can only be assessed convincingly in 10 or 15 years time when the new cohort studies are completed. However, there are two other important gains meanwhile. The first is the right environmental conditions in which to work. The second is a reassuring defence against both the suspect as well as the unknown and unpredictable hazard. The industry's past experience alone fully vindicates the need for this approach.

The means by which long term health and safety can be assured are sound engineering practices, continued epidemiological surveillance and environmental monitoring to ensure compliance with laid-down standards. Health and safety in this context implies freedom from the risk of acute illness, injury or long term disease caused by the work environment.

REFERENCES
1. Winspear GG. *Vanderbilt Rubber Handbook*. New York: RT Vanderbilt Co. Inc., 1968.
2. Stern HJ. *Rubber Natural and Synthetic*. London: Maclaren & Sons Ltd, 1967.
3. Malten KE, Nater JP and van Ketel WG. *Patch Testing Guidelines*. Nijmegen: Dekker & van de Vegt, 1976.

4. Wilson HT. Rubber dermatitis. *Br. J. Dermatol.* 1969; **81**:175–9.
5. Herve-Bazen D, Gradiski D, Duprat P et al. *Industrial Allergic Eczema due to contact with IPPD and DMPPD in Tyres.* Report No. 226/RE. Vandoeuvre-les-Nancy: Institut National de Recherche et de Sécuritée, 1978.
6. Nurse DS. Rubber sensitivity. *Australas. J. Dermatol.* 1979; **20**:31–3.
7. Cronin E. *Contact Dermatitis.* London: Churchill Livingstone, 1980: 714–70.
8. British Rubber Manufacturers' Association. *Toxicity and Safe Handling of Rubber Chemicals—A Code of Practice.* Birmingham: BRMA Health Research Unit, 1978.
9. INRS. Toxicité de produits utilisés dans l'industrie du caoutchouc—revue bibliographique. *Cahiers de Notes Documentaires.* Paris: INRS, 1980; **99**:253–62.
10. Holmberg B and Sjostrom B. *A Toxicological Survey of Chemicals used in the Swedish Rubber Industry.* Investigation Report 19. Stockholm: National Board of Occupational Safety and Health, 1977.
11. Nutt AR. Toxicity of rubber chemicals. *Progress of Rubber Technology* 1979; **42**:141–54. London: The Plastics and Rubber Institute.
12. Hammond CM. Dust control concepts in chemical handling and weighing. *Ann. Occup. Hyg.* 1980; **23**:95–109.
13. British Rubber Manufacturers' Association. *Ventilation Symposium.* London: BRMA, February 1979.
14. Worwood JA. The engineering of dust and fume control systems. *Health and Safety in the Plastics and Rubber Industries.* In: Proceedings of a conference held at Warwick University. London: The Plastics and Rubber Institute, October 1980: Paper No. 22.
15. Case RAM and Hosker ME. Tumour of the urinary bladder as an occupational disease in the rubber industry in England and Wales. *Br. J. Prev. Soc. Med.* 1954; **8**:39–50.
16. Veys CA. A study on the incidence of bladder tumours in rubber workers. MD Thesis 1973, University of Liverpool.
17. O'Connor J. *Judgement: Cassidy and Wright v Dunlop Rubber Co. Ltd and ICI Ltd.* The High Court of Justice, Court No. 21, 1971: 34–6.
18. Veys CA. Bladder cancer in rubber workers: the story reviewed and up-dated. *Health and Safety in the Plastics and Rubber Industries.* In: Proceedings of a conference held at Warwick University. London: The Plastics and Rubber Institute, October 1980: Paper No. 13.
19. The British Rubber Manufacturers' Association. *Health Research Project.* Birmingham: BRMA Health Research Unit, January 1976.
20. Health and Safety Executive. *Mortality in the British Rubber Industries.* London: HMSO, 1967–76.
21. Davies JM. Bladder tumours in the electric cable industry. *Lancet* 1965; **2**:143–6.
22. Fine LJ, Peters JM, Monson RR et al. An industrial epidemiology. *Chemtech.* 1980; **10**:298–301.
23. National Cancer Institute (DHEW). *An Epidemiologic Study of Bladder Cancer among Rubber and Tire Industry Workers.* Occupational Health Studies Group, School of Public Health, University of North Carolina, USA, 1977.
24. McMichael AJ, Andjelkovic DA and Tyroler HA. Cancer mortality among rubber workers: an epidemiologic study. *Ann. NY Acad. Sci.* 1976; **271**:125–37.
25. Tyroler HA, Andjelkovic D, Harris R et al. Chronic diseases in the rubber industry. *Environ. Health Perspect.* 1976; **17**:13–20.
26. McMichael AJ, Spirtas R, Gamble JF et al. Mortality among rubber workers: relationship to specific jobs. *J. Occup. Med.* 1976; **18**:178–85.
27. Parkes HG. Identification and control of occupational bladder cancer. *Strategies for Intervention: Carcinogenic Risks.* IARC Scientific Publ. No. 25. Lyon: WHO/IARC, 1979: 47–58.
28. Office of Population Censuses and Surveys. Occupational Mortality 1970–72 (Decennial Supplement). London: HMSO, 1978.

29. Waterhouse JAH. Current status of cancer risk in the rubber industry. In: *Advances in Medical Oncology, Research and Education,* vol. 3. Oxford: Pergamon Press, 1979: 97–105.
30. Veys CA. Developing opportunities for research in occupational medicine. *North Staffordshire Medical Institute Journal* 1979; **11**:24–36.
31. Baxter PJ and Werner JB. *Mortality in the British Rubber Industries 1967–76.* London: Health and Safety Executive, 1980.
32. Fox AJ, Lindars DC and Owen R. A survey of occupational cancer in the rubber and cablemaking industries: results of five-year analysis, 1967–71. *Br. J. Ind. Med.* 1974; **31**:140–51.
33. Fox AJ and Collier PF. A survey of occupational cancer in the rubber and cablemaking industries: analysis of deaths occurring in 1972–74. *Br. J. Ind. Med.* 1976; **33**:249–64.
34. American Letter. *Rubber Journal* July 1970: p. 22, and August 1971: p. 27.
35. Burgess WA, Peters JM and Monson RR. The rubber workers study at the Harvard School of Public Health. In: Aye A (ed.) *Environmental Aspects of Chemical Use in Rubber Processing Operations.* Washington: Office of Toxic Substances, Environmental Protection Agency, 1975: 413–25.
36. Peters JM, Monson RR, Burgess WA et al. Occupational disease in the rubber industry. *Environ. Health Perspect.* 1976; **17**:31–4.
37. Monson RR and Nakano KK. Mortality among rubber workers. *Am. J. Epidemiol.* 1976; **103**:284–303.
38. Monson RR. Effects of industrial environment on health. *Environmental Law* 1978; **8**:663–700.
39. Monson RR and Fine LJ. Cancer mortality and morbidity among rubber workers. *J. Natl Cancer Inst.* 1978; **61**:1047–53.
40. Fine LJ, Peters JM, Monson RR et al. An industrial epidemiology. *Chemtech.* 1980; **10**:298–301.
41. McMichael AJ, Spirtas R and Kupper LL. An epidemiological study of mortality within a cohort of rubber workers 1964–72. *J. Occup. Med.* 1974; **16**:458–64.
42. McMichael AJ, Spirtas R, Kupper LL et al. Solvent exposure and leukaemia among rubber workers. *J. Occup. Med.* 1975; **17**:234–9.
43. Harris RL. University of North Carolina Occupational Health Studies Program. In: Ayer A (ed.) *Environmental Aspects of Chemical Use in Rubber Processing Operations.* Washington: Office of Toxic Substances, Environmental Protection Agency, 1975:396–412.
44. Andjelkovich D, Taulbee J and Symons M. Mortality experience of a cohort of rubber workers 1964–1973. *J. Occup. Med.* 1976; **18**:387–94.
45. Andjelkovich D, Taulbee J, Symons M et al. Mortality of rubber workers with reference to work experience. *J. Occup. Med.* 1977; **19**:397–405.
46. Andjelkovich D, Taulbee J and Blum S. Mortality of female workers in a rubber manufacturing plant. *J. Occup. Med.* 1978; **20**:409–13.
47. Goldsmith DF, Smith AH and McMichael AJ. A case-control study of prostate cancer within a cohort of rubber and tire workers. *J. Occup. Med.* 1980; **22**:533–44.
48. National Cancer Institute (DHEW). *An Epidemiologic Study of Leukemia among Rubber and Tire Industry Workers.* Occupational Health Studies Group, School of Public Health, University of North Carolina at Chapel Hill, USA, 1977.
49. Mancuso TF. Tumours of the central nervous system — industrial considerations. *Acta Union Internationale contre le Cancer* 1963; **19**:488–9.
50. Mancuso TF and Brennan MJ. Epidemiological considerations of cancer of the gallbladder bile ducts, and salivary glands in the rubber industry. *J. Occup. Med.* 1970; **12**:333–41.
51. Lamperth-Siler E. Urinary tract and brain neoplasms in workers in rubber plants (in German with English summary). *Schweiz. Med. Wochenschr.* 1974; **104**:1655–9.

52. WHO/IARC. *Chemicals and Industrial Processes associated with Cancer in Humans.* IARC Monographs vol. 1–20, Supplement 1. Lyon: International Agency for Research on Cancer, September 1979.

53. Industry Survey – Rubber. *Health and Safety at Work* 1979; March: 23–5, and April: 43–5.

54. Bates RR. Preventing occupational cancer. *Environ. Health Perspect.* 1979; **28:** 303–10.

55. Leading Article. Mortality in the British rubber industries 1967–1976. *Br. Med. J.* 1980; **2:**471–2.

56. Editorial. Getting the facts straight. *J. Soc. Occup. Med.* 1980; **30:**130–1.

57. ASTMS Policy Document. *The Prevention of Occupational Cancer.* London: Association of Scientific, Technical and Managerial Staffs, 1980.

58. OSHA. *Cancer in the Rubber Industry: The Risks and What You Can Do about Them.* Washington: US Department of Labor, 1980: 72.

59. NIOSH. *Criteria for a Recommended Standard — Occupational Exposure to Carbon Black.* DHEW Publication No. 78-204. Washington: US Department of Health, Education and Welfare, 1978.

60. Troitskaya NA, Velichkovsky BT, Kogan FM et al. *Vapr. Onkol.* 1980; **1:**63–7 (in Russian with English summary).

61. Robertson JM and Ingalls TH. A mortality study of carbon black workers in the United States from 1935 to 1974. *Arch. Environ. Health* 1980; **35:**181–6.

62. Stula EF, Barnes JR, Sherman H et al. Urinary bladder tumours in dogs from 4,4′-methylene-bis 2-chloroaniline (MOCA). *J. Environ. Pathol. Toxicol.* 1977; **1:**31–50.

63. Henning HF. Precautions in the use of methylene-bis-o-chloroaniline (MBOCA). *Ann. Occup. Hyg.* 1973; **17:**137–42.

64. Haley TJ. Chloroprene (2-chloro-1,3-butadiene): What is the evidence for its carcinogenicity? *Clin. Toxicol.* 1978; **13:**153–70.

65. WHO/IARC Monographs on the evaluation of the carcinogenic risk of chemicals to humans. *Some Monomers, Plastics, Synthetic Elastomers and Acrolein,* vol. 19. Lyon: International Agency for Research on Cancer, 1979.

66. Ott MG, Kolesar RC, Charnweber HC et al. A mortality survey of employees engaged in the development or manufacture of styrene-based products. *J. Occup. Med.* 1980; **22:**445–60.

67. Bleyer H and Arlon R. Talc: a possible occupational environmental carcinogen. *J. Occup. Med.* 1973; **15:**92–7.

68. Pelfrene A and Shubik P. Le talc, est-il carcinogène (in French with English summary). *Nouv. Presse Méd.* 1975; **4:**801–3.

69. Nutt A. Measurement of some potentially hazardous materials in the atmosphere of rubber factories. *Environ. Health Perspect.* 1976; **17:**117–23.

70. Fajen JM and Carson CA. N-nitrosamines in the rubber and tire industry. *Science* 1979; **205:**1262–4.

71. WHO/IARC *Environmental Aspects of N-Nitroso Compounds.* IARC Scientific Publication No. 19. Lyon: International Agency for Research on Cancer, 1978.

72. Magee PN. N-nitroso compounds and related carcinogens. In: Searle CE (ed.) *Chemical Carcinogens.* Washington DC: American Chemical Society, Monograph 173; 1976: 491–625.

73. Ember LR. Nitrosamines: assessing the relative risk. *Chem. Engng News* 31 March 1980; **13:**20–6.

74. BRMA. *Nitrosamines.* Birmingham: British Rubber Manufacturers' Association, August 1980: Health Bulletin No. 23.

75. BRMA Health Research Project. *A Report on the Environmental Monitoring Study.* Birmingham: British Rubber Manufacturers' Association, 1975.

76. Fraser DA and Rappaport SM. Health aspects of the curing of synthetic rubbers. *Environ. Health Perspect.* 1976; **17:**45–53.

77. Rappaport SM and Fraser DA. Gas chromatographic–mass spectrometric identification of volatiles released from a rubber stock during simulated vulcanization. *Analytical Chemistry* 1976; **48**:476–81.
78. Bebb RL. Chemistry of rubber processing and disposal. *Environ. Health Perspect.* 1976; **17**:95–101.
79. Rappaport SM and Fraser DA. Air sampling and analysis in a rubber vulcanization area. *Am. Ind. Hyg. Assoc. J.* 1977; **38**:205–10.
80. Willoughby BG. Reformulation or process control as a concept for the reduction of rubber curing fume. *Health and Safety in the Plastics and Rubber Industries.* In: Proceedings of a conference held at Warwick University. London: The Plastics and Rubber Institute, October 1980: Paper No. 12.
81. Williams TM, Harris RL, Arp EW et al. Worker exposure to chemical agents in the manufacture of rubber tires and tubes: particulates. *Am. Ind. Hyg. Assoc. J.* 1980; **41**:204–11.
82. Vanert MD, Arp EW, Harris RL et al. Worker exposures to chemical agents in the manufacture of rubber tires: solvent vapor studies. *Am. Ind. Hyg. Assoc. J.* 1980; **41**:212–19.
83. Pagnotto LD, Elkins HB and Brugsch HG. Benzene exposure in the rubber coating industry — a follow-up. *Am. Ind. Hyg. Assoc. J.* 1979; **40**:137–46.
84. Noweir MH, El-Dakhakhny AA and Osman HA. Exposure to chemical agents in the rubber industry. *J. Egypt Public Health Assoc.* 1972; **47**:182–201.
85. Osman HA, Wahdan MH and Noweir MH. Health problems resulting from prolonged exposure to chemical agents in rubber industry. *J. Egypt Public Health Assoc.* 1972; **47**:290–311.
86. Fine LJ and Peters JM. Respiratory morbidity in rubber workers. I. Prevalence of respiratory symptoms and diseases in curing workers. II. Pulmonary function in curing workers. III. Respiratory morbidity in processing workers. *Arch. Environ. Health* 1976; **31**:5–14, 136–40.
87. Fine LJ, Peters JM, Burgess MS et al. Studies of respiratory morbidity in rubber workers. IV. Respiratory morbidity in talc workers. *Arch. Environ. Health* 1976; **31**:195–200.
88. McMichael AJ, Gerger BA, Gamble JF et al. Chronic respiratory symptoms and job type within the rubber industry. *J. Occup. Med.* 1976; **18**:611–17.
89. Lednar WM, Tyroler HA, McMichael AJ et al. The occupational determinants of chronic disabling pulmonary disease in rubber workers. *J. Occup. Med.* 1977; **19**:263–8.
90. Gamble JF, McMichael AJ, Williams T et al. Respiratory function and symptoms: an environmental–epidemiological study of rubber workers exposed to a phenol-formaldehyde type resin. *Am. Ind. Hyg. Assoc. J.* 1976; **37**:499–513.
91. doPico GA, Rankin J, Chosy LW et al. Respiratory tract disease from thermosetting resins: study of an outbreak in rubber tire workers. *Ann. Intern. Med.* 1975; **83**:177–84.
92. Valic F, Beritic-Stahuljak D and Mark B. A follow-up study of functional and radiological lung changes in carbon-black exposures (in English). *Int. Arch. Arbeitsmed.* 1975; **34**:51–63.
93. Crosbie WA. Respiratory function of carbon black workers. *Health and Safety in the Plastics and Rubber Industries.* In: Proceedings of a conference held at Warwick University. London: The Plastics and Rubber Institute, October 1980: Paper No. 16.
94. Falck K, Sorsa M and Vainio H. Mutagenicity in urine of workers in rubber industry. *Mutation Res.* 1980; **79**:45–52.
95. Vainio H, Savolainen H and Kilpikari I. Urinary thioether of employees of a chemical plant. *Br. J. Ind. Med.* 1978; **35**:232–4.

96. van Doorn R, Bos R, Leijdekkers C et al. Thioether concentration and mutagenicity of urine from cigarette smokers. *Int. Arch. Occup. Environ. Health* 1979; **43**:159–66.
97. Kumar S, Taylor G, Wilson P et al. Prognostic importance of specific immunoreactivity in occupational bladder cancer. *Br. Med. J.* 1980; **282**:512–13.

# 2. OCCUPATIONAL HEALTH IN COAL MINING
## *Part 1: NON-RESPIRATORY PROBLEMS*
### *R. M. Archibald*

The outstanding feature in coal mining operations over the past three decades has been the growth of mechanized as opposed to manual coal cutting. This has been particularly marked in the so-called 'developed' countries but less so in the 'under-developed' countries. With increasing mechanization and the general advances in technology there have naturally been changes in the health risks and hazards. Problems associated with manual coal getting, for example, nystagmus, have rapidly declined and have been replaced by problems such as noise related to modern mining techniques.

This chapter deals with current non-respiratory health problems. The continuing importance and magnitude of respiratory diseases in the mining industry merit separate treatment.

**Physical standards**

Although mechanization has removed much of the arduous toil of manual mining, the industry remains one in which physical activity plays a large part, and in which men have to carry out their daily work in a continuously hostile environment. For these reasons all industrial entrants in the United Kingdom are required to have a stringent medical examination, including an assessment of muscular fitness, visual acuity, hearing ability and a chest X-ray. A substantial part of this is implemented by trained nursing staff skilled in technical procedures such as electrocardiography, audiometry and assessment of lung function.

The nature and circumstances of mining operations are such that some medical defects clearly exclude affected individuals from any employment in the industry or for specific occupations within the industry. Men suffering from epilepsy, poor vision or having only one eye, claustrophobia and obstructive lung disease are excluded from mining employment, but in certain circumstances may be considered for surface work. Normal colour vision is essential for men to be employed as electricians or for underground transport operations.

There are, however, some conditions where rejection need not be automatic. Diabetics controlled by diet or small doses of oral drugs may well be found suitable work, whereas the young diabetic controlled by insulin injections would be unacceptable. Severe otorrhea would usually be unacceptable, certainly for underground purposes, but minor degrees of middle ear infection may be considered. This also applies to chronic respiratory disease — severe cases are rejected but mild cases may be found suitable for specific occupations. Of course in cases of doubt the occupational physician has to make the final decision.

The foregoing does not mean that there is no 'light' work in the mining industry. Even underground there are many operations which can be carried out by men unfit for medical or traumatic reasons. Not unreasonably, however, management tends to reserve such jobs for men whose disability arises from their previous service in the industry.

The activity in mining most likely to put severe strain on the fitness and physical endurance of men is that involving rescue operations. Rescue men may be required to undertake extremely hard physical work in a hot, humid and irrespirable atmosphere, requiring continuous use of breathing apparatus. For this reason all rescue men, whether full- or part-time, are subjected to especially severe medical examinations and tests of physical fitness. In the United Kingdom it is a legal requirement that rescue men pass such an examination and test on first appointment and annually thereafter.

Recently the method of measuring fitness has been reviewed with the objective of replacing the present near maximal exertion test with a more sophisticated submaximal test which would give virtually the same information on fitness. In the United Kingdom this would mean changing from carrying a heavy weight whilst stepping on and off a platform for a specified period (the Harvard Pack test) to the use of a tread-mill with continuous electrocardiography shown on an oscilloscope. In other countries bicycle ergometry is the method of choice. Whatever method is chosen, it should be carefully standardized and take place in the presence of a physician, with resuscitation equipment readily available.

In order further to ensure the fitness of rescue workers it is the practice in many countries for rescue men to be retired at ages ranging from 40 to 45. The present tendency for ischaemic heart disease to develop at increasingly earlier ages would make it unlikely that any extension of retirement age would be contemplated.

The potential stress of rescue operations makes it important to examine each individual before every visit underground for rescue operations. The examining physician will be particularly interested to assess an individual rescue man's state of fatigue, to detect any recent respiratory infection and to be satisfied that there has not been a significant alcohol intake since the previous descent. These details are necessary because of the frequency with which breathing apparatus has to be worn during rescue operations. In the United Kingdom it is necessary also to ensure that dentition is adequate as breathing apparatus in this country still requires the use of a mouth piece. In some other countries, however, a full face mask is used.

As in most industries, specific examinations are required for small numbers of workers at potential risk when involved in specialized work, for example, full blood counts on individuals using devices containing a source of ionizing radiations.

## ORGANIZATION OF ACCIDENT AND FIRST AID SERVICES

The structure of the mining industry, involving relatively small units (in the United Kingdom) of on average 1000 men working at coal faces a long way from the pit bottom, means that neither doctor nor nurses can be immediately available to deal with illness or injury, and may not be available for periods of up to 2 hours. For this reason highly trained and dedicated first aiders are an essential part of the mines medical service and can be regarded as the firm base of the pyramidal structure of the service. Clearly a high standard of first aid is essential if adequate primary care is to be given to the injured or ill, and it follows that this can only be achieved by equally high standards in initial and refresher training. Methods of teaching and training are continuously under review to ensure that the most modern and effective methods such as tape slide presentations, films and video tapes are used.

Good first aid demands adequate resources and there are four essentials for this:

1. Small first aid tins, pouches or packs containing a sufficiency of small dressings to treat minor wounds. These should be carried in and out of the mine each shift by first aiders and should be inspected and if necessary replenished daily by the surface medical attendant.

2. Large containers should be stationed at intervals of 1000 m along the mine roadways and proximate to either end of long wall faces. These should contain a supply of triangular bandages and large sterilized dressings, carrying sheets and splints, splint padding, blankets sealed in plastic against dust and humidity and sandbags to be filled with mine dust in the event of a cervical spine injury.

3. Suitable means of transport from the mine bottom for stretcher cases should be available. This may take the form of special trolleys, adaptations to fit mine cars, man-riding trains or motorail systems. Where hand carriage is inescapable, stretcher slings should be provided as stretcher transport is now one of the most physically demanding underground tasks.

4. Good communications: it is vital to alert the surface organizations to the requirements of the emerging casualty, therefore, all first

aiders must be taught the importance of full, accurate and early information. Beside each underground telephone there should be a plastic notice listing the requisite information to be given to the surface and a smaller replica should be in every individual tin or pouch.

Pain relief is of great importance in making the patient more comfortable and reassured. Although increasing safety standards have reduced accident rates significantly, illness causing severe pain, such as coronary ischaemia, still occurs and needs pain relief. In the United Kingdom morphia in injectable syrettes is stored at strategic points underground, but generally this is the exception rather than the rule and in most countries non-morphia substitutes are available. Recently greater use has been made of an oxygen/nitrous oxide mixture (Entonox) as an analgesic, and it has proved possible to store cylinders of this mixture underground and to ensure a high standard of maintenance. It is essential to ensure thorough training and familiarization with such equipment if it is to be used by first aiders with confidence.

First aid organization on the surface will vary with the size of the mine and its proximity to hospital emergency services. The larger and more remote the situation the more essential it is to have a well-equipped medical centre in which resuscitation can be given and in which emergency surgery can be undertaken. When a hospital is within 15 km and road communications are good it is probably advisable to take the casualty directly for treatment without interposing on-surface medical centres.

In severe injuries or in the event of multiple casualties from a colliery disaster, where evacuation may be delayed for some of the injured, the value of transfusion should not be overlooked. It is usual practice in the United Kingdom to stock a high molecular replacement fluid with giving sets which can be taken underground and set up there if required. In addition, it is the practice in some United Kingdom coal fields to arrange liaison with hospital flying squads which can be called out to the mine in the event of a major incident.

From the foregoing the role of the doctor and nurse, both of whom must be trained in emergency resuscitation techniques, will be obvious and it is a *sine qua non* that both must be ready to go underground at any time. Finally, the continuously changing geography of underground operations as coal faces advance or are closed makes it essential that a regularly updated plan of the mine underground should be maintained in the medical centre so that the location of an incident can be clearly identified.


## NOISE

As recently as 25 years ago mines were relatively silent places underground, but increasing mechanization has changed this. Powerful coal cutting machines move along coal faces, other machines are used to prepare face ends and to drive roadways. Ventilation fans are in continuous operation as are conveyor belts. Diesel trains have replaced ponies and compressed air is a frequent source of power for drilling machines. At the mine bottom loading and transporting mine cars add to the noise, and on the surface the operation of coal preparation is virtually continuous and frequently produces noise of more than 90 dBA.

It is essential that systematic noise surveys are conducted in order that working sites with an Leq of 90 dBA can be identified and appropriate hearing conservation measures taken. The underground environment of dust and humidity makes hearing conservation a difficult problem because wearing ear muffs for any length of time is very uncomfortable and it is impossible to isolate the individual from noise. For these reasons a number of countries insist on routine audiometric screening on entry to the industry and periodic re-screening thereafter. Individuals whose serial audiograms indicate undue sensitivity to noise can then be detected and given work in less noisy surroundings.

In West Germany the medical service of the Ruhr mines has developed plastic ear plugs which are individually wrapped so that the plug can be pressed into the auditory meatus directly from its protective envelope without contamination from dust from the hand, thus preventing possible otitis externa. It is estimated that such plugs, which mould easily to the meatus, afford protection of the order of 10 dBA and, therefore, are only effective when noise exposure is to an Leq of 100 dBA or less.

In most countries where compensation for noise-induced hearing loss is given, disability is measured by loss in the conversation frequencies — so-called social disability. When this criterion is applied to United Kingdom coal mining little epidemiological evidence of noise-induced hearing loss has been found. This is because noise is discontinuous apart from ventilation fans underground.

Efforts are increasingly being made to build noise reduction into the design of new machines and coal preparation plants. More efforts are being made to educate management and men in noise problems; for example, a film on noise in coal preparation plants in the United Kingdom has recently been produced for circulation to the sites concerned.


## MINE HYGIENE

It is aesthetically desirable to provide adequate sanitation underground and it is particularly essential in those parts of the world where parasitic and worm infections are common. Underground toilets should be strategically sited at pit bottom and on the travelling roadways where man-riding trains stop or where men congregate at the beginning or end of a shift. In such places a refuge hole should be constructed and rendered private by a door or brattice screen. Most miners prefer to use lavatories before or after the underground shift and, therefore, it is essential to have adequate numbers of toilets in the pit-head baths and to ensure their adequate maintenance. Many types of toilets have been tried underground but the commonest is still probably the chemical closet which is charged with a fixed amount of chemical and emptied regularly. It is possible to obtain self-sealing lids which allow the container to be sealed, replaced by a freshly charged container, loaded on to a mine car and taken out of the mine to be emptied at a suitable drainage point, hosed out and recharged. However, the problems with such systems is that servicing them is an unpopular job for which money will not compensate adequately.

Occasionally it may be possible to install water-flush systems where there is an adequate supply of mine water and old disused workings underneath to which the soil can safely be flushed. Vacuum sewage systems exist and are successfully operated in the iron ore mines at the extreme north of Sweden. They are, of course, more costly to operate.

Over the past decade a number of new sanitation systems have been offered but all seem to founder because of a lack of robustness, an electrical system which would be too costly to make flameproof or the use of an enclosed system of a light mineral oil which does not meet the safety requirements for the use of such materials underground. This is one area in which technology has singularly failed to keep pace with modern mining requirements.


## OCCUPATIONAL DISEASES OF MINERS

The fact that some of these diseases have virtually disappeared from the United Kingdom scene and that the prevalence of others is declining steadily is a sound indication that doctors and mining engineers, working in concert, can effect prevention. The traditional diseases are:


### Weil's disease, leptospirosis (icterohaemorrhagica)

The role of the rat as vector of this condition is too well known to require detailed description, and the primary attack must be on the rat population. To be successful this must be done systematically. The first requirement is to have selected officials or operators in the mine trained by governmental or other agencies specializing in vermin control. Next a systematic survey of the rat population in the mine must be undertaken and an energetic campaign sustained to remove sources of food. At one time it was thought that rats would feed on vegetable oils but not mineral oils and, therefore, mineral oils were widely used to grease underground points. However, it was demonstrated experimentally that rats will eat, and exist on, mineral oils if hungry enough. The pre-baiting and baiting routines should be adhered to and then a further survey made in an attempt to assess the 'kill' rate. In the meantime workmen must be discouraged from discarding food scraps and wrappings underground.

The success of such measures, coupled with the elimination of horses underground, has led to the virtual disappearance of rats from United Kingdom mines and it is now several years since the last authenticated case of Weil's disease was recorded.

## Ankylostomiasis

Where this condition is endemic in the general population good mine hygiene is essential and the provision of stout safety boots together with efficient mine drainage will effect a reduction in the incidence.

## Miner's nystagmus

The introduction of the electric cap lamp and improved standards of underground illumination have virtually eliminated this condition. However, it should be remembered that Wellwood Ferguson[1] demonstrated other factors in causation and, therefore, isolated cases, with or without overt nystagmus, may well be seen.

## Beat conditions

Subcutaneous cellulitis or bursitis occur at the knee, hand or elbow. The commonest of these conditions is beat knee and is most likely to be seen where coal is hand-filled. A series of studies by Sharrard[2] demonstrated that the pressure exerted on about 7 cm$^2$ of the pre-patellar area of the knee could rise to about 130 kg while shovelling coal on to a bottom-loading belt behind the collier. Prevention lies in the provision of adequate knee pads of approved design. Today such cases as do arise are as often due to trauma from crawling along the face amidst the powered supports as to static kneeling. Once a beat knee develops prompt treatment will ensure cure and rapid return to the coal face. When the bursa is infected, broad spectrum antibiotics should be given. The bursa can be aspirated under strict asepsis. Firm bandaging and physiotherapy should follow promptly.

## Dermatitis

The hot, humid and dusty environment of coal mines in all countries is a decisive factor in the causation of occupational dermatitis in miners.

The subject has been well reviewed by Williamson and Vickers,[3] whose comprehensive report should be consulted.

It is important to initiate treatment early to prevent the condition becoming chronic and sensitized to an ever-widening number of primary irritants. New techniques bring fresh hazards to the skin such as the use of cements underground and the ever-increasing use of mineral oils and diesel fuel.

Communal bathing at the end of the working shift provides an ever-present hazard in the form of tinea pedis and the varied flora found on microscopic investigation probably accounts for the comparative failure of anti-fungal treatment.

This is a condition for which carefully controlled clinical studies and adequate epidemiology are the most likely keys to success in prevention.

## SHOT-FIRING

By and large mechanized mining leads to a considerable reduction in the number of explosive shots fired. With all types of shots there will be emissions of the oxides of nitrogen and when there is incomplete detonation, due to failure to explode instantaneously, this emission may be considerable. There is then the danger of pulmonary oedema due to acute inhalation of nitrogen dioxide. Kennedy[4] has claimed that emphysema is a sequel of chronic exposure to a specific type of explosive, Cardox, now no longer used in the United Kingdom, but this has not been substantiated by others.

When any change in shot-firing procedure is contemplated adequate monitoring of the atmosphere before and after explosion is essential so that any change in nitrous oxide levels can be clearly demonstrated.

## HEAT AND COLD

It cannot be assumed that because the miner spends his working shift underground he is immune from the influences of climate. For economic and engineering reasons most countries will develop the shallow and easily accessible seams first, but as these become exhausted (as in most European countries) deeper seams have to be exploited and this inevitably means hotter and frequently more humid working conditions.

The literature on heat stress[5,6] is extensive and many scales or indices have been devised. But to the heat emitted by the strata at depths of 1000 m or more must now be added the heat produced by the complex machine systems of modern mining.

Recent detailed studies by Leamon[7] from the Institute of Occupational Medicine are designed to measure the effects of hot environments on work together with a subjective assessment of the workers and to look at appropriate work-wear. Some experimental refrigeration units are now installed in deep mines in Europe and these can reduce face temperatures by as much as 1 per cent. The process of cooling air involves a refrigerant gas, commonly one of the chloro-fluoro-alkanes, and in confined spaces the possibility of asphyxia, although remote, must be kept in mind.

From what has so far been said it might seem that cold is a relatively minor problem. This is broadly true at the working place, but the effects of rapid changes of temperature require suitable clothing to counter its effects. It is now the practice in the United Kingdom and many other countries to provide a complete set of working clothes which are regularly exchanged and laundered. These must be capable of being discarded at the working place when hot and, yet, must provide protection against the intake air when travelling back to pit bottom and emerging from the shaft in winter time.

## MEDICAL ASPECTS OF MINES RESCUE

Reference has already been made to the standards of physical fitness required of the rescue workers. It is a truism that the safer a mining industry the fewer disasters will occur. The corollary is that rescue workers may train assiduously and realistically for several years before being called to an incident, as is currently the case in the United Kingdom. As the physical and psychological reaction of inexperienced rescue men cannot be forecast, it is important that the doctor should be at the scene of an incident or disaster as quickly as possible to assess the physical and mental fitness of rescue teams. Each team consists of only four or five men and one weakling could easily spell disaster for the others.

By the nature of the industry it is difficult, if not impossible, to discover for some hours the full extent of a disaster, especially in the case of fire or explosion. Indeed the information passed out of the mine by the initial rescue team will be vital in determining the strategy. It is essential for the doctor, and there must be one on duty at all times until the emergency ends, to maintain close contact with the operations room. An underground fire may take days to contain and seal off during which period all available rescue teams will be utilized and most will make five, six or more visits to the affected area of the mine. Physical and psychological fatigue must be carefully monitored and records kept of all examinations conducted during the whole emergency. It is United Kingdom practice to examine each rescue worker before going underground and to see the team briefly as they return at the end of their shift.

However, the rescue team are not the doctor's sole consideration.

Senior management must be advised against excessive hours spent at the emergency, canteen facilities must be adequate and, ideally, there should be some quiet place where the fatigued can rest.

## BACK PAIN

Back problems present one of the most difficult areas in the practice of occupational medicine. The difficulty of making an accurate diagnosis needs no emphasis. Although mechanization has removed much of the severest physical toil, a good deal of strenuous manual handling of heavy and awkward loads is still required in day to day operations. For this reason back injuries form a high proportion of the occupational physician's consultations in mining medicine.

Recent work by Porter[8,9] has suggested that ultrasonic measurement of the thickness of the spinal canal in the lumbar region will enable prediction of the possibility of later back pain to be made with some confidence. He has suggested that the lowest 10 per cent have significantly more back pain, consult their doctors more often, lose more time from work and so on. However, the ethical and practical problems of rejecting an applicant for employment because he might have trouble in the future remains.

There are some advocates of routine radiology of the spine but this finds no favour in the United Kingdom because of the apparent lack of correlation between radiological findings and physical symptoms. For the same reason the use of a supportive belt cannot be justified. In any event, such apparel would be impractical in hot seams.

In the United Kingdom it is the practice to teach manual handling to all entrants, but it must be admitted that there are no controlled studies to demonstrate the effectiveness of this training. In future it is anticipated that the science of ergonomics will make a considerable contribution to spinal problems by advising at the design stage of machine systems and will by scientific study of working methods identify those areas and operations which are likely to produce or aggravate back problems.

When back pain does occur it is essential to ensure early treatment. Treatment will depend on the diagnosis, but where physiotherapy is indicated it will be advantageous to make it readily available as, for example, in the Ruhr mines of West Germany. The relatively restricted skill of the miner makes it socially and economically desirable that back problems are dealt with energetically so that he can return to his former environment rather than have to seek employment outside the industry.

Troup,[10] in detailed studies of the rehabilitation problems of miners, has developed a set of predictive principles which are of value in forecasting the outcome of a back injury. They are as follows:

Restriction of the pain-free range of straight leg raising (SLR) by
15° or more unilaterally or to 45° or less;
Reproduction of the pain caused by SLR by one of the qualifying
tests for root tension;[11]
Incapacity of trunk flexors as shown in the 'sit-up' tests;
Pain, or weakness, of resisted hip flexion, seated; and
Back pain induced on passive flexion of knee, prone (elicited
during the femoral stretch test).

He further lists four likely indications of recurrent back pain:

Absence from work for the current attack of 5 weeks or more;
A history of two or more previous attacks;
Residual pain in the lower back on return to work; and
Falls (of the patient on to buttocks or back) as a cause of back
pain.

In summary, no dramatic advances can be expected but when so
much of the miner's current disability is due to back pain, attention to
the approaches listed above should help to alleviate the problem.


## ERGONOMICS

The influence of ergonomics came comparatively late to mining and it
is only in the past 10 years that detailed studies have been completed in
Eastern Europe. These can be divided into two broad groups; first the
measurement of human response to environmental factors and to
stress, and secondly the study of mining machines and mining
systems. References have already been made to heat and noise. Currently, the problem of vibration is attracting considerable attention
and the potential hazards of whole body vibrations are of special
interest. The vascular effects of hand-operated drills are well known
but those of whole body vibration are less well understood. It has been
suggested that whole body vibration may well be a cause of back pain
by increasing spasm of the lumbar arterioles and consequent deficiency of blood supply and anoxia of the lumbar muscles. A rich field of
research is opening up in the study of vibration (*see* Chapter 9).

Mining machinery design must be a compromise between achieving
the objective for which the machine is designed while at the same time
making the working position of the operator tolerable without producing a machine totally impractical for underground use. Ideally
manufacturers, engineers and ergonomists should collaborate at the
design stage, and certainly not later than the prototype stage. A detailed
ergonomic study will demonstrate faults in control levers, confusing
operations (for example levers running to right and left on similar
machines to put the machine into a particular gear), faults in seating,
difficulties in egress in emergency and, most important, a blind area

diagram which demonstrates those areas all round the machine which the operator cannot see when seated at his controls. These are all ergonomic problems particularly relevant to mining and indicate the need for, and importance of, a team of ergonomists who can play as great a part as any in the achievement of safety.

## MINING HEALTH LEGISLATION

It is not possible to make direct comparisons of details of legislation or even of general principles across national boundaries even in the European Coal and Steel Community. In the United Kingdom the Mines and Quarries Act (1954) is incorporated under the wider umbrella of the Health and Safety at Work Act (1974). The industry has preferred the unambiguous wording of regulations to the more flexible but imprecise wording of codes of practice. The essence of regulations in the United Kingdom is to ensure the examination of juveniles on entering the industry, to lay down standards of first aid both below and above ground, to provide for elimination of pests and vermin, and to ensure the proper provision and maintenance of underground sanitation and the safety of 'wholesome drinking water' on the surface. In 1975 the Respirable Dust Regulations came into force, and for the first time specified the medical measures required for the control of pneumoconiosis.

It is important for all members of the occupational health team to be familiar with health legislation and to be able to comment on the revisions which occur from time to time.

## REHABILITATION

Because of the hazardous and physical nature of their work miners more than any other working group should have ready access to rehabilitation facilities. Rehabilitation should start as soon as possible after injury or the onset of illness. This implies that the facilities and personnel are available at each colliery but this is only practical at units employing two thousand men or more. In such collieries it is desirable to have a qualified physiotherapist who could provide treatment as soon as possible after injury. In many cases this will avoid absence from work altogether but when absence does occur it would permit, in co-operation with the primary care doctor, daily treatment which will accelerate the return to full fitness.

It cannot be too strongly stressed that the delays which occur while awaiting outpatient appointments, admission for treatment, attendance at outpatient physiotherapy departments all militate against the speedy return of the miner to his working environment. Not only do his muscles waste and his joints stiffen with lack of use but he may put on weight and his psychological condition regresses with enforced idleness of uncertain duration.

Many countries favour residential rehabilitation centres where miners can be subjected to an intensive programme which includes simulated mining tasks. Equally important is the fact that in such countries the surgeon in charge and the occupational physician can, together, review progress in individual cases and speed the return of the miner to his familiar working environment (*see also* Chapter 11).

## ALCOHOLISM

The problem of alcoholism among miners is essentially similar to that of other workers with two important exceptions. First, the dangers of working underground are greatly exacerbated by intoxication, however slight, and it is therefore prudent to ban alcohol underground (as is done in the United Kingdom but not in all countries) and to refuse work to any man suspected of excessive drinking prior to starting his shift. Secondly, mining communities still tend to be relatively isolated and socially deprived. It is desirable, therefore, to combine an active health education programme with facilities for recreation, other than drinking clubs.

## USE OF CHEMICALS UNDERGROUND

The days when the armamentarium of the miner consisted solely of a pick and shovel, supports and a cap lamp are gone. Today complex chemicals are used to repair conveyor belts, to plug cavities in the roof and to seal off areas of the mine where spontaneous combustion may occur. There are, in addition, the fire-resistant fluids (such as phosphate esters) which are an integral part of the hydraulic system and the numerous different types of oils in daily use. Hence it is more difficult to produce satisfactory systems of operation underground and to ensure continuous good working practice. On the other hand, most of the procedures listed above are only carried out intermittently and, therefore, individual exposure as measured over the week, month and year is unlikely to be great but peak exposures in excess of the TLV are not uncommon.

It is essential for the occupational physician to withhold his agreement to the introduction of potentially noxious chemicals underground until he is convinced that there is no viable alternative and that, on the balance of probabilities, the proposed procedure carries the least risk. He must then be present at underground trials of the new substances to familiarize himself with the proposed technique and to study the potential risks to the operatives. Next he must attend, with management, a consultative meeting with the workers to explain the health implications and stress the need for such precautions as he may advise. Lastly, he must decide whether routine medical surveillance is called for, what examination is necessary and its periodicity.

REFERENCES
1. Wellwood Ferguson WJ. Miners' nystagmus: A résumé of recent work concerning its causation and treatment. *Trans. Inst. Mining Engineers* 1951; **110**:Part 12.
2. Sharrard WJW and Liddell FDK. Injuries to the semilunar cartilages of the knee in miners. *Br. J. Ind. Med.* 1951; **19**:195–202.
3. Williamson DM and Vickers HR. *Medicine in the Mining Industries.* London: Heinemann, 1972.
4. Kennedy MCS. Nitrous fumes and coalminers with emphysema. *Ann. Occup. Hyg.* 1972; **15**:285–300.
5. Lind AR, Hellon RF, Jones RM et al. Reactions of mines-rescue personnel to work in hot environments. *Med. Res. Memo.* No. 1. London: National Coal Board, 1955.
6. Leithead CS and Lind AR. *Heat Stress and Heat Disorders.* London: Cassell, 1964.
7. Golding DG, Graves RJ, Leamon TB et al. Thermal conditions in mining. Institute of Occupational Medicine Report TM/80/9. 1981 (in the press).
8. Porter RW, Wicks M and Ottewell D. Measurement of the spinal canal by diagnostic ultrasound. *J. Bone Joint Surg.* 1978; **60B**:481–4.
9. Porter RW, Hibbert CS and Wicks M. The spinal canal in symptomatic lumbar disk lesions. *J. Bone Joint Surg.* 1978; **60B**:485–7.
10. Troup JDG. A prospective survey of back pain in the coal-mining industry: a summary of the findings. Royal Liverpool Hospital, 1981 (in the press).
11. Breig A and Troup JDG. Biomechanical considerations in the straight leg raising test: cadaveric and clinical studies of the effect of medial hip rotation. *Spine* 1979; **4**:269–72.

## *Part 2: RESPIRATORY DISEASE*
## *James Dick*

The historical aspects of respiratory problems in the mining industry have been fully documented[1] and although the relative importance of some conditions has altered, pneumoconiosis remains the most important occupational hazard in those countries in which coal extraction is a significant industry. The possible exception to this is South America where at least in one country parasitic infection related to the excessive water present is the principal hazard.[2] As coal mining is extensively carried out in every continent pneumoconiosis is a problem of international dimensions, varying to some extent with the degree of mechanization achieved.

In those countries in which coal extraction is largely hand won it can be anticipated that pneumoconiosis will increase in importance as more and more machine mining is introduced, because dust production usually varies directly with the degree of mechanization. Not all countries publish data on pneumoconiosis, and the basis for estimating prevalence in others is unsatisfactory, for example, because the sample is inadequate. It is, therefore, not possible to comment accurately on the world situation but in general, and particularly in industrialized nations, coalworkers' pneumoconiosis is declining. In the United Kingdom, for example, in the period 1959–79 the prevalence of coalworkers' pneumoconiosis fell from 13·4 to 4·2 per cent of miners employed.[3] Nevertheless, on average 400 individuals are certified for the first time each year, adding to the 35 000 already known to have coalworkers' pneumoconiosis, and each year a large number of men will die with coalworkers' pneumoconiosis as a primary cause of death.

The aetiology of coalworkers' pneumoconiosis is now well understood and is due to the inhalation of excessive quantities of respirable dust; that is, dust in the 1–5 $\mu$m range. It has been clearly shown that as exposure to respirable dust increases in terms of time and quantity, so prevalence of pneumoconiosis also increases.[4] In the British coal mining industry the probability of developing pneumoconiosis related to respirable dust exposure has been established. In 1968 following a massive film reading exercise using films of men whose dust exposure was known, the relationship between dust exposure and subsequent development of pneumoconiosis was determined[5] (*Fig. 2.1*).

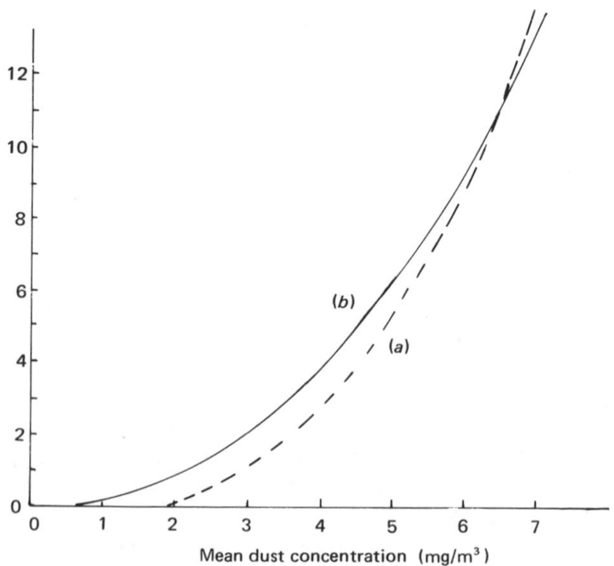


*Fig. 2.1.* Probability (percentage) of developing pneumoconiosis, category 2/1 or more. (*a*) 1968 study; (*b*) 1978 study.

From this information an interim standard of maximum allowable dust exposure was agreed. Subsequently in 1978/79 a further reading exercise with more accurate estimation of dust exposure confirmed the findings of 1968, although demonstrating that the probability of developing pneumoconiosis was slightly under-stated in the original study.[5]

Although it is now established that the mass of respirable dust to which an individual is exposed is a very important factor in the causation of pneumoconiosis, it is not the only one. It has been shown that the quality and carbon content of coal (rank) have an influence on pneumoconiosis.[6] Collieries mining anthracite and high grade steam and coking coal (ranks 100, 200 and 300) had prevalences of up to five times those of collieries mining low grade bituminous coal (rank 900). Furthermore, there are a few collieries which appear to contradict the concept of pneumoconiosis correlating with cumulative respirable dust exposure. These so-called 'anomalous' collieries are known to produce unexpected numbers of pneumoconiotics with low dust exposure on the one hand and few cases when exposure is high on the other. Clearly a factor yet to be found has an important influence on the development of coalworkers' pneumoconiosis and research into this continues.

The factor or factors responsible for the transition from simple to complicated pneumoconiosis remain uncertain. High levels of silica in respirable dust, immunological changes, super-added tuberculosis or other infection, rheumatoid factor and total dust content of the lung have all been investigated but despite intensive research no single cause has been positively identified, so it may be that the transition from simple to complicated pneumoconiosis is multifactorial in its aetiology. It has now been shown that the attack rate of complicated pneumoconiosis increases with increasing category of simple pneumoconiosis and with increasing progression of simple.[22] The attack rate of progressive massive fibrosis (PMF) studied in a sample of 105 490 miners with two serial X-rays taken for each man at the same colliery at intervals of 5 years showed that the probability of PMF increases with increasing degree of pneumoconiosis on the first film, and with increasing progression of simple pneumoconiosis.[7]

In a serological examination of anthracite miners with PMF in the United States it was found that anti-nuclear factor was present in 74 per cent of the men examined.[8] On the other hand, a more recent study of 2455 serum samples of miners, including 420 with PMF, produced little evidence that circulating auto-immune factors, anti-nuclear factors, rheumatoid factor or reticulin antibodies were related to the development of PMF.[9]

The explanation of sudden development of radiological change to dust exposure is even less well understood. Investigation of a recent case in which category 3 appeared 4 years after a succession of three normal films over 20 years in a miner who had worked at the same colliery and the same seam for more than 20 years gave no indication of the reason for the rapid change in the X-ray appearance. All causes other than pneumoconiosis were eliminated. Clinical examination was negative apart from the presence of scleroderma. (Attention has been drawn to an association between coalworkers' pneumoconiosis and scleroderma.[10])

## DIAGNOSIS OF COALWORKERS' PNEUMOCONIOSIS AND X-RAY CLASSIFICATION

The diagnosis of coalworkers' pneumoconiosis in life remains firmly based on radiological appearances together with an appropriate history. It is important in making a diagnosis that the chest X-ray must be of the highest possible quality and should be capable of storage for 20 years or more without deterioration in quality (archival permanence). As regards technique of radiography, there is some difference of opinion as to whether standard technique is more or less efficient than high kV with grid technique. The only published epidemiological investigation into the choice of technique did not show any advantage in relatively high kV with grid films as opposed to standard technique films.[11] Automatic exposure control and automatic film processing are important when large scale surveys with epidemiological implications are involved.

The maintenance of consistency of film reading between individuals and from survey to survey is also of fundamental importance, and doctors involved in epidemiological exercises should have their intra- and inter-observer variability assessed at regular intervals to ensure that these are within acceptable levels.

When the diagnosis is made, the extent of change due to pneumoconiosis is described in terms of the current international classification. Although there have been many classifications of the pneumoconioses, the most important was that following the international meeting in Geneva in 1958. This resulted in the International Labour Office (ILO) (Geneva, 1958) Classification which was accepted by most countries interested in the problem of dust disease.

Briefly, the classification divided pneumoconiosis into two basic groups, simple and complicated — the latter is also known as progressive massive fibrosis (PMF).

Simple pneumoconiosis was that form in which discrete opacities of a maximum size up to 10 mm were recognizable on the radiograph and complicated pneumoconiosis was characterized by larger opacities of 10 mm diameter or greater. This was a natural distinction, as simple pneumoconiosis differs in radiological appearance, pathology and natural history from complicated. The severity of simple pneumoconiosis was assessed by the extent and profusion of small opacities.

The earliest recognizable radiograph appearance was described as category 1, the next as category 2 and the most advanced as category 3. Simple pneumoconiosis was further subdivided by reference to the size of small opacities. Three such groups were recognized: p (pinhead) for opacities up to 1·5 mm in diameter; m (micronodular) for sizes 1·5 to 3 mm in diameter; n (nodular) for sizes 3–10 mm in diameter.* In this way a radiograph characteristic of simple coalworkers' pneumoconiosis might be described as, for example, category 2p or 3n.

Complicated pneumoconiosis was also described as having three stages or categories. Category A represented a large opacity or series of small opacities larger than 10 mm with a combined diameter of up to 5 cm; category B, cases where the opacity was greater than 5 cm in diameter and occupied up to the equivalent of the right upper zone; and category C, where very large masses were present occupying more than the equivalent of the right upper zone.

When complicated pneumoconiosis occurs it does so in the majority of cases on a background of simple pneumoconiosis; thus, the full classification of a radiograph may be 2nA or 1pC.

In recording the classification of a particular film provision was also made for suspect cases of pneumoconiosis and for indicating that some condition other than pneumoconiosis might be present.

In 1968 the classification was again reviewed, principally in order to include asbestosis and also to be able to record smaller degrees of change between one X-ray and a follow-up some years later. At about the same time the American College of Radiology together with the International Union Against Cancer (UICC) was largely concerned with the same objectives. As a result of these discussions a new classification was devised and formally agreed in Bucharest in 1971. The principal changes made were *first* to make it possible to classify asbestosis (by adding irregular opacities — s, t and u), and *secondly* to incorporate the NCB elaboration of the simple categories of pneumoconiosis in order to assess finer degrees of change in serial X-rays of an individual.[12] As chest X-rays may be assessed for clinical purposes only, a short version of the classification was provided; but when used for epidemiological purposes the extended form of the classification is required. This classification was known as the ILO/UC classification 1971 and continued in use until a still further review took place in 1978/79, again under the aegis of the ILO. This is known as the 1980 ILO Classification and is given as Appendix 1 (pp. 57–61).

---

*For phonetic reasons subsequently altered to p, q and r (ILO U/C Classification), *see* Appendix 1.

There is little doubt that the new classification represents a further refinement. For example, the definition of categories is much more precise and thus less liable to misinterpretation, and much greater emphasis is placed on standard films. For the first time international reading trials have been used to select, on a statistical basis, films which by general agreement represent specific categories. Readers from many countries took part in this exercise and the result was selection of radiographs which are likely to lead to much greater uniformity in reading standards between readers and countries.

## NATURAL HISTORY OF COALWORKERS' PNEUMOCONIOSIS

Coalworkers' pneumoconiosis is a slowly developing condition. Radiographic evidence of dust retention is unusual before 15–20 years' exposure. It is thus a condition of the older age group. In 1978 the average age of men certified as having coalworkers' pneumoconiosis in the United Kingdom was 61.[13]

Once developed, it is usual for gradual progression to take place if exposure to respirable dust continues, but progression is not inevitable even with further exposure. A category 1 X-ray may remain unaltered for very many years. Similarly, PMF or complicated pneumoconiosis, although likely to progress, does not invariably do so. For this reason 'complicated pneumoconiosis' is a more accurate term than 'progressive massive fibrosis'.

When coalworkers' pneumoconiosis does progress and reaches the stage of advanced PMF, destruction of lung tissue together with related emphysema may produce anoxia with subsequent right heart hypertrophy and congestive cardiac failure. This is the sequence of events when pneumoconiosis is a primary cause of death.

The significance of the size of small opacities is not clearly established, and for many years workers in the United Kingdom did not record the size of small opacities in epidemiological studies, whereas in Europe the size of a small opacity was thought to be at least as important as category. It has been shown, however, that the smallest opacity p differs from m(q) and n(r),[14] the transfer factor being reduced in the presence of p type opacities and not in the others. It is generally accepted that the largest of the small opacities, n(r), is more likely to be associated with development of complicated pneumoconiosis and with Caplan's syndrome (*see below*). For these reasons, and to conform with European practice, size of small opacities is now recorded in epidemiological surveys in the United Kingdom.

Simple pneumoconiosis does not reduce life expectancy but it has been shown that there is a reduction for men with category B and C complicated pneumoconiosis.[15] A recent study of mortality in the mining industry suggests that this is equally true of category A.

A radiological feature of PMF not generally reported is the tendency for the massive lesion to migrate towards the hilum. This is particularly so when the initial mass is in the right upper zone.

## Pathology

The basic lesion in the lungs of miners with simple coalworkers' pneumoconiosis is the dust macule — an aggregation of dust particles and macrophages surrounded by a greater or lesser degree of emphysema and fibrosis. In a recent post-mortem study of lungs from 500 British coal miners whose dust exposure was known it was found that soft macules, assessed by palpation, were less obvious radiologically than firmer, more fibrotic, macules.[16] In the majority of cases of soft macules, associated chest X-rays were categorized as 0/0, implying absence of radio-opacity in these cases.

Progressive massive fibrosis presents as large black masses macroscopically which on palpation are rubbery in consistency. Histologically the lesion consists of dust, fibrotic tissue and a variable proportion of collagen.

## Symptoms

There are no symptoms specifically related to the simple form of coalworkers' pneumoconiosis. It is not infrequent, however, for bronchitis to be super-added and in that event symptoms present are those related to this condition. In complicated pneumoconiosis in its early stages there are again no specific symptoms, but in the advanced forms (B and C) destruction of lung tissue and accompanying emphysema can cause symptoms of breathlessness. Nevertheless, very advanced radiological appearances with large masses of complicated pneumoconiosis and considerable emphysema are often associated with remarkable absence of symptoms.

## Signs

Physical signs are absent both in simple and the majority of cases of complicated pneumoconiosis. Occasionally when a large mass is close to the periphery dullness on percussion and diminution of breath sounds over the affected area may be detected, and the signs of emphysema may be present in advanced complicated pneumoconiosis.

## Quartz

The role of respirable quartz has long been a contentious topic in the pathogenesis of coalworkers' pneumoconiosis. This is well illustrated by the fact that in many countries (particularly European) lung changes resulting from exposure to coal mine dust are described as 'silicosis'. It seems likely, however, that respirable quartz must reach a level of 10 per cent or more of total respirable mine dust before it has any influence on the development of simple pneumoconiosis and even at this level exposure must be over a considerable period. On the other hand, egg-shell calcification is not an uncommon finding in British miners. In the United States of America this finding is accepted as evidence of exposure to respirable silica.

The precise role of quartz in the causation of complicated pneumoconiosis is by no means established, but in a series of studies on men who had shown unusually rapid progression of simple pneumoconiosis it was found that the individuals involved had been exposed to high levels of respirable quartz. In this series there was a relatively high incidence of progressive massive fibrosis, implying some support for the theory that quartz may play a part in the transition from simple to complicated disease.[32]

## Caplan's syndrome

The radiological appearances of dust retention modified by a rheumatoid process, either overt or subclinical, and known as Caplan's syndrome, were first described in 1953.[17] The syndrome consists of a typical radiological appearance characterized by rounded opacities of varying size scattered over the lung fields with relatively little evidence of simple pneumoconiosis. In addition, the affected individual has either overt rheumatoid changes or a positive rheumatoid arthritis screening test (RAST) and in the course of time would develop clinical evidence of rheumatoid arthritis. Over a period of time the opacities might enlarge or decrease, might cavitate or, rarely, calcify. With the passage of time, however, the classic syndrome has become less precise and there is now a tendency to characterize any radiological evidence of dust retention associated with a positive RAST as Caplan's syndrome.

## Research

There are important problems still to be answered in coalworkers' pneumoconiosis. It is not known what causes the transition from simple to complicated. The explanation for an individual suddenly developing pneumoconiosis after years of exposure is unclear and the basis of individual susceptibility is also not understood. Of these possible factors individual susceptibility is probably the most important when considering the long term elimination of pneumoconiosis in coal mining. Research directed towards the detection of dust-sensitive individuals is being pursued in several countries. Probably the most promising prediction of individual susceptibility involves the removal of macrophages from terminal bronchioles and alveoli by lavage through a fibreoptic bronchoscope. The reaction of the macrophages when exposed to the relevant dust is then observed. This method, currently being developed in France, is not yet applicable in the field.

Until a relatively simple test of susceptibility can be developed, removal to dust-free or very low dust conditions at the earliest detectable radiological change remains the most likely method of preventing susceptible individuals from progressing to the higher categories and to complicated pneumoconiosis. This requires a very high standard of radiography as well as highly competent film readers with demonstrably consistent levels of both inter- and intra-observer reading.

The problems of reading consistency and reproducibility are the subjects of research into computerized reading of chest X-rays for epidemiological purposes. If such a system could be perfected then these problems would be solved, and valuable medical resources released for other activities.[18] There is a good deal of work still to be done in this respect.

Possibly the most important question now to be determined is whether or not linear opacities may be a direct result of exposure to respirable mine dust.

All published data on prevalence and progression of coalworkers' pneumoconiosis are based on chest X-rays showing the classic rounded opacities universally recognized as typical of coalworkers' pneumoconiosis. Linear opacities have, of course, been noted in routine colliery surveys and have been accepted as evidence of ageing, respiratory infection, fibrosing alveolitis or smoking. However, it is the opinion of experienced film readers in at least one country that a large proportion of films are showing fine linear opacities and it may be that changing methods of production have produced dust of a different composition thus producing differing radiological appearances. If this were so, the real prevalence of respirable dust lung changes would be greater than at present believed. An epidemiological investigation to elucidate this problem is required.


**Prevention**

The prevention of pneumoconiosis lies in the hands of mining engineers. Their efforts in the past two decades have contributed very largely to the significant decrease in the problem. However, it is highly unlikely that, given present methods of coal extraction, respirable dust levels can be reduced to a point at which no new cases of pneumoconiosis would be produced, unless some method of detecting the highly susceptible individual can be found. In the absence of such a test the very early detection of dust retention by radiology followed by removal to dust-free conditions is the only method likely to eradicate coalworkers' pneumoconiosis. This implies the use of film readers highly skilled in the technique of diagnosing dust retention at the earliest levels, i.e. 0/1, 1/0 stages of the classification.

Until recently it has been assumed that removal from further significant dust exposure of an individual with simple pneumoconiosis, certainly up to the level of high category 2, would ensure no further progression. However, follow-up studies of men who had left the industry for several years indicate that progression may well take place in simple pneumoconiosis.[19] The impression of experienced readers studying the films of re-entrant miners tends to confirm this view, and this reinforces the need to detect the very earliest evidence of dust retention.

For this reason some method of assessing the effectiveness of dust suppression in reducing pneumoconiosis by monitoring chest radiographs is essential.

While this is possible on a national basis by measuring changes in prevalence, such data are not always of value in the individual colliery where altering populations could reflect dust conditions not at the colliery under consideration but at another from which men have been transferred. To meet this problem in the United Kingdom a method of using serial chest X-rays has been devised to produce a progression index. Serial films of face workers who have been continuously employed at a particular colliery taken at 4-yearly intervals are read side-by-side using the extended ILO classification and the number of steps of progression per 100 face men employed gives the progression index. This has been found to be a useful tool in ranking collieries for their efficiency in dust suppression. The method has been described in detail elsewhere.[20]

The method used is open to criticism because reading films side-by-side must introduce bias, either because the chronological order of films is known (readers are reluctant to read regression) or for reasons of quality. However, extensive film reading trials in the United Kingdom have shown that there is little difference in the amount of progression recorded when side-by-side readings are compared with independent randomized readings on the same film.[5]


## BRONCHITIS (OBSTRUCTIVE AIRWAYS DISEASE)

Since the inception of acceptable epidemiological investigations of chronic bronchitis the agreed diagnosis has been based on a standard questionnaire, usually the Medical Research Council questionnaire. A history of cough and sputum occurring virtually daily is essential to this diagnosis.

The fact that cough and sputum can be present without measurable loss of lung function and that loss of lung function can be demonstrated in the absence of significant sputum has lead to some questioning of the method of diagnosis. It is now probably desirable that an accepted decline in some measurement of lung function, for example $FEV_1$, should be added to the existing definition.[21] That chronic bronchitis as defined by the MRC questionnaire has a high prevalence in the mining as compared to the average population is not seriously questioned, but the reason remains controversial. It has been shown that symptoms of cough and sputum increase with increasing exposure to respirable dust and there is also evidence that lung function as measured by $FEV_1$ declines as dust exposure increases.[23] However, other workers have not confirmed this finding.

All reported investigations of the effect of the relationship between dust and consequential symptoms of bronchitis data have been related to respirable dust (1–5 $\mu$m) — that is to particles liable to reach the smallest bronchioles and associated alveoli. But bronchitis is primarily a disease of the larger bronchi (air passages) and the question arises as to whether dust particles larger than 5 $\mu$m could be more important in inducing bronchitic symptoms. However, from a study of the importance of total inhalable dust in causing upper respiratory disease there appears to be no evidence that estimates of total dust give a better correlation with cough and sputum than respirable dust.[24]

It is universally accepted that the effect of cigarette smoking overwhelms that of respirable dust and in that respect stopping smoking is vastly more important than reduction of dust.

So far as pneumoconiosis is concerned it has been shown that chronic bronchitis does not give protection against development of dust retention.[25]


## EMPHYSEMA

In considering the problem of emphysema one of the major difficulties lies in accurate diagnosis of this condition during life. The disease process has to be far advanced before incontrovertible radiological changes can be detected. It has been shown that of patients with fairly severe emphysema only 41 per cent were diagnosed radiologically, and only 66 per cent of those with the most severe grade of emphysema were diagnosed radiologically.[26] There are no lung function tests specific to emphysema.

As in chronic bronchitis, the precise relationship between dust exposure and emphysema is not established. A post-mortem study showed that cigarette smoking was an important feature in the causation of overall emphysema and levels of emphysema were related to lung dust content.[27]

It has been suggested that nitrous fumes in the underground atmosphere could be a cause of chest conditions[28]. Two forms are described, acute and chronic. The acute type results from high concentrations of nitrous oxide leading to a condition resembling acute pulmonary oedema, and the chronic type from continuous exposure to low concentrations causing emphysema. The latter finding has not been confirmed by investigations carried out by the National Coal Board in the United Kingdom.

It has been shown that in certain forms of severe emphysema alpha-1-antitrypsin is reduced. There is no evidence of reduced alpha-1-antitrypsin level in miners with reduced $FEV_1$.[9]

**Bronchitis and emphysema**

A recent study of coal miners' mortality showed an increase in death due to bronchitis and emphysema with increasing exposure to respirable coal mine dust.[33]

## PULMONARY TUBERCULOSIS

Pulmonary tuberculosis (PTB) is not a problem in the United Kingdom mining industry. A study of 57 000 miners who had a chest X-ray in 1975 gave an incidence of PTB at 0·02 per cent, which is very close to the average for the male population as a whole. However, in other countries, such as France and Germany, where there is a very considerable migrant population working in the coal industry, tuberculosis is indeed a problem of considerable magnitude. In France strenuous efforts have been made to eradicate tuberculosis from the mining industry by BCG campaigns among miners and their children of school age. In addition, every effort is made to detect and to treat PTB found in routine examination of miners from foreign countries supplying migrant labour.

## CARCINOMA OF THE BRONCHUS ('LUNG CANCER')

It is generally accepted that carcinoma of the bronchus is one serious illness in which the mining industry fares rather better than the average adult male population.[29] There is a clinical impression that the prognosis in established cases of carcinoma of the bronchus in miners is rather better than for non-miners. If this is correct, the explanation is not clear. A study of German miners showed no correlation of carcinoma of the bronchus with pneumoconiosis.[30] A more recent study of coal miners' mortality provided no obvious evidence that coal dust exposure or the presence of simple pneumoconiosis affected lung cancer risks.

On the other hand, early indications from yet another incomplete mortality study indicate an increase in mortality from bronchial carcinoma in miners with complicated pneumoconiosis. Accurate differential diagnosis of carcinoma of the bronchus from PMF is of considerable importance. Surgical resection of a mass of complicated pneumoconiosis thought to be carcinoma frequently transforms a mobile patient into a respiratory cripple.

## CORONARY HEART DISEASE

As in the general population there has been an increase in the incidence of coronary heart disease in the mining population. There is, however, no relationship between this and dust exposure as measured by pneumoconiosis.[31]


REFERENCES
1. Meikeljohn A. History of lung disease of coal miners in Great Britain. *Br. J. Ind. Med.* 1951; **8**:127; 1952; **9**:93; 1952; **9**:208.
2. ILO. Programme of Industrial Activities. Coal Mines Committee. 10th Session. Geneva, 1976. Health and Safety in Coal Mines. 110.
3. NCB Medical Service Annual Reports 1959 and 1979.
4. Jacobsen M. Paper to seminar on epidemiology and technical and medical prevention of coal miners' pneumoconiosis. Commission of the European Communities, 1979.
5. Hurley JF, Copeland E, Dodgson J et al. Simple pneumoconiosis and exposure to respirable dust. Institute of Occupational Medicine TM/79/13. UDC 616.24-003.6.
6. Bennett JG, Dick JA, Kaplan YS et al. The relationship between coal rank and the prevalence of pneumoconiosis. *Br. J. Ind. Med.* 1979; **36**:206–10.
7. McLintock JS, Rae S and Jacobsen M. The attack rate of progressive massive fibrosis in British coal miners. In: Walton WH (ed.) *Inhaled Particles III*. Woking, Surrey: Unwin, 1971: 933–52.
8. Lippman M, Eckert HL, Hahon W et al. The presence of circulating antinuclear and rheumatoid factors in United States coal miners. *Ann. Intern. Med.* 1973; **79**:807.
9. Boyd JE, Robertson MD and Davis JMG. The examination of serum samples from coalminers for the presence of autoantibodies that might be involved in the development of progressive massive fibrosis. Institute of Occupational Medicine TM/79/12. CEC Contract 6244-00/8/104. UDC 616.24-003. 6.616-097.
10. Rodnan GP, Benedek TB, Medsger TA et al. The association of progressive systemic sclerosis (scleroderma) with coal miners' pneumoconiosis and other forms of silicosis. *Ann. Intern. Med.* 1967; **66**:323–34.
11. Washington JS, Dick JA, Jacobsen M et al. A comparison of conventional and grid techniques for chest radiography in field surveys. *Br. J. Ind. Med.* 1973; **30**:365.
12. Liddell FDK and May JD. *Assessing the Radiological Progression of Simple Pneumoconiosis.* London: NCB Medical Service, 1966.
13. NCB Medical Service Annual Report 1977/78: 22.
14. Cotes JE and Field GB. Transfer factor reduced in 'p' type coal workers' pneumoconiosis. *Br. J. Ind. Med.* 1972; **29**:268–73.
15. Cochrane AL. Relationship between radiographic categories of pneumoconiosis and expectation of life. *Br. Med. J.* 1973; **2**:532.

16. Davis JMG, Chapman J, Collings P et al. Autopsy studies of coalminers' lungs. Institute of Occupational Medicine TM/79/9. CEC Contract 6244-00/9/103.

17. Caplan A. Certain unusual appearances in the chest film of miners suffering from pneumoconiosis. *Thorax* 1953; **8**:29.

18. Jagoe JR and Paton KA. Reading chest radiographs for pneumoconiosis by computer. *Br. J. Ind. Med.* 1975; **32**:267.

19. Melville AWT, Paris I, Hurley JF et al. Pneumoconiosis, lung function and exposure to airborne dust: epidemiological research to compare responses of working coalminers with responses of ex-miners. Institute of Occupational Medicine TM/79/11. CEC Contract 6244-00/8/106. UDC 616.24-003.6.

20. Liddell FDK and Lindars DC. An elaboration of the ILO classification of simple pneumoconiosis. *Br. J. Ind. Med.* 1969; **26**:89.

21. Fletcher C, Peto R, Tinker C et al. *The Natural History of Chronic Bronchitis and Emphysema.* Oxford: Oxford University Press, 1976.

22. Jacobsen M, Rae S, Walton WH et al. The relationship between pneumoconiosis and dust exposure in British coal mines. In: Walton WH (ed.) *Inhaled Particles III.* Woking, Surrey: Unwin, 1971: 903–19.

23. Rogan JM, Attfield MD, Jacobsen M et al. Role of dust in the working environment in the development of chronic bronchitis in British coal miners. *Br. J. Ind. Med.* 1973; **30**:217–26.

24. Cowie A, Miller B and Dodgson J. A study of the importance of total inhalable dust in causing upper respiratory disease. Institute of Occupational Medicine. CEC Contract 7246-16/8/003. 1980.

25. Muir DCF, Burns J, Jacobsen M et al. Pneumoconiosis and chronic bronchitis. *Br. Med. J.* 1977; **2**:424–7.

26. Thurlbeck WM and Simon G. Radiographic appearance of the chest in emphysema. *AJR* 1978; **130**:429–40.

27. Davis JMG, Chapman J, Collings P et al. Autopsy studies of coalminers' lungs. Institute of Occupational Medicine TM/79/9. CEC Contract 6244-00/8/103. UDC 616.24-003.6.

28. Kennedy MCS. Nitrous fumes and coalminers with emphysema. *Ann. Occup. Hyg.* 1972; **15**:285–300.

29. Goldman KP. Mortality of coal miners from carcinoma of the lung. *Br. J. Ind. Med.* 1965; **22**:72–7.

30. Schimanski P and Rosmarith J. Contribution to the study of the correlation between bronchial carcinoma and anthracosilicosis. *Beitr. Silikose-Forsch.* 1974; **26**:59–110.

31. Lindars DC, Rooke GB, Dempsey AN et al. Pneumoconiosis and death from coronary heart disease. *J. Pathol.* 1972; **108**:249–59.

32. Seaton A, Dick JA, Dodgson J et al. Quartz and pneumoconiosis in coalminers. *Lancet* 1981; 1272–5.

33. Miller BG, Jacobsen M. Coalminers' mortality in relation to radiological category, lung function and exposure to airborne dust. CEC Contract 7246-16/9/001. Institute of Occupational Medicine.

# APPENDIX 1

ILO 1980 International Classification of Radiographs of the Pneumoconioses:
Summary of Details of the Classification (reproduced by permission of ILO)

| Features | | Codes | Definitions |
|---|---|---|---|
| Technical quality | | 1 | Good |
| | | 2 | Acceptable, with no technical defect likely to impair classification of the radiograph for pneumoconiosis |
| | | 3 | Poor, with some technical defect but still acceptable for classification purposes |
| | | 4 | Unacceptable |
| Parenchymal abnormalities | | | |
| *Small opacities* | Profusion | | The category of profusion is based on assessment of the concentration of opacities by comparison with the *standard radiographs* |
| | | 0/- 0/0 0/1<br>1/0 1/1 1/2<br>2/1 2/2 2/3<br>3/2 3/3 3/+ | Category 0 — small opacities absent or less profuse than the lower limit of category 1<br>Categories 1, 2 and 3 — increasing profusion of small opacities as defined by the corresponding standard radiographs |
| | Extent | RU RM RL<br>LU LM LL | The zones in which the opacities are seen are recorded. The right (R) and left (L) thorax are both divided into three zones — upper (U), middle (M) and lower (L)<br>The category of profusion is determined by considering the profusion as a whole over the affected zones of the lung and by comparing this with the standard radiographs |
| | Shape and size | | |
| | Rounded | p/p q/q r/r | The letters p, q and r denote the presence of small rounded opacities. Three sizes are defined by the appearances on *standard radiographs*<br>p = diameter up to about 1·5 mm<br>q = diameter exceeding about 1·5 mm and up to about 3 mm<br>r = diameter exceeding about 3 mm and up to about 10 mm |

| Features | | Codes | Definitions |
| --- | --- | --- | --- |
| *Small opacities (cont.)* | Irregular | s/s t/t u/u | The letters s, t and u denote the presence of small irregular opacities. Three sizes are defined by the appearances on standard radiographs<br>s = width up to about 1·5 mm<br>t = width exceeding about 1·5 mm and up to about 3 mm<br>u = width exceeding about 3 mm and up to about 10 mm |
| | Mixed | p/s p/t p/u p/q p/r<br>q/s q/t q/u q/p q/r<br>r/s r/t r/u r/p r/q<br>s/p s/q s/r s/t s/u<br>t/p t/q t/r t/s t/u<br>u/p u/q u/r u/s u/t | For mixed shapes (or sizes) of small opacities the predominant shape and size is recorded first. The presence of a significant number of another shape and size is recorded after the oblique stroke |
| *Large opacities* | | A B C | The categories are defined in terms of the *dimensions* of the opacities<br>Category A — an opacity having a greatest diameter exceeding about 10 mm and up to and including 50 mm, or several opacities each greater than about 10 mm, the sum of whose greatest diameters does not exceed about 50 mm<br>Category B — one or more opacities larger or more numerous than those in category A whose combined area does not exceed the equivalent of the right upper zone<br>Category C — one or more opacities whose combined area exceeds the equivalent of the right upper zone |
| Pleural abnormalities<br>*Pleural thickening*<br>Chest wall | Type | | Two types of pleural thickening of the chest wall are recognized: circumscribed (plaques) and diffuse. Both types may occur together |
| | Site | R L | Pleural thickening of the chest wall is recorded separately for the right (R) and left (L) thorax |

| Features | | Codes | Definitions |
|---|---|---|---|
| | Width | a b c | For pleural thickening seen along the lateral chest wall the measurement of *maximum width* is made from the inner line of the chest wall to the inner margin of the shadow seen most sharply at the parenchymal–pleural boundary. The maximum width usually occurs at the inner margin of the rib shadow at its outermost point<br>a = maximum width up to about 5 mm<br>b = maximum width over about 5 mm and up to about 10 mm<br>c = maximum width over about 10 mm |
| | Face-on | Y N | The presence of pleural thickening seen face-on is recorded even if it can be seen also in profile. If pleural thickening is seen face-on only, width cannot usually be measured |
| | Extent | 1 2 3 | Extent of pleural thickening is defined in terms of the *maximum* length of pleural involvement, or as the sum of maximum lengths, whether seen in profile or face-on<br>1 = total length equivalent up to one-quarter of the projection of the lateral chest wall<br>2 = total length exceeding one-quarter but not one-half of the projection of the lateral chest wall<br>3 = total length exceeding one-half of the projection of the lateral chest wall |
| Diaphragm | Presence | Y N | A plaque involving the diaphragmatic pleura is recorded as present (Y) or absent (N), separately for the right (R) and left (L) thorax |
| | Site | R L | |
| Costophrenic angle obliteration | Presence | Y N | The presence (Y) or absence (N) of costophrenic angle obliteration is recorded separately from thickening over other areas, for the right (R) and left (L) thorax. The lower limit for this obliteration is defined by a *standard radiograph* |
| | Site | R L | If the thickening extends up the chest wall then both costophrenic angle obliteration and pleural thickening should be recorded |

| Features | Codes | Definitions |
| --- | --- | --- |
| *Pleural calcification* | | The site and extent of pleural calcification are recorded separately for the two lungs, and the extent defined in terms of *dimensions* |
| Site | | |
| Chest wall | R L | |
| Diaphragm | R L | |
| Other | R L | 'Other' includes calcification of the mediastinal and pericardial pleura |
| Extent | 1 2 3 | 1 = an area of calcified pleura with greatest diameter up to about 20 mm, or a number of such areas the sum of whose greatest diameters does not exceed about 20 mm<br>2 = an area of calcified pleura with greatest diameter exceeding about 20 mm and up to about 100 mm, or a number of such areas the sum of whose greatest diameters exceeds about 20 mm but does not exceed about 100 mm<br>3 = an area of calcified pleura with greatest diameter exceeding about 100 mm, or a number of such areas whose sum of greatest diameters exceeds about 100 mm |
| Symbols | | It is to be taken that the definition of each of the symbols is preceded by an appropriate word or phrase such as 'suspect', 'changes suggestive of', or 'opacities suggestive of', etc. |
| | ax | coalescence of small pneumoconiotic opacities |
| | bu | bulla(e) |
| | ca | cancer of lung or pleura |
| | cn | calcification in small pneumoconiotic opacities |
| | co | abnormality of cardiac size or shape |
| | cp | cor pulmonale |
| | cv | cavity |
| | di | marked distortion of the intrathoracic organs |
| | ef | effusion |
| | em | definite emphysema |
| | es | eggshell calcification of hilar or mediastinal lymph nodes |

| Features | | Codes | Definitions |
|---|---|---|---|
| | | fr | fractured rib(s) |
| | | hi | enlargement of hilar or mediastinal lymph nodes |
| | | ho | honeycomb lung |
| | | id | ill-defined diaphragm |
| | | ih | ill-defined heart outline |
| | | kl | septal (Kerley) lines |
| | | od | other significant abnormality |
| | | pi | pleural thickening in the interlobar fissure or mediastinum |
| | | px | pneumothorax |
| | | rp | rheumatoid pneumoconiosis |
| | | tb | tuberculosis |
| Comments | Presence | Y N | Comments should be recorded pertaining to the classification of the radiograph, particularly if some other cause is thought to be responsible for a shadow which could be thought by others to have been due to pneumoconiosis; also to identify radiographs for which the technical quality may have affected the reading materially |

# 3. OCCUPATIONAL HEALTH IN THE AGRICULTURAL INDUSTRY
*John E. Burgess*

The time when pre-historic man took his first steps towards becoming truly civilized was when he became a grower of crops and a tender of animals. These basic components of agricultural practice have changed little over the centuries but their development has proceeded at variable rates in different parts of the world.

The health of those engaged in the agricultural industry has been studied only recently, for it had long been considered that the active outdoor life of the farm worker was a health-promoting one. There was some justification for this attitude, for during the years 1949–53 the standardized mortality ratio for agricultural workers was only 174.[1] Indeed, the pursuit of certain occupations in the industry led at the turn of the century to some workers being provided with their own vaccination programmes; a fact quickly seized upon by Jenner whose original work has contributed to the eradication of smallpox.

The agricultural industry has undergone dramatic changes since those days. There has been increasing mechanization and widespread use of numerous chemical agents. Alongside these changes there has been a drastic reduction in total numbers employed in the industry. In 1841 the total population engaged in the agricultural industries in Great Britain was over 1·5 millions. This figure had been reduced to 1·2 millions by 1951 and 10 years later had been reduced to under a million.[2,3] This decline in numbers has continued over the past few years and the number employed in agriculture in 1979 was a mere 557 000.

It is often assumed that pesticides are the only chemicals which pose a risk to the health of the agricultural worker. However, modern farming practices often require the use of a great diversity of materials to which the worker may be exposed, such as solvents, oils, paints, detergents and creosote. The well-known hazards of these need no elaboration.

Over the past decade there has also been a tendency for greater specialization in both agricultural and horticultural practices. The days of really 'mixed farming' seem to be numbered and we now hear of intensive pig and poultry units, exclusively arable farms and an extension of greenhouse acreage to cope with the public demand for a wider choice of products throughout the year. In some ways this progress has meant that one has been able to define more accurately the likely occupational health problems in a particular group of employees. However, one must be aware that exposure potentials may be relatively greater with more intensive specialized work. This is evident if one considers that the agricultural worker of 20 years ago may have been exposed to pesticides for perhaps a few days of his working year, but regular spray operators of today have a potential for continuous exposure. The control of such hazards should be accomplished more easily in such a specialized group by utilization of education and training programmes and by the experience gained from regular work in controlled conditions. This evolution of specialized agriculture has meant that employees in the industry may be variably exposed to more easily defined groups of hazards. These groups may be conveniently classified as hazards from:

mobile equipment and work practices
physical agents
chemical agents
infectious diseases.


## HAZARDS FROM MOBILE EQUIPMENT AND WORK PRACTICES

Accidents on farms leading either to injury or to death have for long been the major concern of various government departments. Reported non-fatal and fatal injuries and diseases from 1970 to 1979 are shown in *Tables 3.1* and *3.2*.

There has been a 40 per cent reduction in total non-fatal accidents from 1971 to 1979 with a corresponding reduction in the population at risk of 11 per cent. Falls represent 23 per cent of total non-fatal injuries, whilst injuries from animals account for 13 per cent. It is perhaps more important to note that poisonings represented 3·5 per cent of the total in 1971 but 4·2 per cent in 1979. Whilst the numbers reported are small, one wonders to what extent unrecognized chronic effects of chemicals may be present in agricultural workers.

The 23 per cent reduction in total fatal injuries from 1971 to 1979 has not been quite as dramatic as the reduction in the non-fatal group but, nevertheless, does indicate overall improvement. Twenty-eight per cent of fatal injuries are still caused by self-propelled machinery. It is very apparent that the farm is a hazardous area for children, for of the total fatal injuries in 1973, 23 per cent were in children under the age of 16, whilst in 1979, 25 per cent of children are included in the total. It seems that the massive publicity campaign to farmers and agricultural establishments generally to keep children away from

**Table 3.1.** Non-fatal injuries and diseases in agriculture in the UK 1970–9

| Type of injury/disease | 1970 | 1971 | 1972 | 1973 | 1974 | 1975 | 1976 | 1977 | 1978 | 1979 |
|---|---|---|---|---|---|---|---|---|---|---|
| *Injuries caused by:* | | | | | | | | | | |
| Self-propelled machines | 532 | 559 | 508 | 437 | 427 | 400 | 437 | 401 | 410 | 319 |
| Other field machines | 642 | 652 | 613 | 574 | 618 | 509 | 543 | 487 | 470 | 431 |
| Stationary machinery | 324 | 326 | 289 | 247 | 267 | 234 | 203 | 225 | 217 | 182 |
| Electrical | 7 | 10 | 20 | 9 | 10 | 11 | 18 | 29 | 20 | 11 |
| Falls | 1601 | 1485 | 1521 | 1424 | 1300 | 1186 | 1146 | 1022 | 990 | 916 |
| Drowning | – | – | – | – | – | – | – | – | – | – |
| Animals | 833 | 745 | 805 | 785 | 713 | 701 | 658 | 614 | 593 | 518 |
| Poisoning | 31 | 24 | 29 | 46 | 33 | 23 | 11 | 34 | 34 | 17 |
| Other injuries | 3355 | 2878 | 2922 | 2604 | 2341 | 2132 | 2190 | 1974 | 1795 | 1630 |
| Total injuries | 7325 | 6679 | 6707 | 6126 | 5709 | 5196 | 5206 | 4786 | 4529 | 4024 |
| *Diseases:* | | | | | | | | | | |
| Ringworm | 15 | 10 | 8 | 7 | 3 | 3 | 2 | 2 | 1 | 2 |
| Weil's disease | 7 | 6 | 7 | 3 | 9 | 3 | 4 | 2 | 3 | – |
| Tetanus | 1 | 1 | 1 | 1 | – | – | 1 | – | – | – |
| Non-infective dermatitis | 48 | 44 | 29 | 27 | 20 | 26 | 20 | 8 | 18 | 18 |
| Tenosynovitis | 101 | 74 | 92 | 60 | 54 | 38 | 34 | 36 | 53 | 36 |
| Farmers' lung | 18 | 14 | 10 | 4 | 10 | 7 | 5 | 1 | 2 | 5 |
| Total diseases | 190 | 149 | 147 | 102 | 96 | 77 | 66 | 49 | 77 | 61 |
| Grand total | 7515 | 6828 | 6854 | 6288 | 5805 | 5273 | 5272 | 4835 | 4606 | 4085 |

*These figures are based on information supplied by DHSS about industrial injuries and *prescribed* diseases accepted by that Department under the National Insurance (Industrial Injuries) Act, now superseded by the Social Security Act 1975. They relate therefore only to workers, and include only those injuries resulting in 3 or more days' absence from work.

No statistics are available for children, those over pensionable age or anyone between those stages who cannot or does not claim.

**Table 3.2.** Fatalities from accidents and diseases in agriculture in the UK 1970–9

| Type of injury/disease | 1970 | 1971 | 1972 | 1973 | 1974 | 1975 | 1976 | 1977 | 1978 | 1979 |
|---|---|---|---|---|---|---|---|---|---|---|
| *Injuries caused by:* | | | | | | | | | | |
| Self-propelled machines | 54(13) | 53 (7) | 36 (7) | 43(10) | 32 (6) | 43(12) | 44 (4) | 42 (8) | 25 (4) | 26 (6) |
| Other field machines | 16 (6) | 12 (4) | 15 (1) | 13 (1) | 16 (6) | 9 (2) | 18 (6) | 14 (2) | 10 (2) | 12 (1) |
| Stationary machinery | 2 (1) | 6 (2) | 2 (1) | 4 | 3 | 4 | 2 | 2 | 3 (2) | 3 (1) |
| Electrical | 4 | 3 | 7 | 6 | 1 | 5 (1) | 11 (1) | 6 (2) | 5 | 3 (1) |
| Falls | 18 (3) | 16 | 15 (3) | 20 (8) | 19 (5) | 20 (4) | 12 (4) | 13 (4) | 8 (1) | 15 (3) |
| Drowning | 3 (2) | 13(10) | 5 (5) | 6 (5) | 6 (6) | 1 (1) | – | 10 (6) | 3 (3) | 5 (2) |
| Animals | 5 | 7 | 7 | 6 | 5 | 10 (1) | 3 | 1 | 4. | 5 (1) |
| Poisoning | 3 (1) | 2 (1) | – | 4 (1) | – | – | – | – | 1 | 2 |
| Other injuries | 21 (8) | 29 (6) | 13 (3) | 16 (3) | 19 (7) | 7 (2) | 16 (6) | 14 (3) | 13 (4) | 23 (9) |
| Total injuries | 126(34) | 141(30) | 100(20) | 118(28) | 101(30) | 100(23) | 106(21) | 102(25) | 72(16) | 94(24) |
| *Diseases:* | | | | | | | | | | |
| Weil's disease | 1 | – | 3 | – | – | 1 | – | 1 | – | – |
| Tetanus | 1 | 1 | 1 | 1 | – | – | 1 | 1 | – | – |
| Farmers' lung | 2 | 2 | 3 | 1 | 3 | – | 1 | 1 | 1 | – |
| Total diseases | 4 | 3 | 7 | 2 | 3 | 1 | 2 | 3 | 1 | – |
| Grand total | 130(34) | 144(30) | 107(20) | 120(28) | 104(30) | 101(23) | 108(21) | 105(25) | 73(16) | 94(24) |

*The above figures relate to deaths occurring as a result of agricultural operations, so they include not only workers, but farmers, members of their families, members of the public and others.

The figures in brackets show children under 16 (15 before 1973) included in the totals.

hazards has had no great effect on improving the proportion of child fatalities. Indeed, whilst total injury fatalities fell by 20 per cent from 1973 to 1979, the rate for children fell by only 14 per cent.

One would have expected that the development of intensive methods of husbandry would have reduced the number of injuries involving animals but there has been little decline in the proportion of fatal and non-fatal injuries in this group between 1971 and 1979. Undoubtedly, some of these may relate to accidents occurring during leisure pursuits rather than normal working procedures but the latest figures should give cause for concern and should prompt a closer look at ways to reduce this particular hazard.

Agricultural work has, in the past, been generally of a heavy manual type. This continuous activity, whilst beneficial in some ways, has led to problems. It was not uncommon, even 20 years ago, for agricultural workers regularly to lift and carry weights in excess of 120 kg. Sacks of grain were carried on the back often over cluttered barn floors and even up ladders. It was not surprising that backache was a common complaint amongst these workers. Whilst modern farming practices have almost eliminated the necessity to handle such loads, there is still an element of heavy work in the industry.

It is seen from *Table 3.2* that tenosynovitis is still one of the most important occupational diseases in agriculture. It was not unusual to have epidemics of tenosynovitis during the sugar beet harvesting when 'topping' was carried out by hand, but the continuous repetitive hand movements found in many farm jobs have now mostly disappeared. However, present farming methods relying on more mechanization do not appear to have eliminated all aetiological factors in this disease.

The ergonomic problems associated with long spells of work on mobile equipment are considerable and back pain related to ligamentous or muscular damage is still common. The simple introduction of a swivel-type tractor seat has been shown to improve some of the problems associated with long spells of tractor driving.[4] Much driving has to be undertaken over rough and uneven terrain and attention to correct seat stability and shock-absorbing facilities on farm vehicles is essential.


## HAZARDS FROM PHYSICAL AGENTS
### Heat and cold

Prolonged manual work in areas such as greenhouses, grain storage areas or mushroom growing farms might give rise to problems related to heat exposure. Additionally, the heat stress occurring in cockpits of planes used by aerial sprayers has been well recognized.[5] It is not uncommon for some poultry farms to provide their own packing, freezing and storage units. If cold-store working is undertaken on farms, then the same stringent safety precautions and work practices as are in operation in other industries should be adhered to. It is unlikely that exposure to the weather in our temperate climes could lead to hazard but there has been some improvement in comfort standards for employees. There is now universal use of tractor cabs which afford protection from the elements but which were designed and introduced primarily for safety reasons.

## Radiation

It is well known that agricultural workers in Australasia have a greater incidence of skin neoplasia than the general population because of their continuous exposure to sunlight. Although this particular form of ultraviolet radiation is unlikely to be hazardous in Britain one must also remember that many farm workers do experience rather more radiation from certain rock formations than does the general population.

## Noise and vibration

There was recently much publicity in East Anglia when a village resident urged the local environmental health officer to restrict the dawn crowing of a neighbour's rooster. For a short while a restriction order was placed upon the bird! There has been an increasing awareness of the total noise exposure experienced by the modern agricultural worker. Noise sources from tractors, corn mills and other machinery have always been well recognized. Modern farming practices requiring continuously operating corn driers, the development of intensive livestock husbandry and the utilization of more effective saws in forestry have all led to potentially more hazardous noise sources.

The dangers of chain saws are not confined to their noise output. They have been repeatedly investigated for their potential to cause vibration white finger. They are also notoriously difficult to guard effectively and extensive facial and eye injuries resulting from their malfunction have been described.[6]

## Dusts

Respiratory disease occurring as a result of dust inhalation has been investigated in many work situations but until recently the only contribution from the agricultural industry remained the classic 'farmers' lung', first described nearly 50 years ago. It is now apparent that inhalation of dusts causes a variety of different responses of the respiratory tract, such as occupational asthma, extrinsic allergic alveolitis, chronic bronchitis and non-specific disease. The agricultural industry has been associated with dusty occupations for many years but it has often been difficult to study groups of workers who have experienced respiratory diseases. Exposure to known fungi transported in dust has given us classic pictures of aspergillosis and actinomycosis. Extrinsic allergic alveolitis occurring after inhalation of mushroom spores has been termed 'mushroom workers' lung'. More recent studies on grain dust have shown a variable picture. Davies[7] and his co-workers described the occurrence of nocturnal asthma in grain handlers, and identified the grain mite *Glycyphagus destructor* as the sensitizing agent. 'Harvest workers' lung' has been related to sensitivity to various fungi on cereals. Grain dust has also been shown to be heavily contaminated with bacteria, particularly *Erwinia herbicola*. Further, it has been suggested that this organism may be a causative agent of some respiratory disorders in grain workers.[8,9]

It seems likely that inhalation of any cereal dust with its attendant contaminants may give rise to respiratory symptoms or disease. There is probably a wide range of aetiological agents responsible and these may vary according to geographical, seasonal and crop strain differences, as well as depending upon differing pesticide treatments. It is certain that prevention of exposure to dusts in the agricultural setting is of prime importance.

### Gases

Another classic occupational disease described in the industry is 'silo fillers' disease', which is the result of inhalation of nitrogen dioxide. The introduction of intensive livestock production has led to investigation of other toxic gases to which workers may be exposed. Donham[10] found that in swine confinement buildings the levels of carbon dioxide, hydrogen sulphide, carbon monoxide and ammonia all exceeded threshold limit values, often for long periods. Again, one must be aware of the possibility of danger in such areas and protect employees carefully from any toxic fumes.

## HAZARDS FROM CHEMICAL AGENTS

The possible effect of chemicals on agricultural workers has received more attention in recent years than any other aspect of occupational health in the group. However, when one notes the actual incidence and prevalence of disease caused by chemicals it seems to be a minor problem. Reported diseases or fatalities are nearly always the result of acute exposure. There are few statistics which relate to possible chronic exposures and long term effects. Several factors should be considered when assessing actual risks to agricultural workers: (*a*) the overall acute and chronic toxicity of the material; (*b*) the methods of mixing and application; (*c*) the degree and effectiveness of protective equipment; (*d*) the biodegradability of the material in plants, soil and water; (*e*) the possibility of synergistic effects; (*f*) the attitude of the worker to handling potentially harmful materials.

## Toxicity

There is a tendency for chemical manufacturers to produce pesticides with ever-increasing specificity. This is necessary to ensure not only effective eradication of the pest but also to eliminate any hazard to other species. It seems unlikely, however, that the degree of specificity will ever be such that a chemical will affect only a single species. In the United Kingdom two schemes exist which have been designed to protect workers in the field from exposures to toxic materials.

### *Pesticides Safety Precaution Scheme*

Under this scheme chemical manufacturers have undertaken not to market a product containing any new chemical for use in agriculture, horticulture or food storage, or to introduce a new formulation, until recommendations for safe use have been agreed with the government departments concerned.

### *Agricultural Chemicals Approval Scheme*

This is a voluntary scheme under which proprietary brands of agricultural chemicals can be officially approved. The purpose of the scheme is to enable users to select, and advisers to recommend, efficient and appropriate crop protection chemicals and to discourage the use of unsatisfactory products. The chemicals included in the scheme are those used for the control of plant pests and diseases, for the destruction of weeds, for growth regulation and certain other crop protection purposes and for the control of insect and mite pests of farm-stored grain. The scheme does not deal directly with the operator and consumer safety requirements for crop protection chemicals, but approval cannot be given to a product containing a new chemical or to a new use of an existing chemical until it has been considered and cleared under the Pesticides Safety Precautions Scheme.

In addition to these two schemes, the Health and Safety (Agriculture) (Poisonous Substances) Regulations 1975[11] place a statutory duty on users to observe certain precautions when handling chemicals listed in the regulations. Most of these precautions relate to the wearing of appropriate protective equipment during handling of these chemicals. The majority of substances listed are those possessing acutely toxic properties such as organophosphorus, carbamate and some nitrophenolic compounds. The Ministry of Agriculture, Fisheries and Food has produced a most useful guide[12] for the users of pesticides and it is to be hoped that the publication will be afforded the attention it deserves.

## Mixing and application

The majority of reported incidents involving pesticides are the result of inadequate notice being taken of instructions for their use. The increasing utilization of aerial application techniques may lead to more widespread contamination if correct procedures are not followed. Many formulations are of a wettable powder variety which require tank mixing prior to spraying; others are concentrates in various solvents which require dilution. Conventional spraying apparatus is usually tractor-drawn but in some circumstances hand-held spray units may be used, particularly in greenhouses.

Protection of workers during the stages of mixing and application is important and instructions as to proper use are given not only on product labels but often by technical staff associated with the distributor or manufacturer. This so-called 'product stewardship' function cannot be overemphasized for there often appears to be little attention paid to manufacturer's label instructions. I have seen workers mixing known skin allergens with a bare hand and forearm and this after considerable effort has gone into making the user aware of the hazard. When considering potential exposure to pesticides during application it is necessary to take account of meteorological conditions, the actual position of the operator, the physical effort required during the work, the length of working time and, above all, the attitude of the operator. Standards differ greatly from country to country but more efforts should be made to reduce exposures generally in the field situation.

## Protective equipment

In spite of the tendency to use less toxic materials there will always remain some chemicals which are hazardous to man, and his protection from them is vital. The Health and Safety (Agriculture) (Poisonous Substances) Regulations 1975 specify the types of clothing to be worn when handling chemicals, primarily to avoid their acute toxic effects. Many substances, for example some organophosphorus compounds or arsenicals, which may not have acute toxicity potential in low dosage, may still have the ability to cause long term effects and protection from them will depend, to a large extent, upon the information supplied and the technical assistance given by the manufacturer. Most emulsifiable concentrates contain a solvent, such as xylene, and clothing has to provide protection from this as well as from the active ingredient. Correct forms of clothing for use with certain materials are described in the Ministry of Agriculture, Fisheries and Foods Approved List.[12] Wettable powders are notorious for their ability to form clouds of dust during mixing processes, so it is essential that working areas are kept clean and free of dust deposits. Manufacturers are becoming more aware of this problem and the trend now is to provide packs in smaller and more easily handled units. It is to be hoped that the days of 'scooping' from a 45 kg drum of powder are numbered.

## Biodegradability

Many pesticides which persist in the environment have been used in the past, thus placing agricultural workers at risk from regular potential exposure. Some are retained in animal or plant tissues but many remain in the soil. Disturbance of the soil by subsequent tillage operations might lead to release of such materials long after their original use. The persistence of the organochlorine herbicides in the environment has received much publicity. Polychlorinated biphenyls, although not normally used as pesticides, have been shown to be present in many wild life samples. Fortunately, there is a trend nowadays to use materials which rapidly degrade after use.

## Synergism

It is not uncommon to find several active ingredients combined in a single formulation but there is often little in the way of assessment of possible effects in man. The problems in the field may arise in two ways. First, users may themselves combine two or more chemicals in the same spray tank and secondly, operators may be using a whole cocktail of pesticides over a short period whilst not actually mixing them. Information about possible synergistic effects is often limited because the products concerned may be manufactured by different companies. Most technical advisers will be able to give guidance as to which compounds may safely be combined. It would seem sensible to suggest that no mixing of different products should take place, but in practice it is necessary to spray according to a specific programme which is dependent upon variable weather conditions. Such mixing will, therefore, take place and in these circumstances where an unknown hazard may exist one should urge operators to take utmost care to avoid exposure.

## Attitudes

The application of toxic chemicals often requires a high level of technical expertise and demands the use of sophisticated equipment. Understandably, the small farmer or horticulturist may not always have access to such equipment but may still handle dangerous chemicals. There is an increasing need for manufacturers, distributors and suppliers to become involved with field usage of their products. Education of operators about the hazards of chemicals, the protective devices required and the proper and safe techniques for handling is not easily accomplished. Reliance upon the attention given to label instructions is hardly adequate. More emphasis on product stewardship by manufacturers would be only too welcome, although some still maintain that if adequate instructions are given on the label the responsibilities of the manufacturer or distributor are ended.

## Pesticides

The problems of providing adequate food supplies to an ever-expanding global population will be the constant concern of economists, agriculturalists and health experts alike over the next 50 years. Food production in the future will be dependent upon the better utilization of land for crops, improved irrigation schemes and the use of newer and more productive strains of plants and cereals. The yield of crops will depend not only upon soil quality and climatic conditions but also on our ability to control the numerous pests which may attack crops of all kinds. In order to fulfil this need for pest control there will be an increasing dependence upon pesticides. The use of such chemicals, however, must be carefully controlled so that their usefulness far outweighs their attendant problems and dangers.

A useful definition of a pesticide has been given in the US Federal Environmental Pesticide Control Act: 'A pesticide may be defined as any substance or mixture of substances intended for preventing, destroying, repelling or mitigating any pest (insect, rodent, nematode, fungus, weed, other forms of terrestrial or aquatic plant or animal life or viruses, bacteria or other micro-organisms, except viruses, bacteria or other micro-organisms on or in living man or other animals) and any substance or mixture of substances intended for use as a plant regulator, defoliant or desiccant.'

This somewhat lengthy definition serves to indicate the extent of this group of materials and there have been many different classifications of pesticides related either to their intended use, that is insecticides, nematocides or herbicides, or to their chemical groupings. Materials with the same general chemical structure may be used against one or more groups of pests. As the potential toxic effects may be similar within such a chemical grouping it is felt that some classification by this method is preferable.

*Inorganic and organic metal compounds*

These are, in the main, a group of pesticides of historic interest. Odysseus utilized the known properties of sulphur when fumigating his house following the elimination of his wife's suitors. A whole range of materials has been used at some time or other, and this includes barium carbonate (rodenticide), ferrous sulphate (herbicide), sodium fluoride (insecticide and herbicide), thallium sulphate (rodenticide) and sodium and potassium chlorate (herbicides). Other members of this class of compounds will be described in a little more detail.

*Arsenic*

Inorganic arsenicals (arsenic trioxide, sodium arsenite and lead arsenate) have been used as rodenticides and herbicides for many years but organic arsenicals have been developed more recently for use on lawns. They are not as widely used in the United Kingdom as elsewhere in the world but the toxicology of arsenic is such that it merits special mention. Although the organic pentavalent arsenical compounds are generally less toxic than the inorganic trivalent members, arsenic-containing substances should be regarded as serious threats to life and health. Inorganic arsenicals are usually absorbed after ingestion and injure the splanchnic vasculature causing abdominal colic and diarrhoea. Once absorbed, however, they bind sulphydryl-containing enzymes in tissues and produce toxic injury to the liver, bone-marrow, brain and peripheral nerves. These effects may occur after a single dose and may be immediate or delayed, or as a result of repeated small doses. They are manifest by cirrhosis, hypoplastic bone-marrow, acute and chronic renal failure and peripheral neuropathy. Furthermore, skin contact with arsenicals has resulted in the production of neoplasia. It is possible to confirm a diagnosis of arsenic poisoning by analysis of blood, urine or other tissues. Treatment is generally supportive but dimercaprol may be useful.

*Copper*

The fungicidal properties of copper salts were first discovered in 1807 and copper sulphate has been used as a seed dressing and as a selective herbicide. Several inorganic copper compounds are still manufactured and utilized to control mildew on hops and many common leaf and fruit diseases of fungal origin. Copper compounds are of low toxicity to mammals but may be harmful to fish.

*Mercury*

Inorganic mercury compounds have been used as fungicides since 1891 but the highly toxic mercuric chloride has been gradually replaced by mercurous chloride, still employed as a fungicide and selective herbicide. The use of organic mercury compounds is usually confined to seed treatments and has on occasion led to poisoning on an epidemic scale when contaminated grain has been consumed, for example over 6000 cases of poisoning were reported in Iraq after pesticide-treated seed had been used as food.[13] The highly toxic ethyl and methyl mercury have been largely superseded by rather less toxic materials, but great care has still to be exercised when handling them. Organic mercurials cause skin irritation and vesiculation. Their systemic effects are usually related to central nervous system damage.

*Tin*

Organo-tin compounds usually contain either tricyclohexyl tin which is used as a miticide, or triphenyl-tin acetate which is employed as a fungicide. They are sometimes combined with dithiocarbamates. Overexposure may therefore result in a combination of symptoms and signs related to the proportions of the various ingredients. The organo-tins are of moderate toxicity and may be absorbed through the gastrointestinal tract, the skin and, in the case of powders, by inhalation. They may produce marked skin irritation. After absorption they affect the central nervous system and produce generalized weakness, depression and occasionally paralyses. Penicillamine may be used in cases of poisoning, but otherwise treatment is symptomatic.


## Organochlorines

The organochlorine compounds have undergone more research and have been involved in more public discussion than, perhaps, any other group of pesticides. They have been widely used in agricultural and public health programmes and are outstanding in retaining their potency for long periods. Whilst this property makes them valuable for some purposes it results in unwanted environmental contamination. When it was suggested that enzyme induction was an important factor influencing carcinogenesis, the organochlorines came in for even closer scrutiny, as many of them exhibit potent enzyme induction properties. 1,1,1-trichloro-2,2 di(chlorophenyl) ethane (DDT) has been outlawed as a result of public opinion. However, other members of the cyclodiene group of insecticides, such as aldrin, dieldrin and endosulfan, are still used, although such use is limited under the Pesticides Safety Precautions Scheme. Acute poisoning with organochlorines is not common but if adequate dosage is received they interfere with axionic transmission of nerve impulses, particularly in the brain. This results in behavioural changes, sensory and equilibrium disturbances, involuntary muscle activity and depression of the respiratory centres. Of greater importance are their possible long term effects. Organochlorines are retained in body fat for long periods, but although much research has postulated their likely effects, there is little conclusive evidence of their having done so. Jager[14] carried out an epidemiological survey on pesticide workers handling various organochlorine materials for periods of up to 15 years and found no adverse effects. It is, of course, possible that a longer period of medical surveillance will be required to ensure that such effects do not subsequently arise.

## Organophosphates

The first organophosphorus material to be used as an insecticide was the compound tetra ethyl pyrophosphate. This was marketed in Germany in 1944 but was originally prepared by de Clermond in 1854 and not recognized as an insecticide until 1938. No group of synthetic chemicals has aroused such interest in the agricultural field particularly since the demise of the organochlorines. They are used as potent insecticides, fungicides, plant foliants and plant desiccants. Their relatively non-persistent nature has led to greatly increased production and to the development of a variety of compounds. The general formula is:

$$R_1O \diagdown \overset{\overset{\textstyle O(S)}{\|}}{\underset{}{>}} P - OR_3$$
$$R_2O \diagup$$

The $R_1$ and $R_2$ groups may be of a widely variable nature but the normal substituent is an alkyl group or an amine. The $R_3$ group is usually an acidic group. In 1941, organophosphorus compounds were first shown to depress activity of the enzyme cholinesterase. At that time the state of knowledge of neuromuscular physiology was such that direct correlations were made with respect to toxic effects of the organophosphates in relation to their cholinesterase-inhibiting effects. We are now discovering that neuromuscular transmission, transmitter release mechanisms and transmitter–receptor interactions are all highly complex functions. It would, therefore, seem reasonable to expect varied effects upon enzymes not necessarily confined to cholinesterase systems. The basic inhibition concept will be described which applies to *in vivo* cholinergic synapses such as at the neuromuscular junctions or at parasympathetic nerve endings.

At these synapses and junctions the impulses are initiated by the chemical transmitter, acetylcholine. The secretion of this substance at nerve endings results in electrical changes which culminate in the production of an action potential. In order to evoke such a response, the action of acetylcholine is required for only about two milliseconds. It is then hydrolysed by the enzyme acetylcholinesterase which is also present in nerve endings. If this enzyme is inhibited then disturbance of normal nerve function will follow. Many organophosphorus compounds interact with acetylcholinesterase forming an enzyme inhibitor complex which becomes progressively phosphorylated. This process is known as 'ageing' and the degree to which it occurs is dependent upon

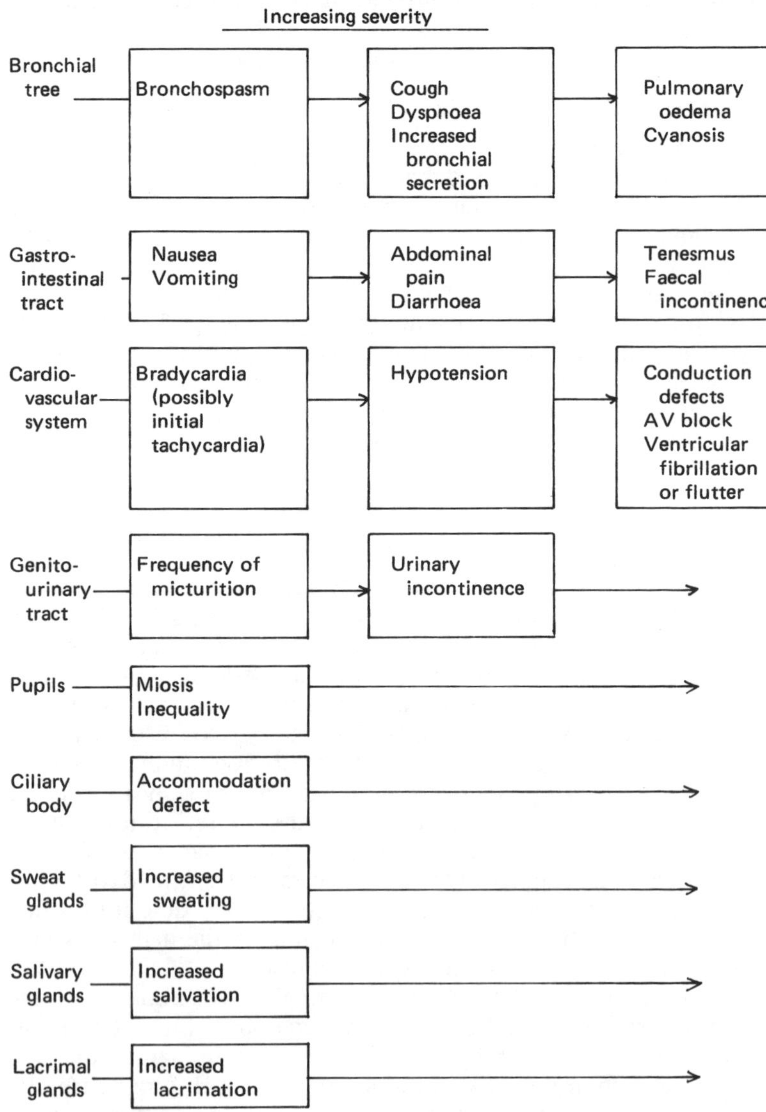


*Fig. 3.1.* Muscarinic effects.

the particular compound involved. Further, once phosphorylation has taken place no effective release of the enzyme can then occur. Atropine, given parenterally in relatively large doses, effectively blocks the continuing activity of acetylcholine but such treatment must be continued until such time as adequate quantities of acetylcholinesterase have been released or synthesized. Pralidoxime acts as a competitive inhibitor and, if given early in cases of poisoning, may 'protect' acetylcholinesterase by forming a relatively loose link with the enzyme which can subsequently be released. It must be remembered however that the oximes themselves are, therefore, enzyme inhibitors.

Acute poisoning with organophosphorus compounds may be recognized by classic symptoms related to muscarinic, nicotinic and central nervous system effects, as shown in *Figs. 3.1–3.3*.

In acute organophosphorus poisoning nerve ending cholinesterases will be markedly inhibited. This inhibition may be reflected in both plasma (butyrylcholinesterase) and erythrocyte (acetylcholinesterase) enzymes. For many years measurements of either blood enzyme have been made to determine exposure to and absorption of organophosphorus compounds. There are several points to be considered when assessing this method of monitoring:

1. *Are blood enzyme levels truly reflective of nervous system enzymes?* If not, could some form of neurological damage be taking place in the presence of normal blood levels? It is well known that peripheral neuropathies caused by some organophosphorus com-

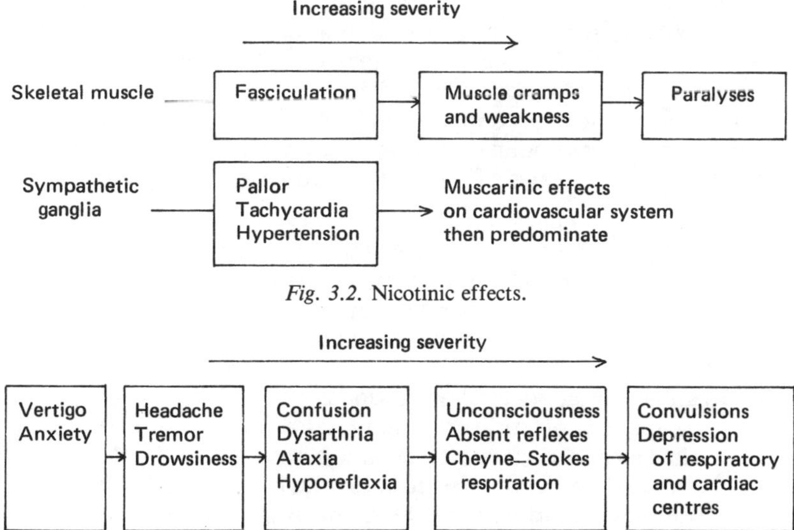


*Fig. 3.2.* Nicotinic effects.



*Fig. 3.3.* Central nervous system effects.

pounds are the result of inhibition of 'neurotoxic esterase' described by Johnson,[15] but no inhibition of plasma or erythrocyte cholinesterase takes place. It has also been shown that some organophosphorus compounds have effects upon neuromuscular end-plates whilst blood enzymes remain normal and other compounds have an effect upon blood cholinesterases in the presence of normal neuromuscular function. Recovery of blood enzymes to normal levels, after inhibition, takes place at a variable rate. The discovery of a single depressed blood cholinesterase level will not indicate whether recovery or further inhibition is occurring. When abnormally low blood levels are found a repeat test should be performed, within 3 days, to determine progress.

2. *How variable are the blood cholinesterases and is it possible to correlate changes of these with low level exposures?* Variability of plasma and erythrocyte cholinesterase is dependent upon individual genetic factors, the presence of disease or the taking of drugs, especially alcohol. There are many different techniques employed to measure the enzymes ranging from Michel's electrometric method[16] to the rather more acceptable spectrophotometric method of Ellman.[17] For monitoring purposes it is essential that an individual's baseline plasma and erythrocyte cholinesterase be determined. At least three separate analyses are required for this. It is also important for repeatability that analyses are carried out in the same laboratory, using the same techniques and if possible utilizing the services of the same analyst.

3. *Can adaptation to inhibition take place?* If long term organophosphorus exposure leads to repeated minor inhibitions of blood enzymes it is conceivable that some adaptation may take place in the form of increased synthesis of the enzyme or increased erythrocyte production. Such adaption would effectively negate the use of blood enzyme measurements as a monitoring tool. In any form of monitoring programme these points must be considered.

The following schemes of medical surveillance are suggested:

(i) Occasional single contacts:
   complete protection from material as far as is practicable;
   no form of medical monitoring of real value;
   previously obtained baseline levels of plasma and erythrocyte cholinesterases may be helpful in the event of an accidental exposure.

(ii) Regular contact over short periods:
   complete protection from material as far as is practicable;
   plasma and erythrocyte cholinesterase levels should be obtained and compared to individuals' baseline levels;
   samples should be taken at frequent intervals initially to ensure that hygiene standards are sufficient to prevent exposure.

It will be impossible to eliminate contact altogether and one has to fix an arbitrary level of cholinesterase depression which should prompt further action. The recommendation of the Fifth International Workshop of the Scientific Committee of Pesticides of the International Association on Occupational Health was that a 30 per cent depression of erythrocyte cholinesterase activity should lead to withdrawal from work and suspension should continue until the erythrocyte cholinesterase value has returned to 80 per cent of the normal level. Normal levels are based on the three pre-exposure tests. Rather less emphasis was placed on plasma cholinesterase levels. They do reflect exposure and so close attention to their values is helpful when assessing hygiene standards. A depression of plasma cholinesterase by 50 per cent (based on individual normal level) should also prompt temporary removal from potential exposure.

(iii)  Regular contact over long periods:
   complete protection from material as far as is practicable;
   plasma and erythrocyte cholinesterase measurements, as indicated in (ii) above;
   assessment of neuromuscular transmission and nerve conduction velocities, employing an electroneuromyographic technique,[18,19] on a regular basis.

The organophosphorus pesticides are of great importance throughout the world and it is appreciated that the scheme mentioned above may not always be practicable. There is, however, no doubt that efforts should be made to improve hygiene standards and to ensure that at least some form of medical surveillance is undertaken where these materials are used.


## Carbamates

This group of insecticides and fungicides can be conveniently divided into two major divisions dependent upon the ability to affect cholinesterases.

### Cholinesterase-inhibiting carbamates

This class includes materials such as aldicarb, carbaryl, methomyl and proxopur. They are of variable toxicity and several are controlled by statutory regulations. They are able to cause reversible carbamylation of acetylcholinesterase and, consequently, may produce clinical pictures similar to the organophosphorus compounds. However, the enzyme-carbamyl combination dissociates more readily than phosphorylated enzyme and this ability tends to mitigate the acute toxicity of carbamates. It also limits the usefulness of blood cholinesterase measurements in biological monitoring programmes. Treatment of acute poisoning relies largely upon the use of parenterally administered atropine but pralidoxime is of no value and its use is contraindicated.

### Dithiocarbamates

The majority of commercially used preparations of this series include a metallic group in their structure. The fungicides maneb, mancozeb and zineb are included in this class of chemicals. Some compounds with this type of structure act as potent irritants and allergens to the skin and they may exacerbate pre-existing allergic skin or respiratory disease. They are metabolized in a manner similar to disulfiram and their systemic toxic effects may be related to the toxicant alone or to the combination of the toxicant with alcohol. Functional and anatomical central nervous system damage has been demonstrated in rats on long term, high dietary intake of some dithiocarbamates and peripheral neuropathy and psychotic reactions have occurred in individuals taking disulfiram alone. Absorption of dithiocarbamates followed by alcohol intake results in illness due primarily to inhibition of liver enzymes necessary for oxidation of acetaldehyde to acetic acid.

Both groups of carbamates should be handled with the greatest of care and although no monitoring techniques of value are available, some form of medical surveillance should be undertaken on those potentially exposed.

### Organo-nitrogen

Most of this group of pesticides are used for their herbicidal properties and they may be derivatives of urea (linuron, monolinuron), uracil (bromacil), acetaniline (propachlor) or picolinic acid (picloram). Compounds of other groups such as acetamides (allidochlor), s-triazines (atrazine, simazine) and carbamates (chloropropham) are also included in the organo-nitrogens. The various compounds are of low general toxicity and adverse effects have only rarely occurred in man. Certain of these chemicals are potent irritants to skin and mucous membranes (uracil, acetanilide, acetamide and picolinic acid compounds) and some cause allergic responses (acetanilides and acetamides). No real difficulties should be encountered with their use in properly controlled situations but awareness of their adverse effects should be borne in mind when they are handled in the field.

### Chlorophenols

Chlorinated phenols are employed as wood preservatives, fungicides, herbicides and insecticides in their own right as well as being intermediate in the production of the phenoxyacids.

#### Pentachlorophenol and sodium pentachlorophenate

These have limited use in the agricultural industry and are more often used for their potency as fungicides and bactericides in products such as adhesives, leather and cosmetics. They are, however, still utilized for their properties as wood preservatives and agricultural or forestry workers may be exposed to them. Pentachlorophenol is a potent irritant of the skin and mucous membranes and is also well absorbed through skin, lung or gastrointestinal lining. Like nitro-phenolic compounds it stimulates oxidative metabolism in tissue cells and in common with other phenols it is toxic to the liver, kidneys and central nervous system. It is likely that chloracne described as a result of pentachlorophenol exposure is related to the inclusion of dioxin impurities in the technical formulation. Mutagenic and carcinogenic properties of these materials have been looked at carefully, especially in view of the possible dioxin contamination. Further epidemiological surveys and research will be necessary to resolve the whole question and, in the meantime, these materials should be handled with caution.

*2,4,5-Trichlorophenol and its derivatives*

There has been great public debate about the use of 2,4-dichloro-phenoxyacetic acid (2,4 D) and 2,4,5-trichlorophenoxyacetic acid (2,4,5 T), which are both manufactured from 2,4,5-trichlorophenol. Reports of teratogenic and carcinogenic effects and chloracne in populations exposed to them in Vietnam and Seveso have prompted some agricultural workers' representatives to call for a complete ban on their use. It is again likely that highly toxic contaminants such as 2,3,7,8-tetrachlorodibenzo-*p*-dioxin (TCDD) have been responsible for these effects but there is no epidemiological evidence that the use of the compounds in properly controlled situations has caused ill health. It would seem reasonable, however, to suggest that such epidemiological studies should be undertaken and that in the mean-time only highly purified compounds should be used and only then according to strictly enforced hygienc standards.


## Nitrophenols

These range from highly toxic herbicides DNOC (dinitro-ortho-cresol) and dinoseb (2-sec-butyl-4,6-dinitrophenol) to the relatively less toxic fungicide binapacryl. All of the materials should be regarded as toxic to man. Most are well absorbed through the skin, gastrointestinal tract and by the lung if fine droplets are inhaled. They usually produce a yellow skin stain and like other phenols they are toxic to the liver, kidneys and nervous system. The basic mechanism of their toxicity is a stimulation of oxidative metabolism in cell mitochondria, by inter-ference with the normal coupling of carbohydrate oxidation to phos-phorylation reactions. Increased oxidative metabolism depletes body carbohydrate and fat stores and leads to pyrexia, tachycardia and dehydration. Treatment of poisoning is supportive. Some compounds are subject to statutory regulations when used in the field.

*Dipyridyls*

Paraquat and diquat are both used as contact herbicides and their toxicology has been related more to accidental or suicidal ingestion rather than to overexposure in their agricultural usage. They are capable of injuring epithelial tissues of the skin, eyes, nose, mouth and respiratory and gastrointestinal tracts and paraquat appears to be relatively more toxic than diquat. After ingestion there may be a latent period of up to 14 days before a diffuse toxic pneumonitis occurs which frequently results in death. Treatment of poisoning is supportive but forced diuresis and administration of corticosteroid or immunosuppressive drugs may help. A regimen of azothioprine and potassium aminobenzoate has been described by Laithwaite[20] and this has met with some success.

*Halocarbons and sulphuryls*

This group consists, in the main, of gases and highly volatile liquids. They are used for soil fumigation processes often within the confines of greenhouses. In the UK their use is carefully controlled because they are highly toxic. Several compounds have been used in the past, such as chloroform and carbon tetrachloride, but methyl bromide has superseded these. Paradichlorobenzene and sulphuryl fluoride are also marketed as fumigants.

Methyl bromide is often manufactured with the addition of a warning gas (chloropicrin) and is highly toxic to man. It penetrates skin and rubber and makes protection difficult. It is an irritant of skin, eyes and respiratory tract and has been known to cause severe vesiculation if left on the skin. In addition to its toxic effect on the lung, where it is capable of producing pulmonary oedema and haemorrhage, it also acts on the central nervous system. It is essentially a depressant but early changes may be of behavioural or of a non-specific neurological nature. These effects may progress to epileptiform convulsions and coma. All of these manifestations of poisoning may be delayed for up to 48 hours after contact. It follows, therefore, that prevention of exposure is vitally important. Small exposures can be judged by comparing blood bromide levels with baseline readings, always bearing in mind that many medicaments contain bromides and may have an effect on such results.

Dibromochloropropane is used as a soil fumigant because of its nematocidal properties. Not only does it possess high general toxicity but it has been implicated in the causation of reduced testicular function.[21]

Paradichlorobenzene is normally used in solid form and is not particularly toxic, although it is known to be a skin allergen.

*Miscellaneous pesticides*

Warfarin has been used for many years as a rodenticide and its anti-coagulant properties should pose no problem to the user. Warfarin concentrates are often added to sugar and poor labelling and storage of the resulting mixture could, conceivably, lead to mistaken ingestion.

Hexachlorobenzene is used as a fumigant and is well known for its ability to induce porphyria.

Derivatives of pyrethrum have been known to have marked insecticidal properties. The development and preparation of synthetic pyrethroids has increased greatly in recent years and it is likely that their use will become more widespread in the foreseeable future. Many animal toxicology studies have been carried out and there has been medical monitoring of those involved in field usage. There is, as yet, no conclusive evidence about their neuropathic potential but they do produce unusual effects on the skin. Complaints of paraesthesiae, burning and erythema of skin have been made by those exposed. The symptoms disappear within 24 hours and are promptly relieved by application of a steroid preparation. It has been suggested that the paraesthesiae represent sensory nerve ending irritation. There is also some evidence that clinical features may exhibit the pattern of an allergenic type reaction.


## INFECTIOUS DISEASES

Many diseases transmitted to man by animals have been described but most are of historical interest only. Donald Hunter gives vivid accounts of anthrax, glanders, orf, Weil's disease, louping ill, Q-fever, erisipeloid and ringworm.[22] *Tables 3.1* and *3.2* show that these diseases play only a minor role in the overall picture of occupational diseases in agricultural workers. The improvement of animal health has contributed greatly to reduced incidence of these diseases but one must be constantly aware of them and of other more recently described conditions such as toxoplasmosis and leptospirosis.[23] Brucellosis and tetanus, although not causing excessive ill health, are at present real hazards to many agricultural workers, but it is hoped that intensive vaccination programmes for employees and livestock, and attention to hygiene standards, will lead to a further reduction in illness.


## CONCLUSIONS

The occupational health problems of the agricultural industry as a whole have received scant attention. There is a definite need to provide occupational health care for employees in the many branches of the industry. At present there are three major requirements. First, the necessity to introduce a scheme which will enable basic health data to be collected, which can then be used for more effective surveys and epidemiological studies. There does remain the problem that the industry is often divided into numerous small working units but some form of standardization could be introduced through government departments, employee or employer organizations. The second requirement is to provide occupational health expertise at field level in order to assess the nature and extent of possible health hazards. Thirdly, programmes based on sound preventive medicine practice will be required, especially for those who work in specialized units. The likely areas of concern are those related to ergonomic design, dust inhalation and respiratory disease, possible long term effects of a multitude of chemicals and improvement of safety, particularly amongst children and animal handlers.

## Acknowledgement

I acknowledge gratefully the help given by Mr J. C. Weeks, Director of Agricultural Health and Safety of the Health and Safety Executive, in providing the data for *Tables 3.1* and *3.2*.


REFERENCES
1. Schilling RSF (ed.) *Occupational Health Practice.* London: Butterworths, 1973: 173.
2. Mitchell BR. *Abstract of British Historical Statistics.* Cambridge: Cambridge University Press, 1962: 60.
3. Central Statistical Office. *Annual Abstract of Statistics.* London: HMSO, 1962: 99–104.
4. Bottoms DJ and Barber TS. A swivelling seat to improve tractor drivers' posture. *Appl. Ergon.* 1978; **9:**77–84.
5. Richter ED, Gribetz B, Krasna M et al. Heat stress in aerial spray pilots. In: Tordoir WF and van Heemstra-Lequin EAH (ed.) *Field Worker Exposure during Pesticide Application.* Amsterdam: Elsevier, 1980.
6. Rigg BM. Chain saw facial injuries. *Can. J. Surg.* 1979; **22:**149–51.
7. Davies RJ, Green MMcC and Schofield N. Recurrent nocturnal asthma after exposure to grain dust. *Am. Rev. Resp. Dis.* 1976; **114:**1011–19.
8. Dutkiewicz J. Exposure to dust-borne bacteria in agriculture, I. Environmental studies. *Arch. Environ. Health* 1978; **33:**250–9.
9. Dutkiewicz J. Exposure to dust-borne bacteria in agriculture, II. Immunological survey. *Arch. Environ. Health* 1978; **33:**260–70.
10. Donham KJ, Rubino M, Thedell TD et al. Potential health hazards to agricultural workers in swine confinement buildings. *J. Occup. Med.* 1977; **19:**383–7.
11. *The Health & Safety (Agriculture) (Poisonous Substances) Regulations 1975.* London: HMSO, 1975.
12. Ministry of Agriculture, Fisheries and Food. *1980 List of Approved Products and their Uses for Farmers and Growers.* London: HMSO, 1980.
13. Bakir F, Damluji SF, Amin-Zaki L et al. Methyl mercury poisoning in Iraq. *Science* 1973; **181:**230–40.

14. Jager KW. *Aldrin, Dieldrin, Endrin and Telodrin. An epidemiological and toxico-logical study of long term occupational exposure.* Amsterdam: Elsevier, 1970.
15. Johnson MK. Organophosphorus and other inhibitors of brain neurotoxic esterase and the development of delayed neurotoxicity in them. *Biochem. J.* 1970; **120**:523–31.
16. Michel OH. An electrometric method for the determination of red blood cell and plasma cholinesterase activity. *J. Lab. Clin. Med.* 1949; **34**:1564.
17. Ellman GL, Courtney D, Andres V et al. A new and rapid colorimetric determination of acetylcholinesterase activity. *Biochem. Pharmacol.* 1961; **7**:88–98.
18. Roberts DV and Wilson A. Electromyography in diagnosis and treatment. In: Greene R (ed.) *Myasthenia Gravis.* London: Heinemann, 1969: 29–42.
19. Jager KW, Roberts DV and Wilson A. Neuromuscular function in agricultural workers using pesticides. *Arch. Environ. Health* 1970; **25**:395–8.
20. Laithwaite JA. Paraquat poisoning. *Br. J. Clin. Pract.* 1976; **30**:71–3.
21. Sandifer SH, Wilkins RT, Loadholt CB et al. Spermatogenesis in agricultural workers exposed to dibromochloropropane (DBCP). *Environ. Contam. Toxicol.* 1979; **23**:703–10.
22. Hunter D. *The Diseases of Occupations,* 5th ed. London: English Universities Press, 1974.
23. Crawford SM and Miles DW. Leptospira hebdomada associated with an outbreak of illness on a farm in North Yorkshire. *Br. J. Ind. Med.* 1980; **37**:397–8.

# 4. OCCUPATIONAL HEALTH IN COMMERCE
## *W. M. Dixon*

The practice of occupational medicine in commerce does not differ in principle from that in manufacturing industry. Although there are fewer obvious specific occupational hazards, the psychosocial problems which occupy the time and attention of occupational physicians the world over loom larger. This is not because hazards are more common in these situations but because the more esoteric problems of, for instance, the motor manufacturing or the petrochemical industry, do not predominate. This enables the occupational health team to concentrate on the many personal concerns which beset all people from time to time. However it cannot be assumed that offices, shops and warehouses are free from physical and chemical hazards of which the occupational health service must be aware.

In recent years the introduction of sophisticated electronic equipment has increasingly taken the place of the pen and cash box, so that the secretary and the shop assistant are surrounded by and supported by word processors, calculators and computers. The visual display unit (VDU) has partly taken over from paper documentation and the television telephone may soon take over from the conference table.

In shops the ready sale of potentially hazardous chemicals, for example herbicides and pesticides, and the increased vigilance of consumer protection groups, for example concerning the carcinogenicity of hairdyes, have of recent years brought the occupational health department into a new role supporting the buyer who frequently calls for expert interpretation of conflicting scientific and pseudo-scientific evidence about the products he purchases.

Food poisoning is still one of the few increasingly common infectious diseases of western civilization; it is prevalent all over the world. In supermarkets, restaurants and hotels the importance of strict food hygiene has brought the occupational health team to the assistance of management. By the use of bacteriological techniques, a better scientific assessment of the environment and of food workers' equipment as well as the health of food handlers can be achieved.[1]

It therefore follows from these examples that the training and professional expertise of occupational physicians and nurses in this field must be no less wide than that of their colleagues in other areas of industry. Each must have a broad enough overall view of occupational health to be able to spot the unexpected and to delve into the previously unexplored.

Occupational medicine is practised in an increasing variety of non-industrial circumstances in many countries. Indeed, even in an apparently *industrial* situation, such as the control room of an oil refinery or an electricity generating station, the nature of the work is strikingly similar to the *commercial* world of the computer operator in a department store headquarters.

## SHOPS

The health and safety of workers in shops is the subject of legislation in the UK in the Shops Act[2] and the Offices, Shops and Railway Premises Act,[3] later consolidated into the Health and Safety at Work Act.[4] Statutory inspection is usually carried out by environmental health officers (EHOs), but one of the principles of the latest Act[4] is to encourage self-regulation by the establishment of health and safety policies in each company under the guidance of a senior executive. Each shop manager, whether he be a singlehanded corner-store owner or the managing director of a large department store, is responsible for the health and safety of all employees and contractors and of customers whilst on the premises. The Consumer Protection Act[5] and the Dangerous Goods Act[6] place responsibility for the safety of goods sold on to the retailer whereas previously he could pass the blame for untoward effects on to the supplier. Good commercial sense provides an over-riding stimulus to sell products which are safe and effective, but the retailer now has also a statutory responsibility.

There is an almost infinite variety of shops even within the UK, and overseas the variety increases. However, for simplicity, they can be divided into three main classes.

1. *Department stores,* housed in large multi-storey buildings in city centres, or more recently in suburban shopping precincts or malls. These shops sell everything from a 'pin to an elephant' but concentrate on household goods in the widest sense, including clothing and personal services such as hairdressing. The majority of department stores in the USA and in Europe form chains owned by one company, often with central offices controlling development and research, as well as commercial, purchasing, personnel, financial, property and credit policies. The degree of central control varies from company to company but the advantages of scale and the introduction of computers and better communications have led to increasing centralization both in Western Europe and in North America. Central buying brings the advantages of scale to the strength of the buyer in the market place.

2. *Supermarkets and hypermarkets:* In Europe and the USA these shops are primarily in the food business, although most companies sell some non-food lines partly because of the higher profit margins. Self-service with payment at a terminal check-out marks them out as different from other shops, although minimarkets and superstores at either end of the size spectrum merge with small shops and department stores respectively. In the USA, France and Canada there is a stronger trend towards hypermarkets selling a wider assortment usually on one floor but still with self-service and check-outs. Chains of supermarkets with central administration are increasingly replacing the small family business.

3. *Neighbourhood shops:* These are commonly grouped in clusters in a town or city centre or in a suburb and are often owner-operated. They each specialize in a relatively narrow assortment of goods such as men's clothing, meat or hairdressing, and may have an element of self-service, particularly in grocery, do-it-yourself or bookselling. However, most continue the tradition of over-the-counter selling.


**Distribution**

The need for storage of goods behind or above the selling area of a shop itself varies from the high value/low volume stock of the jeweller to the other extreme of the low value/high volume stock in grocery. The business of buying, collecting into a central warehouse and distributing to a large number of scattered supermarkets or department stores is often computerized. Stock control is an important element for profitability in the retail business. Electronic cash registers, which automatically feed information from the point of sale to a central computer, have been installed in many stores. The objective is to maintain stock at the minimum level so as to avoid money being tied up unnecessarily but simultaneously to ensure that the sales assistant has an adequate supply of goods at the right place at the right time. Large warehouses situated centrally with satellites in or near the bigger cities and stockrooms in the shops form a chain of distribution linked by a transport system run on tight schedules. Every item purchased in a shop has been moved several times, and on each occasion people lift and handle it either mechanically or by hand. Large numbers of retail employees are therefore never seen by the shopper but their health and safety is still the concern of the occupational health service.


**Occupational health services**

Provision of occupational health and safety services to such a wide range of human activity varies from the most advanced to the minimum permitted by legislation. The minimum UK legal requirement is the appointment of a trained first aider if the number of employees exceeds 150 at any one time. However, the vast majority of shops employ less than twenty people. Food retailers must comply with relevant legislation[7] which is monitored by EHOs whose primary responsibility is to the consumer rather than to the retail employees.

Department store groups in the UK often have well-established occupational health services and rely on nursing staff to provide the mainstay of the department with either full- or part-time doctors visiting at intervals.

Supermarkets vary; some have excellent medical advice available to management and employees, but the majority still lack any sort of occupational health service. Each supermarket employs around 100–150 staff, many of whom are part-time so that effective deployment of doctors and nurses is difficult. The best administration is for a group of say 8–10 supermarkets to share one itinerant full-time nurse who has access to a doctor for medical advice. Another arrangement is for a part-time doctor to visit each shop regularly.

Expansion of occupational health services is likely with increasing emphasis on food hygiene and consumer protection and also on the well-established value of occupational health services to the employer and his employees.


## Physical Environment

Working conditions in small shops especially food, fruit and vegetables and in open-air markets are often very poor with excessive heat and humidity in summer and cold, wet, draughty conditions in winter. Many self-employed people in small family retail businesses work in conditions which would provoke industrial and perhaps even legal action in a factory. The profit incentive, the self-reliance of the small business community and the entrepreneurial attitude bring about a camaraderie and a cheerful stoicism which can accept the disadvantages of cold for the warmth of human contact and the constant stimulus to sell. At the other extreme the air-conditioned department store or office with fitted carpets throughout and a totally controlled physical environment gives optimum conditions for work at all times of the year.

Selling is a demanding occupation, requiring the talents of an actor, a banker, a counsellor and sometimes of a philosopher. Failing to make a sale is demoralizing but the rewards of a satisfactory transaction, especially in departments selling large and expensive items such as televisions or furniture, are not purely financial, but personal as well. The shop assistant, in contrast to the factory hand, sees instant success or failure recorded on the cash register. Personal contact with the customer replaces the distant relationship with the factory's consumer. Feedback to the employees of branch and company sales figures often at weekly or even daily intervals ensures close personal involvement. This instant feedback is a two-edged sword: in good times bad working conditions are overlooked, but when sales figures fall and boredom develops even good working conditions become the subject of criticism.

Teaching of sales assistants, porters, drivers and so on in good lifting techniques is an essential part of occupational health work in business. A film such as 'Mind Your Back',[8] which includes some retail scenes depicting shop assistants lifting rolls of cloth, is a useful training aid. Constant reminders are necessary because the effect of teaching on the incidence of back injury lasts only 2–3 years. The practical problems of carrying heavy furniture up narrow staircases in customers' houses make teaching of delivery men difficult.

## OFFICES

'An office is a room with a desk and two chairs', a definition which becomes inadequate if one gives a moment's thought to the wide variety of offices within every adult's personal experience. A branch bank or a building society office are really shops selling money and a betting shop sells dreams. A major computer control centre, the dealing floor of the Stock Exchange and the room at Lloyds are all offices. They and the offices of a mail order company may more closely resemble factories than the traditional concept of an office. All have one thing in common — they are places where decisions are communicated. The invention of the telephone and its subsequent developments into telex, VDUs and data transmission from computer to computer without the intervention of human language, have revolutionized the nature of office work. An office now contains a range of electro-mechanical equipment; some, such as photocopiers, using complex chemicals with their attendant occupational hazards — occupational dermatitis has taken the place of writer's cramp. The introduction of open-plan offices in the 1950s and the landscaping of large offices, particularly in the USA, brought many problems. The trend is now for a return to individual offices, particularly for managers who need privacy and lack of interruption and distraction. The occupational health problems in offices stem from two main causes: the physical environment and the stress induced by human relationships.

### Occupational health services

The provision of occupational health services in offices varies even more widely than in shops. Some of the major banks, insurance companies and head offices of large manufacturing companies have sophisticated occupational health provision but the vast majority have little or none. Statutory requirements are similar to those for shops[3] but the appointment and training of first aiders is often neglected. A major bank, for example, has a full-time medical staff with nursing assistance at its head office, but a large proportion of the 50 000 employees have no medical advice in the branch offices scattered all over the UK unless some crisis develops. An attempt was made in the early 1960s to set up a group occupational health service in the City of London but it failed for lack of support, despite active help from local government.

## Physical environment

The physical environment is, in modern offices, often controlled by air conditioning and in nearly all American and British offices central heating is provided. This should give the ideal thermal environment but, human nature being what it is, people vary in their needs. This gives rise to some conflict, especially when first introduced, and numerous complaints of sore throats, headache, dry skin and so on find their way to the medical department. Humidification, although expensive, solves many of these problems but inevitably there will be some workers who cannot accept imposed thermal environments. Humidifier fever is a possible hazard in offices and shops as it is in printing works,[9] but so far no case of this disease or of Legionnaires' disease has been recorded in a retail shop. Fitted carpets have provided suitable conditions for the growth of various mites, particularly the house-dust mite which causes allergic reactions in many people. Some carpet-tiles manufactured from hog's-hair produce allergenic dusts. The build-up of static electricity in carpeted, heated offices is usually a minor nuisance but can become a major problem at times.

Provision of toilet facilities is controlled by UK legislation[3] which gives minimal requirements. However, the good employer will exceed these minima. It is wise to establish with the company architects an agreed set of criteria for accommodation in offices. Canteens are now commonplace in large offices. (Medical departments are frequently asked to advise both on nutritional content of what for many office workers is their main meal of the day and on food hygiene.) The objective of the planners should be to provide an environment of which the users are largely unconscious — neither too hot nor too cold, too dry nor too moist, with effective, unobtrusive lighting, free of draughts and noise. Good lighting to an appropriate standard[10] is advisable and avoidance of glare essential.

Recently VDUs have been the subject of much controversy. They replaced the punch cards which were used to feed data into a computer and the printer on which results came back to the operator. A VDU consists of a cathode ray tube display and a keyboard. Complaints of eye discomfort, excessive lacrimation and of blurring or double vision occur in a proportion of operators, varying from 10 to 30 per cent. The causes of these complaints are complex, but are often related to luminance of the characters on the screen, their persistence and sharpness and to the seated posture of the operator. Reading[11] has edited a series of authoritative articles on the subject. A more recent publication by Grandjean and Vigliani[12] gives a comprehensive review of the literature and stresses the importance of psychological factors, including the fear felt by some employees towards new technology. Provided sound ergonomic principles are followed in the design of these units they can be used for prolonged periods without any harmful effect on the health or eyesight of the operator.[13]

## Psychosocial environment

Much as jargon is to be deprecated, it is difficult to think of a better phrase than 'psychosocial environment' to describe the mélange of influences which surround office workers and affect their daily life, hopes and fears, and which contribute to, or detract from, physical and mental health. Everyone is subject to influences of this kind in human interpersonal relationships. The psychosocial environment may be defined as the summation of human factors in our surroundings. Adaptation to change is a characteristic of modern man. Inflexibility produces inner stress which results in ill health.

The psychosocial environment, as opposed to purely physical surroundings, has an important bearing on the health of all workers, particularly those in offices. In the hierarchical structure of management the pecking order becomes of great importance to some people, and stress may be caused by real or imagined slights due to insensitive handling of these problems by senior managers.

Stress has been written about by so many medical authors[14,15] and by psychologists[16] and social scientists on both sides of the Atlantic that any bibliography would, if exhaustive, be exhausting! Whether the effects of stress are greatest for the managed or the manager is a matter of opinion, but probably the first-line manager is the person who suffers most because pressures come from above and below. Once a man or woman has reached the top, stress often lessens because he now makes his own decisions, sets his own pace and is answerable only to the shareholders. The foreman, the junior manager on the struggle up the ladder of promotion, the senior manager who has reached or passed his zenith, the specialist who is frustrated because he can advance no further are all liable to greater stress than either the rank and file or the managing director.

These people frequently become casualties of the commercial battlefield and present in the medical departments with physical and psychological symptoms. The particular manifestation will vary according to genetic inheritance, to upbringing and education, and to the level of intelligence and insight of the individual. Some present with overt anxiety states, some with depression occasionally complicated by alcoholism, some with coronary heart disease, peptic ulceration or with irritable colon. Some will have bronchial asthma, chronic skin disorders or frequent absences for a succession of trivial injuries or illnesses. More often than not breakdown occurs when pressures at work cannot be discussed at home. Threats of demotion coupled with thoughts of divorce form a recipe for personal disaster. In these situations the patient's manager is usually only too anxious to be helpful. Especially with alcoholics, he may become over-protective and indulge in cover-up, thus delaying the inevitable crisis and its resolution by condoning late arrivals, long lunch breaks and frequent short absences because of personal liking for the patient. Alcoholics are often highly skilful manipulators and may gain sympathy from the most hard-hearted bosses. Grant[17] stresses the importance of avoiding involvement in this manipulative process by spurious promises of abstinence in return for discretion.

An important part of the treatment of mental illness is an explanation of the situation to the senior manager by the occupational physician. Depression frequently causes a gradual fall-off in work performance and comes to light only when a manager asks whether the patient should be relieved of responsibility or retired on medical grounds. These patients are often suspicious of company doctors and, claiming that their own doctor is treating them, refuse to discuss their problems. It is important to recognize withdrawal as a symptom of the underlying depression and to persist in attempts to help. Managers find it difficult to appreciate the long time span of change in depressive illness and the frequency of relapse, often extending over periods of years rather than months. The over-conscientious obsessional often makes a good middle manager until some life crisis projects him into depression or, in either sex, the menopause provides added physical, social and psychological stress which precipitates a depression.

Counselling in these situations may achieve results unobtainable by the most potent drugs. If a man or woman can be brought to an understanding that they can continue to contribute to the business, although their ambition may be at an end, then they may adjust to a new lifestyle. Often these crises coincide with the departure of children from home, the menopause and the realization that the chairman's desk is finally beyond reach. Ten or more years until retirement stretches ahead and uncomfortable facts may have to be faced. Most people can cope with a crisis at work or at home, but when the two occur simultaneously coping mechanisms fail and mental illness results.

Occupational medicine is a clinical specialty with strong links to preventive medicine. The prevention of stress-induced illness, whether physical or mental, is a complex process. The doctor should know the management structure of his company, should have some understanding of the internal politics of the business and should be able to share the problems whilst remaining on the sidelines of the organization. Fortunately he is well placed because he is precluded from promotion in line management. He shares this quality with the nursing staff who are seen to be part of the company but without entering into the rat race of promotion. Often the opportunity to talk through a problem with an interested, informed and independent counsellor is sufficient to enable the patient's own coping mechanisms to come into play and prevent further trouble. Experienced, sympathetic and well-motivated nursing staff can counsel successfully but the younger nurse needs special training in counselling techniques. The medical and nursing professions in hospital are so accustomed to giving didactic instructions, under the guise of advice, that they find it difficult to adjust to the listening role. A period in general practice is helpful for the doctor but few occupational health nurses have worked outside the hospital environment. Active listening is a technique which has to be learnt. The doctor or the nurse should persuade the patient to work out his own solution. In the long run this is less time-consuming because coping mechanisms will have been brought to a state of readiness to deal with the next crisis.

Many of these problems present to the medical department as trivial physical illnesses, for example recurrent headaches, frequent short absences for a variety of minor ailments, attacks of dyspepsia, acute anxiety states, pain in the chest and so on. The alert nurse will find time to go into the problems more deeply, grasping an opportunity when it arises. A planned consultation is often resisted by the patient, who perversely seems willing to talk only when the waiting room is full of patients at the end of a busy day. It is often better to persuade the patient to talk over his problems with his immediate manager, but if the manager is antipathetic an offer to intercede may be welcomed.

**Executive health programmes**

The practice in some companies of offering regular screening to senior executives provides an opportunity to discuss health problems in the wider context of the career of the individual, especially when this is done by the occupational physician himself and not by an outside consultant. However, the evidence in favour of executive health programmes in the discovery of physical disease is equivocal. D'Souza[18] is sceptical of their value, while Wright[19] is enthusiastic. Early diagnosis of coronary heart disease, of carcinoma or diabetes or of other life-threatening diseases is no better from routine screening than from prompt investigation of early symptoms. Useful screening programmes may be conducted within the company's own occupational health service but such schemes must be genuinely voluntary and this may be more difficult to achieve than one imagines, especially if the chairman regards the examinations with enthusiasm. Secondly, they must be genuinely confidential and thirdly, the results must be fully discussed with the individual. The yield of significant abnormal physical findings is likely to be low except for obesity, systemic hypertension, obstructive lung disease due to tobacco and general flabbiness due to lack of exercise.

## ORGANIZATION OF OCCUPATIONAL HEALTH SERVICES: GENERAL

The provision of an occupational health service to a commercial or retail organization is in principle the same as for manufacturing industry. An administrative unit consisting of a chief medical officer (CMO) and a chief nursing officer (CNO) or adviser supported by appropriate secretarial staff is recommended for any company employing more than say 10 000 people. The CMO will respond to the chief executive of the company or at least have direct access to him when necessary. The CMO's first task is to establish the aims and objectives of the service and to agree them with the chairman. This rule applies whether in a new service or in taking over from a retiring predecessor. Next, agreement must be sought on a budget, in terms both of money and of manpower, to fulfil these aims over a period of time, with appropriate priorities. These policies should be examined critically at least annually. Third, suitable medical, nursing and other staff are recruited or if already in post are professionally supervised. It is common practice in commerce for the local or branch medical and nursing staff to be responsible to local management, as are the accountants, personnel managers, engineers and so on, with strong links to the relevant head office department. This system functions satisfactorily provided that there is good communication between line managers and professional heads of departments. Medical conferences are an invaluable means of establishing personal contacts, of explaining new policies and practices, of seeking new ideas and of reaching common high standards of work.

Professional training of nursing staff is the second priority of the CNO, the first being recruitment of nurses. The CNO should have a plan of training for each nurse in the company tailored to her or his individual needs and reviewed annually. This will include at one extreme attendance at day release courses for the Occupational Health Nursing Certificate (OHNC) — usually over a period of 18 months — to, at the other extreme, the occasional meeting of a local occupational health group. Training within the company on commercial matters gives a better understanding of the business, enables occupational health staff to identify with the enterprise and to share and appreciate the hopes and fears of their colleagues.

Occupational health medical staff, some of whom will be part-time general practitioners selected by the CMO in collaboration with local management, and some full-time appointed centrally, should attend some or all of the medical conferences, but their professional supervision and training are clearly a priority for the CMO. Only through them and the nursing staff is he able to achieve his objectives. A common administrative arrangement is a nucleus of full-time accredited specialists in occupational health working from head office, to provide professional advice to management within a reasonable distance and for part-time doctors to be appointed further away or for small units. Increasingly occupational health trained nurses are taking over the administration of distant or isolated units and in these situations the part-time doctor advises on clinical problems leaving the environmental aspects of his work to the central medical staff. It is difficult for a part-timer, especially in commerce, to gain enough experience in occupational health to provide expert advice without reference to the better trained full-time doctors at head office. Where there is a specific health problem, for example noise-induced hearing loss in a warehouse or shift working in computer control, then the part-timer can become very knowledgeable but in a limited field. The well-trained occupational health nurse is better able than the part-time doctor to recognize occupational hazards and to supervise the health of workers in general. In these cases the nurse should be in charge of the department, under local management, and the part-time doctor should act as a visiting consultant.

It is important for the full-time doctors and the CNO to establish their differing roles in professional supervision of nursing staff. There is an obvious dividing line when dealing with individual patients. The doctor, whether full-time or part-time, should clearly be in clinical charge. However, in dealing with groups of employees, for example in a food hygiene training programme, the nurse will organize the arrangements, the doctor providing some lectures. Cases of premature retirement on health grounds will be handled by the doctor whereas rehabilitation after minor illness or injury would be in the nurse's province. Between these extremes, each will contribute.

The CNO is responsible for supervision of all nursing staff. She or he will seek opinions from line management and from the company doctors about the performance of the nurses including their professional competence, level of knowledge and application of that knowledge. She will observe standards of appearance, methods of work, organizing ability, counselling skills and judge their effectiveness. It is her job to criticize, to praise, to intervene when difficulties arise, to advise on salary adjustments, to arrange training, to correct faults and to consider promotional opportunities within the company. In recruitment she works with the appropriate personnel manager, reviewing applications, interviewing candidates and putting forward a short list to the line manager with recommendations. She will then arrange a suitable induction course, preferably in another branch, before the nurse takes up her new appointment, and supervise the first few months' work with particular care. She should be responsible to the CMO for these functions.

## ANCILLARY SERVICES

In commerce it is rarely necessary to employ full-time occupational hygienists because specific toxicological or environmental problems are rare. Consultant hygienists are available for advice when necessary.

Physiotherapists are invaluable in offices, shops and warehouses, both in treatment of musculoskeletal injuries and in health education. Part-time appointments of one session per week for up to 1000 people would seem appropriate. Physiotherapy can reduce sickness absence and prevent back injuries which are common in lifting and handling tasks. The physiotherapist must be responsible to the company doctor for her work and, except in an emergency, all cases should be seen by a doctor.

In a large office or shop the nurse is responsible for organizing first aid, particularly for major emergencies such as fire or explosion. City centre department stores unfortunately are a frequent target for terrorists, so plans should be drawn up with management for handling large scale casualties. Explosions and fires can cause large numbers of minor injuries from broken glass and masonry and many people may suffer from nervous shock. First aiders should be trained to deal with these cases, leaving nurses and doctors free to cope with the major casualties and with triage. First aiders can be trained by the St John Ambulance Brigade, Red Cross or by St Andrew's Ambulance Brigade or within the company to certificate standard, but the nurse should arrange brief refresher training sessions every few months and give responsibility to first aiders in her absence in order to maintain their skills and to retain their interest.

Chiropodists are usually appointed in shops because many men and women who stand for long periods at work develop foot ailments. This is a worth-while service, often paid for by the patients but subsidized by the employer.

All these ancillary services should be under the general management of the CMO.

The CMO should lay down criteria for pre-employment medical examinations. Nurses conduct the majority of these tests referring doubtful cases to a doctor, but the nurse should have the authority to reject on medical grounds applicants considered unfit for the job. Subsequent routine health checks are offered in some companies or for specific groups — for example drivers and store detectives — or at certain ages — for example when approaching retirement. The CMO's responsibility is to set appropriate standards and to ensure that these standards are consistently reached and applied throughout the company.

The majority of employees will work through to normal retirement age, but the medical service will offer advice to managers and to the pensions department on the criteria for premature retirement on medical grounds. The CMO should see all the papers in these cases so that consistent standards are applied. Advice is given on sick pay schemes in general and specifically about difficult or unusual cases.

In retail companies advice is often sought by buyers on the safety of the products sold. International standards are now recognized for labelling of such common items as pesticides, cleansing agents and cosmetics, but occasionally a new product will prove quite complicated to investigate. Exaggerated accounts of supposed dangers appear frequently in the media. For example, alleged carcinogenicity of hairdyes following the finding of a positive Ames test caused alarm amongst hairdressers.[20] More recently advice was sought about the risk of toxic shock syndrome from tampons.[21]

In summary, the job content of a CMO in commerce is more varied than many realize — clinical investigation of a 40-year-old manager with chest pain; the analysis of 100 audiometry results in a textile plant; answering a buyer's inquiry about the toxicity of a new paint stripper; advising a nurse about the placement of an epileptic; recommending a premature pension for a case of terminal cancer. All may occur in a single day, and one never knows what the next day will bring. Although occupational physicians increasingly rely on nurses and are constantly adding to the responsibilities placed upon them, a commercial organization needs both professions working together.

REFERENCES
1. Kearns JL. Standards of food hygiene. In: Ward Gardner A (ed.) *Occupational Medicine.* Bristol: Wright, 1979.
2. Shops Act 1950. London: HMSO.
3. Offices, Shops and Railway Premises Act 1963. London: HMSO.
4. Health and Safety at Work etc. Act 1974. London: HMSO.
5. Consumer Protection Act 1961. London: HMSO.
6. Dangerous Goods Act 1978. London: HMSO.
7. The Food Hygiene (General) Regulations 1970. London: HMSO.
8. 'Mind Your Back.' London: Millbank Films, 1980.

9. Parrott WF and Blyth W. Another causal factor in production of humidifier fever. *J. Soc. Occup. Med.* 1980; **30**:63–8.
10. Lighting in Offices, Shops and Railway Premises 1969. London: HMSO.
11. Reading M. *Visual Aspects and Ergonomics of Visual Display Units.* London: Institute of Ophthalmology, 1978.
12. Grandjean E and Vigliani E. *Ergonomic Aspects of Visual Display Terminals.* London: Taylor & Francis, 1980.
13. Health and Safety Executive. *Human Factors Aspects of Visual Display Operation.* London: HMSO, 1980.
14. McLean A. *Work Stress.* Reading, Mass.: Addison–Wesley, 1979.
15. Hollingworth A. Stress and the individual. In: Ward Gardner A. (ed.) *Current Approaches to Occupational Medicine.* Bristol: Wright, 1979.
16. Marshall J and Cooper CL. *Executives under Pressure.* London: Macmillan, 1979.
17. Grant M. Alcohol-related disabilities. In: Ward Gardner A. (ed.) *Current Approaches to Occupational Medicine.* Bristol: Wright, 1979.
18. D'Souza M. Screening for all: excellence or extravagance? In: Ward Gardner A. (ed.) *Current Approaches to Occupational Medicine.* Bristol: Wright, 1979.
19. Wright HB. *Executive Ease and Dis-ease.* London: Gower, 1975.
20. Leading Article. Hairdyes and cancer. *Lancet* 1975; **2**:201.
21. Leading Article. Toxic shock and tampons. *Br. Med. J.* 1980; **281**:1161–2.

# 5. OCCUPATIONAL HEALTH PROBLEMS IN DIVING
## R. R. Pearson

**INTRODUCTION AND HISTORICAL BACKGROUND**

The ability of man to carry out useful work underwater dates back to pre-Christian times and, according to Herodotus 460 BC, divers are said to have been used by Xerxes in an offensive role and to recover treasure from sunken Persian ships. Certainly, Alexander the Great employed divers in 332 BC to destroy the boom defences of Tyre. No clear evidence exists to describe how these divers operated and it is probable that they were breath-hold divers, although Aristotle in 360 BC described a form of appliance to allow breathing underwater. By description, Aristotle's appliance appears to have been a bell, which, together with breath-hold diving, remained the only practical way for man to venture underwater until comparatively recent times. Breath-hold diving has continued until modern times in the form of the sponge divers of the Eastern Mediterranean and the pearl divers of Korea, Japan and other Pacific regions.

The Korean and Japanese pearl divers, called amas, reached depths of up to 30 m. These breath-hold divers showed considerable and interesting physiological adaptation to underwater work but have virtually been eliminated by the availability of cheap and reliable underwater breathing apparatus. To non-experts, breath-hold diving has many hazards, particularly where hyperventilation is employed to extend breath-hold endurance by washing out carbon dioxide and delaying the so-called 'break-point' at which the ventilatory stimulus caused by a rising arterial partial pressure of carbon dioxide ($Pa_{CO_2}$) becomes impossible to resist. With such hyperventilation, the poor respiratory stimulus derived from a falling arterial $Pa_{O_2}$ may lead to unconsciousness underwater due to hypoxia without the diver being aware of any impending problem.

Diving bells with open bottoms — not to be confused with the modern counterpart to be described later — allowed a diver to take down his own air supply, albeit limited by the size of the bell. The first recorded successful use of a diving bell was in 1531 when Lorena used one to remain underwater for 1 h during an attempt to raise one of Caligula's pleasure galleys which had sunk in the lake of Nemi. From then on, there were many different bells used with varying degrees of success for a variety of underwater tasks. Papin, in 1689, proposed a method of supplying fresh air to the submerged bell thereby greatly increasing the endurance of the divers, although his suggestion does not appear to have been used until 1788 when Smeaton, who had earlier achieved fame as the constructor of the third Eddystone lighthouse, used a force-pump mounted on the roof of the bell to provide a stream of fresh air to the occupants.

An offshoot of the diving bell was the caisson, of which the first recorded example is that designed by Williams in 1692. A caisson is essentially a tube with an open bottom end reaching from the surface to a work site, either underwater or in a water-logged stratum. The work site is kept free of water by filling the tube with compressed air and workers gain access to the work site through air-locks in the upper end of the tube.

The basic concept of the caisson has remained essentially unchanged and it is still an invaluable means of carrying out civil engineering work in relatively shallow waters or marshy ground.

Despite the many ingenious devices to assist man in his underwater work, the major goal has always remained the provision of relatively unlimited breathing gas to the individual diver. This clearly meant supplying gas from the surface or providing the diver with his own portable supply of breathing mixture. The goal of much greater flexibility to be gained from such methods led to a great number of ill-fated or impractical inventions. It is now generally acknowledged that practical success in this area was first achieved by Augustus Siebe with his closed diving dress.

Whereas some success had been achieved by reducing the concept of a diving bell to a helmet applied to the diver's head with a compressed air supply coming from the surface by hose, such systems were open and air vented freely out of the bottom of the helmet. They had the disadvantage that the helmet could flood if air pressure was not maintained or if the diver bent forward. Siebe had himself designed such a helmet in 1819 and it was used in the salvage and demolition work on the wreck of HMS *Royal George* at Spithead in the Solent between 1834 and 1844. Siebe's 1837 closed dress enclosed the diver in an air-tight suit which was fastened to, and continuous with, the helmet. As before, compressed air was fed from the surface to the helmet but control of supply and venting became necessary to avoid overinflation and excessive buoyancy. This control was achieved by inlet and exhaust valves on the helmet.

The closed dress was most effective during the later part of the work on the *Royal George*.

The Siebe closed dress, with little change, continued into the 1950s as the main underwater breathing apparatus for both commercial and certain types of military divers. Called the Standard Rig, it is still in limited use in shallow waters where the diver needs a firm basis from which to carry out heavy work, and in this respect the lead boots, which form an integral part of the outfit, are invaluable.

The truly self-contained diver, as opposed to the surface diver, probably stems from an 1825 invention of W. H. James whereby the diver carried his compressed air in a metal belt round the waist. No attempt was made to eliminate carbon dioxide and the diver's endurance was inevitably very limited. The concept was developed to include regeneration of breathing gas by caustic potash to remove carbon dioxide with the diver breathing from a counter-lung, which could be periodically refreshed with air or, in later developments, oxygen. Again, use of a closed circuit breathing apparatus of this basic type has continued into modern times although it is now virtually limited to military diving operations.

As evidence that rarely is anything truly innovative, very recent years have seen the development of extremely complex closed circuit deep diving sets where economy in the use of helium is of major importance.

Finally, military operations by frogmen in World War II highlighted the flexibility of the self-contained diver and the development of the Cousteau–Gagnan aqualung in 1943 represented another major advance. The Cousteau–Gagnan development was significant in that it used a refinement of the 1866 invention by Rouquayrol of the 'demand regulator'. This regulator allowed automatic self-regulating delivery of a breathing mixture to a diver.

The aqualung led to a veritable explosion in diving activity and the demand regulator, or demand valve as it is now called, is an essential part of what is now commonly known as a SCUBA set, which stands for self-contained underwater breathing apparatus. Perhaps the main beneficiaries of this development have been sports divers, and diving is now a major recreational activity on a world-wide scale.

For a full historical account of the development of diving equipment and techniques through the ages, the reader is referred to the classic work of Sir Robert H. Davis *Deep Diving and Submarine Operations*.[2] Although currently out of print, editions of this work exist in most good libraries.


## LIMITATIONS OF THE DIVER

In all the developments referred to, the diver is exposed to the pressure of sea water either directly or indirectly and must breathe air or some other gas at a pressure equivalent to his depth in the water. While the physiological implications of this are profound and extremely complex, it is necessary to describe briefly the problems associated with the breathing gases for they exert a major influence on the depth to which divers may go.

## Air

Inevitably, air formed the basis of all underwater breathing gas until relatively recent times. Unfortunately, the 79 per cent nitrogen content of air begins to exert a narcotic effect as the diver descends. Each 10 m of sea water (msw) involves an increase in pressure of 1 atmosphere (or 1 bar) and air supplied to a diver at 30 msw needs to be at a total pressure of 4 atmospheres absolute. The term 'absolute' indicates an overall pressure which includes the one-atmosphere environment from which the diver began his descent from the surface. At a pressure of 4 atmospheres, the partial pressure of nitrogen in compressed air would be 3·16 atmospheres (that is, 0·79 × 4, in accordance with Dalton's law) and it is roughly at this level that the effects of nitrogen narcosis become noticeable and disabling to a diver. Although nitrogen is an inert gas, its narcotic potential is quite high, particularly in comparison with some other inert gases, and the diver becomes increasingly handicapped in both manipulative and cognitive skills as he goes deeper. The narcotic process is much akin to anaesthesia and is accompanied by euphoria in its early stages. Although dives to 90 msw and beyond have been carried out on air, the risks are high and the divers ineffective. Commercial diving on compressed air has been limited to 50 msw by United Kingdom legislation. With increasing depth, the partial pressure of the 21 per cent of oxygen in air rises to toxic levels and the very density of air makes it difficult for the diver to breathe from his set. However, neither of these factors is a problem to the diver at depths of less than 50 msw.

## Oxygen

Theoretically, the ideal gas for divers to breathe, pure oxygen, is toxic to the diver in the short term sense at pressures in excess of 1·2 atmospheres. This acute toxicity manifests itself mainly on the central nervous system and is sometimes referred to as the Paul Bert effect. Epileptiform convulsions may occur with little in the way of premonitory signs. The threshold for such acute oxygen toxicity is a time–pressure multiple, and in military diving divers are limited to 8 msw breathing pure oxygen, a depth at which endurances of up to 75 minutes are possible. Pure oxygen is rarely, if ever, employed in commercial diving.

Breathing mixtures containing a partial pressure of oxygen in excess of 0·6 atmospheres may give rise to toxic problems in the lung if breathed for relatively long periods. To avoid pulmonary oxygen toxicity, which is also called the Lorraine-Smith effect, breathing mixtures in saturation diving, where prolonged periods are spent under pressure, do not normally exceed 0·5 atmospheres oxygen partial pressure for any great length of time.

## Helium

In an attempt to overcome the depth limitations imposed on the air diver by nitrogen narcosis and gas density, a light inert gas with a low narcotic potency is necessary to act as a carrier gas, or diluent gas, for the oxygen which must be supplied to the diver. Helium, being non-narcotic, is ideal for such a purpose and so-called heliox diving had become a standard procedure in the United States Navy by the late 1930s. Virtually all commercial and military diving to depths in excess of 50 msw is now carried out with oxygen–helium mixtures. Helium does possess disadvantages in that great speech distortion occurs and speech processors become necessary to maintain communications with divers. A further disadvantage is that heliox mixtures greatly accelerate heat loss from divers, particularly respiratory heat loss. For this reason, it is necessary to actively heat divers for all but the shortest and shallowest dives on heliox mixtures. United Kingdom regulations require suit heating for all dives on heliox below 50 m with the addition of breathing gas heating for dives below 150 m.

Other inert gases have been tried as diluent gases but are unsuitable for reasons of narcotic properties or, in the case of hydrogen, instability when mixed with oxygen. A small amount of diving has been done with neon to depths of 180 m and the use of mixtures of two inert gases is in growing use. For the foreseeable future however, helium, despite its high cost, will continue to be used as the principal inert gas for all diving below 50 msw.

## OTHER LIMITING FACTORS

Without doubt, the underwater environment is singularly hostile and diver performance is limited by a whole host of factors which usually act in combination to affect performance adversely. Gas density alone ensures that the limit to physical exertion underwater is of a respiratory nature in contrast to the cardiovascular limitation which normally applies at the surface. It is only possible to list some of the many other hostile features such as the density of water, cold, limited vision (at best), visual distortion, tidal conditions, weightlessness and the cardiovascular responses to immersion. Also, the need to supply breathing mixtures at increased pressures leads to increased uptake of inert gases by body tissues which may require a gradual return to the surface and normal pressure if the inert gases are to be released in an orderly fashion without initiating decompression sickness. The need for controlled decompression is fundamental to the selection and viability of many of the techniques in use in commercial and military diving.

## DIVING TECHNIQUES

Until the advent in the 1960s of saturation diving as a practical procedure, all diving was limited in depth and time by the decompression techniques necessary to avoid decompression sickness. Tables exist to inform divers as to when decompression is necessary and at what depth decompression stops have to be carried out in the water before return to surface. In general, the technique of bounce diving, a term in general use for non-saturation diving, only allows relatively short working periods underwater. A typical example for an air diver would be an 85 minute decompression for 45 minutes work at 50 msw and for a heliox dive to 150 msw, something over 6 hours' decompression would be required for 15 minutes work. Therefore, in commercial diving, the use of bounce diving techniques below 50 msw is very limited.

Although it is uncertain as to where the credit lies for envisaging saturation diving techniques as a means of getting much longer underwater working periods for divers, the SEALAB and TEKTITE series of dives, carried out by the US Navy and US Department of the Interior respectively, showed that it was possible to keep men under pressure for prolonged periods during which time their ability to work was not interrupted by the need for decompression. The SEALAB and TEKTITE series which took place in the 1960–70 period used air, oxygen–nitrogen and oxygen–helium mixtures for breathing gases in underwater habitats which acted as a home for the divers between in-water working periods.

The term 'saturation' applies to an equilibrium reached by body tissues in relation to the extra inert gas, be it nitrogen or helium, taken up by tissues in the hyperbaric state. This uptake of inert gas is controlled by the rate at which various body tissues take up the inert gas. There are differences between the rates of uptake for nitrogen and helium, but after a period usually taken to be 6–8 hours, the body tissues are regarded as saturated with inert gas. Of course, the term saturation applies to exposure to a constant pressure and any increase or decrease in pressure will initiate further uptake or release of inert gas. A further important feature of saturation diving is that once a diver is saturated with inert gas, the decompression obligation, although increasingly long as depth increases, will at least be constant and unaltered by the time the diver stays at pressure. Therefore a diver who spends 20 days at a steady pressure will not require any longer decompression than a diver who has spent one day. Representative decompression times for saturation dives to 100 msw, 200 msw and 300 msw would be in the order of 4, 7 and 10 days.

Following the early experience with underwater habitats, saturation diving as a technique has become widely used for both commercial and military diving. The habitat concept has been replaced by deck chambers on diving vessels or platforms and these deck chambers act as the living quarters for the divers during the time spent under pressure. A saturation dive is initiated by compressing the divers in the deck chambers to a pressure equivalent to their intended working depth. Once at the intended pressure, and recovered from any side effects of compression, the divers are transferred under pressure to a bell, sometimes termed a PTC (personnel transfer capsule) or SCC (submersible compression chamber), which is mated to one of the deck chambers. The bell is then sealed, detached from the deck chamber and lowered into the water over the side or through the hull of the ship or platform. At the intended working depth, the pressure within the bell will be equal to that of the surrounding water and the bell door can be opened to allow the diver or divers to commence in-water work. Invariably, the diver's breathing gas will be supplied from the bell via a gas umbilical hose to which is usually attached the hot water hose for his suit and breathing gas heating. The diver carries a limited supply of breathing gas in emergency bottles to enable him to reach the bell in the event of a failure of his normal breathing gas supply. The bell is always manned by an attendant who may alternate with the in-water diver to allow a longer period of work to be carried out. In practice, the bell may be deployed for as long as 10 hours, although periods of 6–8 hours are more common. Once work is completed, the bell is closed at depth, recovered on board and re-mated to the deck chamber allowing the divers to return to their pressurized living quarters. Most saturation diving systems are large enough to allow a fresh team of divers to be deployed immediately, thus achieving what is virtually uninterrupted work.

Life support systems enable the gas within the deck chambers and diving bell to be monitored, circulated and purified by removal of carbon dioxide and odours. A very accurate check must be kept on the oxygen content of the chambers and the addition of oxygen for the metabolic needs of the divers must be carefully regulated. The life support system must also be able to control chamber temperature to a high degree of accuracy and also regulate humidity within the chamber.

The modern saturation diving system is indeed a complex affair with control panels that, in their multiplicity of gauges, dials and meters, reflect the need to monitor and control all phases of the dive. However, it is this close control, which includes communications and television monitoring, that makes saturation diving safer in many ways than bounce diving.

Although an operational dive has been carried out to 450 msw, saturation dives in North Sea and other commercial diving operations are rarely in excess of 180 msw. In the southern sector of the North Sea — an area off East Anglia which contains rich deposits of natural gas — diving is almost exclusively on air to depths shallower than 50 msw.

Research into diving at great depths continues in many parts of the world and a dry experimental dive, confined to compression chambers, has been carried out by the British Ministry of Defence to a depth equivalent to 660 msw. Such dives have revealed a variety of problems which indicate that it will yet be some time before a diver can work in the water at such depths. Indeed, a considerable body of opinion would suggest that 450 msw will remain the limit for practical diving operations.

For detailed descriptions of the physiological and human aspects of diving, the reader is referred to *The Underwater Handbook* of Shilling, Werts and Schandelmeier[7] and *The Physiology and Medicine of Diving and Compressed Air Work* by Bennett and Elliott.[1]


## NUMBERS AND USE OF DIVERS

There are currently approximately 1000 commercial divers working offshore in the UK sector of the North Sea. This includes divers working in both the northern and southern sectors of the North Sea which are respectively associated with oil and natural gas production. The diving activity in the North Sea and other areas of the UK continental shelf reaches a peak during the summer months and is strictly limited by weather conditions during the winter. However, diving is an international activity and the offshore exploration of petrochemical reserves is a major industry in the South China Sea, New Zealand, Western Australia, West Africa, the Persian Gulf and several areas in North and South America to name only some of the areas involved. In the offshore industry, divers are principally involved with construction and maintenance of offshore structures and in underwater pipe laying. As exploration for natural resources moves into deeper waters, divers are becoming less and less involved with this aspect of offshore work.

The attention received by the offshore industry tends to disguise the fact that in the UK there are perhaps as many as 3000 or more divers employed in other diving activities. These divers are largely involved in shallow diving in association with civil engineering projects, salvage and underwater demolition, harbour and canal maintenance, and a very varied range of scientific projects. Additionally, divers are employed by a variety of organizations such as fire brigades, police forces and several government departments. Military diving tends to be of a highly specialized nature using a range of breathing mixtures and techniques which have no counterpart in commercial diving.

## LEGISLATION

The veritable explosion of world-wide commercial diving activity has led to a plethora of international legislation. In the case of the UK, however, legislation to regulate the conduct and safety of diving has evolved in a deliberate and controlled way, with measures which have, on the whole, been welcomed by the diving industry and, without doubt, have been used world wide as the basis for other national legislation. UK legislation is entirely in harmony with Norwegian legislation but detailed differences do exist with legislation in other parts of the world. The Diving Operations Special Regulations 1960, made under the Factories Act, were the first specific attempt at controlling commercial diving in the UK. As new diving techniques extended the sphere of diving activities, new legislation evolved as follows: The Offshore Installations (Diving Operations) Regulations 1974; The Merchant Shipping (Diving Operations) Regulations 1975; The Submarine Pipe-lines (Diving Operations) Regulations 1976.

As the name of each set of regulations implies, diving from fixed offshore installations and ships is covered as is pipeline work in coastal waters.

In 1981 the Health and Safety Commission introduced a set of harmonized regulations after some 3 years of consultation with employers of divers and their customers. The Health and Safety Diving Operations at Work Regulations replaced all but the Merchant Shipping (Diving Operations) Regulations of 1975 and, in the case of diving operations covered by both sets of regulations, the Health and Safety at Work Regulations take precedence.

The new regulations have one feature of great importance in that they include for the first time a definition of what constitutes a commercial diver. They are quite specific in their application to 'all diving operations carried on in the course of or in connection with any profit or not'. This definition has had far-reaching effects in bringing groups of divers such as scientific divers within the legislative control of government.

The new harmonized regulations certainly represent a great improvement on the Factories Act Regulations in their application to the standards and conduct of non-offshore commercial diving.

## HEALTH STANDARDS FOR DIVERS

The Health and Safety Diving Operations at Work Regulations, introduced in July 1981, require that divers who are subject to this legislation will comply with a set of health standards to be decided by the Health and Safety Commission's Medical Advisory Committee. These Health Standards, finalized after a widely ranging consultative process also became operative on 1 July 1981. They represent an attempt to subject divers to a screening process which will require expert opinion on any deviation from a clearly defined set of rigid standards. An appeal process already exists to arbitrate in cases where divers disagree with an adverse medical opinion on fitness to dive. Since the advent of the 1974 Offshore Installations (Diving Operations) Regulations, suitably experienced doctors have been formally approved to examine offshore commercial divers for fitness to dive. The 1981 regulations added doctors appointed under the 1960 Factories Act regulations to the approved doctors list making a larger body of doctors available to examine divers covered by the new regulations.

Approval to examine divers does not recognize any expertise in the area of treating diving accidents. However, a significant degree of knowledge of diving techniques and potential health problems in divers is essential if decisions on fitness to dive are to be reliable and it is now required that approved doctors have had a recognized course in the medical aspects of diving. Further information on the qualifications necessary for medical practitioners to be granted the status of an approved doctor may be obtained from the Employment Medical Advisory Service, Health and Safety Executive, 25 Chapel Street, London NW1 5DT.

## OCCUPATIONAL HAZARDS AND OTHER HEALTH PROBLEMS IN DIVING

It is conventional to classify diving-related health problems into short term and long term aspects of the compression, bottom time, which is the time spent at depth before decompression begins, and decompression phases of diving.

### Compression barotrauma

Although body tissues can, for practical purposes, be regarded as fluid and incompressible, the gas-containing spaces in the body may be unable to communicate freely enough to prevent tissue damage as the effect of pressure causes the volume of contained gas to contract in accordance with Boyle's law. Typical of such damage is barotrauma to the sinuses and tympanic membranes, which may occur if sinuses or the eustachian tubes are blocked as a diver descends through the water or is compressed in a chamber. Sinus barotrauma most frequently affects the frontal or maxillary sinuses and is easily diagnosed by intense localized pain with occasional bloodstained nasal discharge. Treatment is entirely conventional, although repeated sinus barotrauma should indicate an end to diving.

Compression barotrauma to the tympanic membrane may range from minor haemorrhage to frank perforation. Pain is again the characteristic presentation with accompaniment of aural bleeding in most cases where perforation has occurred. Treatment is again conventional and prophylactic antibiotics are sometimes recommended where frank perforation has occurred.

One quite rare but serious form of compression barotrauma may occur to the middle ear, with rupture of the round window. This may be the result of overvigorous attempts at ear clearing or may be an alternative manifestation of eustachian tube blockage. The presentation of middle ear barotrauma is usually dramatic with sudden, extreme vertigo, severe unilateral deafness and tinnitus. Early diagnosis and corrective surgery may avert permanent hearing loss and the association of such symptoms with compression is virtually diagnostic.

Noise-induced hearing losses may occur in divers who are exposed to excessive noise from gas inlets in compression chambers. Hearing protection should be provided in compression chambers.

If the external ear canal is blocked by wax or sealed by a tightly fitted diving hood, compression may give rise to a condition known as 'reversed ears'. This somewhat misleading term applies to damage to the lining of the canal as tissues try to compensate for the contracting gas. Typically, large haemorrhagic blisters form which may rupture to give frank bleeding. This condition is peculiarly free from pain and does not require specific treatment.

Compression barotrauma to the lungs is rare and only occurs where a diver is being supplied with gas from the surface and, for a variety of reasons, the gas pressure falls below that required for the depth of the diver. If the diver survives such an insult, which is often termed 'lung squeeze', treatment is similar to other pulmonary conditions where pulmonary haemorrhage and oedema occur.

Diving equipment may give problems during compression and suit squeeze and face mask squeeze may occur with results that are more discomforting than disabling.


**Other compression problems**

With the advent of deeper diving it became clear that certain adverse responses to compression placed a limit on the rate at which men could be compressed. Compression arthralgia is a common accompaniment of compression to depths in excess of 150 msw. It is manifest by aching of the larger joints and, although a wide variation exists in individual tolerance to compression rates, it is rarely disabling if compression rates of 1 m/min are used and divers exercise during their compression in the deck chambers. Compression arthralgia is fortunately self-limiting and disappears in a few hours once compression ceases. The exact cause of this condition is a matter for speculation but an osmotically induced dehydration of the joint surface seems a reasonable explanation for what often feels like dry joints.

A much more limiting feature to compression rates is a complex central nervous system response which is collectively known as the high pressure nervous syndrome (HPNS). Again, whilst there is a wide variation in individual responses, over-rapid compression to depths in excess of 150 msw may result in coarse muscular tremor, vertigo, nausea and vomiting. In more severe cases, a peculiar form of somnolence called micro-sleep may occur and characteristic EEG disturbances have been observed. Animal work indicates that generalized epileptiform convulsions will accompany the most severe forms of HPNS. If slow compression rates of the order of 1 m/min are used, HPNS should not normally be disabling down to depths of 300 msw. However, compression rates below 300 msw need to be slower still or have interruptions at fairly frequent intervals. It has also become clear that HPNS is not solely linked to compression — a relationship to pressure alone has been demonstrated in deep experimental dives. Various recent attempts have been made to suppress HPNS. The most successful have used the addition of 8–10 per cent of nitrogen to the oxygen/helium breathing mixture. It appears that the narcotic effect of the nitrogen acts at cell membrane level to counteract the effect of pressure. However, much research continues into HPNS. It is rarely disabling at the depths achieved by commercial diving.

## Problems of working and living at depths

The diver in the water suffers from many handicaps which combine to limit his performance severely. Reference has already been made to the density of the surrounding medium. The virtually weightless state of the diver requires special torque-free tools or a firm working platform if work is to be effective. Poor visibility and visual distortion combine with weightlessness and relative silence to produce sensory deprivation that may result in disorientation.

The air diver below 30 msw is increasingly handicapped by narcosis and breathing gas density, the latter problem even becoming manifest in heliox mixtures supplied to divers in the water at depths in excess of 300 msw. For very deep diving, modern equipment gives both inspiratory and expiratory assistance to overcome gas density problems. For such equipment, the term push–pull is commonly used.

However, at all depths, perhaps the greatest enemy of the diver is cold. The effects of cold on manual dexterity and strength are well known but the distracting effect of cold, and the impairment of judgement and awareness that accompany the early stages of hypothermia, are less well researched. In general, the targets for thermal protection of divers may be summed up as preventing core or deep body temperature falling more than 1°C in 1 hour and maintaining deep body temperature above 36°C and all other skin temperatures in excess of 25°C.

The diver breathing shallow air must rely on passive thermal protection in the form of dry suits with thermally insulating undergarments or wet suits made of neoprene. However, because of the compressibility of the gas in the cells of neoprene suits, these suits become too thin to afford much thermal insultation below depths of 30 msw. Dry suits are much favoured for shallow diving in cold water and new materials for undergarments ensure good insulation even if the outer suit integrity is lost and the undergarment becomes wet. Gloves and hoods for shallow diving are invariably made of neoprene. There must always be a compromise between the dexterity afforded by gloves with separate fingers and the lesser heat loss from mittens.

For deeper diving, where divers breathing heliox mixtures are quite unable to compensate for respiratory and conductive heat loss, both breathing gas and suit heating must be employed. The most commonly used method is a free-flooding hot water system whereby hot water is supplied from the surface to the bell and thence to the diver where a heat exchanger is used to heat breathing gas and a small bore heating circuit is built into the diver's suit with exit points at the hands and feet. This system, which aims to supply water to the diver at approximately 40°C, is relatively simple but effective and has the particular merit of ensuring a free flow of hot water over the vulnerable hands and feet. Electrically heated undergarments and gas heaters are being developed but will have to overcome the bad reputation for safety earned by earlier attempts at electrical heating of divers.

Loss of heat to a diver or to a bell is a critical situation if recovery cannot be rapidly effected. Two tragedies have occurred in the North Sea where divers have died in bells which have lost heating and could not be recovered for some hours. Much research has gone into providing divers in such situations with passive thermally protective garments and simple regenerative respiratory heat exchangers which use the exothermic reaction of carbon dioxide absorption by an absorbent material to warm inhaled gas. Also, emergency self-contained heaters are now being mounted on diving bells.


**Speech distortion and diver communications**

The most effective way of monitoring divers is by direct communication. For shallow divers breathing air, this is relatively easily achieved by a direct link or by through-water transmission, although the latter has a limited range.

The substitution of helium as the inert gas causes very severe speech distortion. Speech processors, often called unscramblers, are necessary to ensure good communications between the divers themselves and their supervisors on the surface. Within compression chambers divers quite rapidly learn to understand each other, but still require to have their speech processed for intelligibility outside the chamber.

## Chamber environmental problems

A whole range of discomforts and potential hazards face divers during their stay in the deck chambers used for saturation diving.

Thermal balance still remains a problem and the ambient temperature within chambers that allows divers to remain in comfort and in thermal balance rises with increasing pressure. At the equivalent of 100 msw the ambient temperature must be approximately 29°C ± 1°C whereas at 300 msw it needs to be about 31·5°C ± 0·5°C. The temperature zone within which divers feel comfortable narrows with depth and deep dives require most accurate temperature control.

The partial pressure of oxygen in the chamber atmosphere should be kept below 0·5 atmospheres to avoid pulmonary oxygen toxicity. However, during their in-water working periods, higher oxygen partial pressures are customary and 1·0 atmosphere is a commonly used level. Such intermittent exposure to relatively high oxygen partial pressures is tolerable. During a saturation dive, to keep the chamber atmosphere within prescribed limits for oxygen partial pressures requires a gas mixture which varies with depth. To get an oxygen partial pressure of 0·4 atmospheres at 300 msw, only 1·24 per cent oxygen with a balance of helium is required. The same oxygen partial pressure at 100 msw requires 4·4 per cent oxygen.

Ideally, humidity within chambers must be kept fairly high and this, together with the raised ambient temperatures, ensures that within 36 hours of the start of a saturation dive the bacteriological content of chambers is flourishing and hostile. Unfortunately, *Ps. pyocyanea* is an invariable contaminant and, predictably, otitis externa is a common cause of disablement in hyperbaric conditions. Rigorous domestic cleaning routines should be carried out on a daily basis using soft soap and water or proprietary antiseptic solutions which have been cleared for use in such closed environments. Personal hygiene is equally important and regular showers with daily changes of clothing are the rule. Aural hygiene is of the greatest importance and twice daily prophylactic use of 10 per cent aluminium acetate ear drops, to which 1 per cent acetic acid is sometimes added, has proved effective in reducing greatly the incidence of otitis externa. Each ear should be flooded with the drops for at least 5 minutes. If otitis externa does occur, preparations containing gentamicin and polymixin are frequently used but are not always successful and systemic broad spectrum antibiotics may be necessary. A further problem is that strains of *B. pseudomonas* which are resistant to gentamicin may develop rapidly and, in general, this is a malady where prevention is easier than cure.

Bacterial and fungal skin infections are also relatively common and require early, rigorous treatment. Certainly, any chronic skin condition of an eczematous or infective nature is an absolute contraindication to saturation diving.

The dietary requirements of divers during a saturation dive involve a high calorie intake and the weight loss encountered in certain very deep experimental dives may be a reflection of the greater metabolic heat production necessary to ensure thermal balance. A seemingly inevitable accompaniment of saturation diving is some blunting of taste, to which is added the very rapid cooling of hot food in a heliox environment.

While there is no apparent theoretical limit to the duration of saturation dives, 30 days seems a sensible upper limit for a complete dive, although some experimental dives have taken 40 days to complete. For such long dives, in-chamber entertainment systems are essential and much work remains to be done on habitability of chambers.

## DECOMPRESSION

For both the shallow and the saturation diver, decompression is a potentially hazardous time. Decompression sickness, almost universally known as 'the bends', may occur at some stage after surfacing from a shallow air dive or saturation dive, but may also occur during decompression from a saturation dive. Although universally described in terms of mild (fatigue, cutaneous manifestations and joint pain) or serious (central nervous system, vestibular or cardiorespiratory manifestations) presentations, decompression sickness may present in such a variety of fashions that it should be a rule that any unusual sign or symptom arising within 36 hours of diving should be regarded as a decompression illness until proved otherwise.

Another serious hazard of decompression is pulmonary barotrauma due to trapping and expansion of breathing gases within the lung. The immediate result of pulmonary barotrauma is pulmonary interstitial emphysema which may be followed by mediastinal emphysema with tracking of gas into the neck or rupture in the pleural cavities. Gas may sometimes pass from the ruptured alveoli into the pulmonary veins and into the left heart to give arterial gas embolism. This most serious complication usually causes cerebral arterial involvement with potentially disastrous results. Coronary artery embolization is a relatively infrequent result. The differential diagnosis of arterial gas embolism from decompression sickness may be very difficult and, indeed, both may coexist. In general the presentation of arterial gas embolism involves profound and dramatic cerebrally initiated symptoms during or within 5 minutes of the completion of decompression. The gas trapping that initiates pulmonary barotrauma during decompression may be the result of the diver's failure to exhale, a precipitate and uncontrolled ascent or some pulmonary pathology, such as bullae or cysts, that either allows gas trapping at depth or lessens lung compliancy and the ability to withstand sudden decompression.

The treatment of these decompression illnesses is highly specialized and almost invariably involves recompression as the major therapeutic procedure. The provision of on-site compression chambers for therapeutic purposes is required by law for all but the shallowest of commercial diving operations. Decompression sickness occurring during decompression in a saturation dive is treated by recompression either to the depth of relief of signs and symptoms, or to an arbitrary limit. Breathing mixtures which contain higher partial pressures of oxygen than the chamber environment are frequently used on an intermittent basis. For a fuller description of therapeutic procedures for decompression illnesses, the reader should consult either the Royal Navy Diving Manual (BR2806)[6] or the US Navy Diving Manual.[9] Perhaps the major facet in the treatment of decompression illnesses is the avoidance of any unnecessary delay in obtaining expert advice and specific therapy.

### Decompression barotrauma

Both the sinuses and middle ear may become blocked at depth and the reverse process to compression barotrauma may cause sinus pain or damage to the tympanic membrane, although such events are rare.

### Dental problems

Pockets of gas which are trapped in fillings may be a source of pain during pressure changes of compression or decompression and so a high standard of dental fitness is essential for divers.

## DIVING ACCIDENTS

While the initiating cause of most diving accidents is either human error or equipment failure, the final result is often a complex interaction of several factors. Of particular importance is the possibility of partial drowning as a complicating factor both in the presentation and treatment of diving accidents.

Traumatic accidents or intercurrent illnesses occurring during saturation diving are a source of particular concern in that treatment has to be carried out under pressure and the remoteness of most commercial offshore diving ensures that some hours would elapse before skilled medical help could be introduced into the chamber. Further, in deep diving, slow compression rates would have to be used when compressing medical personnel. Thus, much of what is desirable for medical care of acute trauma or illness is either impossible or needs to be modified in hyperbaric conditions and the cramped surroundings of the majority of compression chambers. Even anaesthesia presents severe technical problems. It is well known that pressure may enhance or diminish the effects of many drugs. A surgical and anaesthetic team is available in Aberdeen to deal with North Sea emergencies occurring under pressure and air transportable titanium chambers are also available which would allow transfer under pressure of a patient and attendant to a larger chamber complex in Aberdeen where definitive treatment could be carried out in relatively favourable conditions. It is considered essential that at least some of the divers in a saturation dive have had training in resuscitation procedures. In particular they should be capable of cardiopulmonary resuscitation, establishing and maintaining airways and establishing intravenous fluid therapy. Experience in dealing with accidents and acute illnesses under pressure is very limited but considerable thought has been given to the problems that might be anticipated.

## LONG TERM HEALTH HAZARDS ASSOCIATED WITH DIVING

Despite a great deal of evidence to show considerable alteration in the physiology and function of many bodily systems in response to all phases of diving, it seems that those responses are of a transient nature and do not give rise to permanent changes. Nevertheless, deep saturation diving is a comparatively recent technique and it would be imprudent if a constant watch was not kept on the long term significance of some of the changes observed in such diving.

Perhaps the only recognized long term hazard of diving is dysbaric osteonecrosis, also called aseptic bone necrosis. Dysbaric osteonecrosis is a result of infarcts occurring in either the marrow or cortical bone of the long bones of the legs and upper arms. The reasons for circulatory impairment and infarction are not known but are linked to the decompression phase of a dive and are not always associated with other overt manifestations of decompression-induced illness. Until recently, dysbaric osteonecrosis has mainly been diagnosed by X-ray. The X-ray changes which indicate repair processes at work may not be apparent for months or even years after the initiating infarct. Recent work by Pearson[10] has established that, with the use of radioisotope labelled bone-seeking substances, it is possible to identify bone infarcts within a few days of the completion of a causative decompression. Some of these infarcts do not proceed to X-ray changes. It is almost certain that infarcts in the shafts of the femur, tibia and humerus have little long term significance but lesions of the femur and humerus adjacent to the articular surfaces of the hip and shoulder joints, known as juxta-articular lesions, are much more significant and may lead to joint collapse with crippling arthritic changes. All commercial divers must have bone and joint X-rays before taking up a diving career and thereafter X-ray screening is carried out at intervals decided by the depth to which they dive. For diving in the 30–50 msw range, X-rays at 3-yearly intervals are required with annual X-rays being reserved for divers who go deeper than 50 msw. Currently, the overall incidence of dysbaric osteonecrosis in commercial divers is 4·8 per cent for definite radiographic lesions and only 1·2 per cent for definite juxta-articular lesions. However, from figures produced by the Medical Research Council Registry in Newcastle upon Tyne, where X-rays and records for over 4000 commercial divers are held, it is possible that the incidence of dysbaric osteonecrosis may be rising and may involve over 15 per cent of divers who have dived deeper than 200 msw.

## FUTURE TRENDS IN UNDERWATER WORK

It is certain that for some tasks divers will be replaced by unmanned, remotely controlled vehicles or manned underwater vehicles where the operators remain at normal atmospheric pressure and thus avoid any decompression obligation. Despite great advances in the development and deployment of submersibles and atmospheric diving systems, the manipulative capabilities of such devices leave a good deal to be desired and fall short of the diver's manual dexterity. Also, most developments in this area can only carry out work on the sea-bed or when positively attached to underwater structures. It is impossible to predict the ways in which commercial considerations will dictate future employment of divers but it is likely that the present advantages of submersibles and atmospheric diving systems for underwater inspections and photography will lead to an extension of their use.

REFERENCES

The following books are recommended for both background reading and as a source of specialized information. They include those mentioned in the text.

1. Bennett PB and Elliott DH. *The Physiology and Medicine of Compressed Air Working and Diving.* London: Bailliere, Tindall & Cassell, 1975.
2. Davis RH. *Deep Diving and Submarine Operations.* Surrey: Siebe, Gorman & Co., 1969.

3. Drew EA, Lythgoe JN and Woods JD. *Underwater Research.* New York: Academic Press, 1976.
4. Hills BA. *Decompression Sickness,* vol. 1. Chichester: Wiley, 1977.
5. Miles S and Mackay DE. *Underwater Medicine.* St Albans: Granada, 1976.
6. *Royal Navy Diving Manual* (BR2806). London: HMSO, 1980.
7. Shilling CW, Werts MF and Schandelmeier NR. *The Underwater Handbook.* Chichester: Wiley, 1976.
8. Strauss RH. *Diving Medicine.* New York: Grune & Stratton, 1976.
9. *US Navy Diving Manual,* vol. 1. Navsea 0994-LP-001-9010. Washington DC: US Government Printing Office, 1980.

For more precise information concerning specific areas of research and development the reader is referred to:

10. Pearson RR and MacLeod MA. Bone scintigraphy with $99^m$ technetium methylene diphosphonate (MDP) as an aid to the early diagnosis of dysbaric osteonecrosis. Royal Navy Clinical Research Working Party Report No. 2/76. Alverstoke: Institute of Naval Medicine, 1981.
11. Werts MF and Shilling CW. *Underwater Medicine and Related Sciences. A Guide to the Literature,* vol. 1–4. Bethesda, Maryland: Undersea Medical Society Inc., 1979.

# 6. PROTECTION AGAINST IONIZING RADIATION
## *A. N. B. Stott*

In 1896 the British Association for the Advancement of Science was holding its annual meeting at Harrogate. Its President that year was Joseph Lister, not yet ennobled, but already the foremost physician of his era. In his Presidential Address there occurred a passage which is startling in its prescience. He stated, 'If the skin is long exposed to their action it becomes very much more irritated, affected with a sort of *aggravating sunburning*. This suggests that their transmission through the human body may not be altogether a matter of indifference to internal organs.' Lister was referring to the phenomenon which had caused the greatest excitement in the scientific world since its discovery only a few months before in 1895 by Wilhelm Röntgen — the existence of a 'new kind of ray' which Röntgen termed 'X' because of its unknown nature.

The 'aggravating sunburning' (or erythema) to which Lister alluded had even then become a recognizable result of exposure to these X-rays. The subsequent development of blood disorders and malignancies in the early workers with X-rays, and soon after with radium, led to the creation in Britain in 1921 of an X-ray and Radium Protection Committee to try to establish safer conditions for such work, which was then confined mainly to the medical field. In 1925 there took place the first International Congress of Radiology (ICR) during which the need for standardized radiation units was met by the formation of the International Commission on Radiological Units (ICRU). That same year came the first recommendation for a maximum permissible dose of radiation of one-tenth of an erythema dose per year. At the second meeting of the ICR in 1928 there was established, on the pattern of the British model, the International X-ray and Radium Protection Committee which published the first set of recommendations for protection from ionizing radiation, using the röntgen as the unit of exposure. This committee continued to give guidance on radiation protection but in the post-war period its status became of greater significance because of the rapid developments in the nuclear field and the widespread use of radioisotopes in medical, research and industrial applications. In 1950 it became the International Commission on Radiological Protection (ICRP) and is now regarded as the appropriate body to provide international guidance on standards in the

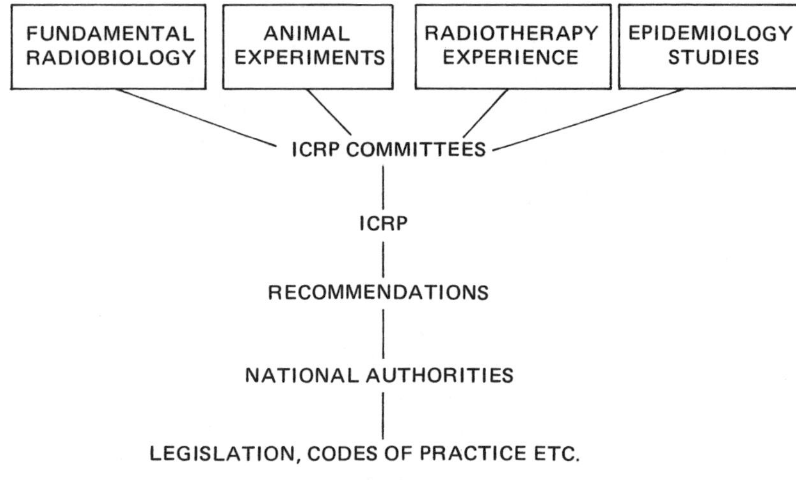


*Fig. 6.1.*

field of radiological protection. This guidance, which is almost universally adopted for the formation of national regulations, is founded on a continuum of information and assessment as illustrated in *Fig. 6.1.*

In order to understand the current philosophy of radiation protection it is necessary to review at least some of the information on which the ICRP bases its recommendations and also to remind the reader of some of the properties of ionizing radiation.

## PROPERTIES OF IONIZING RADIATION

The term 'ionizing radiation' is a generic one which encompasses:

Directly ionizing radiation, such as charged particles (alpha particles, beta particles, fission products, protons and so on).

Indirectly ionizing radiation, such as gamma rays, X-rays and neutrons.

The X-rays discovered by Röntgen are electromagnetic radiations of very short wavelength. An X-ray is produced by the same process as a photon of visible light through the excitation of one of the electrons from the cloud of electrons surrounding the nucleus of an atom. The disturbance or ejection of the excited electron leads to the emission of a photon with an energy corresponding to the energy change in the electron. To produce an X-ray the tightly bound inner electrons have to be disturbed in this way.

Most other radiations of interest arise from excitation of the nucleus of the atom rather than the surrounding electrons. This disturbance or fission of the nucleus can be induced in a nuclear reactor or occur 'spontaneously' in naturally occurring radioactive elements (*Table 6.1*).

**Table 6.1.** Main types of ionizing radiation

| | |
|---|---|
| Alpha particles | The nucleus of a helium atom |
| Beta particles | Positive or negative electrons |
| X-rays | Photons of electromagnetic radiation |
| Gamma rays | Photons of electromagnetic radiation of shorter wavelength than X-rays |
| Neutrons | Particles of neutral charge arising from the nucleus |
| Protons | Particles of positive charge emitted from nucleus or produced by recoil from neutron collisions |
| Fission fragments | The nuclei of atoms produced when an unstable nucleus ruptures into two or more lighter nuclei |

Several types of radiation may be produced in the decay of a single excited nucleus. Thus gamma radiation invariably accompanies the emission of an alpha or beta particle. Fission produces gamma radiation and neutrons as well as fission fragments. In addition, the decay of an unstable nucleus through the emission of radiation may result in another unstable nucleus. There may be a series of such steps before a stable nucleus is reached.

## Radioactivity

All excited nuclei have a finite lifetime which delays the appearance of the reaction products. For some the delay is very short so that for all practical purposes the nuclear reaction can be regarded as instantaneous; an example of this is the fission reaction in uranium. For many excited nuclei the decay products appear an appreciable time after the initial excitation; for example, this is invariably the case with alpha and beta decay. Elements with such nuclei are described as 'radioactive' and are referred to as radioisotopes or radionuclides.

Each radionuclide can be characterized by a 'half-life', the time taken for half the nuclei present to decay. In two half-lives the radioactivity is reduced to a quarter of its original level, and in ten half-lives to about one-thousandth. Over 1000 radioisotopes are known and their half-lives vary from fractions of a second to millions of years. Examples of some common radionuclides and their half-lives are given in *Table 6.2*.

**Table 6.2.** Some important radioisotopes

| Isotope | Half-life |
| --- | --- |
| *Uranium-238 | $4.5 \times 10^9$ years |
| *Radium-226 | 1622 years |
| *Radon-222 | 3·8 days |
| *Lead-210 | 22 years |
| *Carbon-14 | 5730 years |
| *Potassium-40 | $1.36 \times 10^9$ years |
| Cobalt-60 | 5·27 years |
| Tritium | 12·3 years |
| Iodine-131 | 8·04 days |
| Plutonium-239 | $24 \times 10^3$ years |

*It is important to note that these naturally occurring radionuclides have been and will be present always, thus contributing to natural background radiation.

## The absorption of radiation

All directly ionizing radiations (alpha and beta particles, protons and fission products) rapidly lose their energy in transition through any matter. This means that the heavier and more highly charged alpha particles are stopped by the outer layer of the skin and even the lighter beta particles will not penetrate beyond the dermis.

The indirectly ionizing radiation (gamma and X-rays and neutrons) lose energy through random interactions with electrons or nuclei to produce secondary radiation which dissipates by ionization. Since single interactions of this sort only absorb part of the total initial energy the penetration of such radiation is considerably greater. It is this property of course which makes possible their medical uses in radiology or radiotherapy but also imposes a need for greater thicknesses of protective shielding to avoid unintended human irradiation. The required thicknesses of such shielding vary from a few inches of water or lead for laboratory gamma or neutron sources to about ten feet of concrete for the higher intensities emitted from a nuclear reactor.

## BIOLOGICAL EFFECTS

Because of the comparative ease with which experiments can be mounted a great deal of useful knowledge about radiation effects has been gleaned from cellular radiobiology. Two considerations amongst others confirm the validity of such knowledge. Whereas 40 years ago the observable harmful effects of radiation on man and experimental animals were attributed to complex biochemical changes at organ level, it is now considered that most of these effects can be attributed to the loss of proliferative ability, 'cell death', in a high enough number of cells in an irradiated tissue.[1] Furthermore, although the initial mechanisms of carcinogenesis are still far from unravelled, there is increasing evidence that one prerequisite is an effect on the hereditary macromolecule, DNA. In the field of *chemical* carcinogenesis such effects, demonstrating mutational changes in cells treated with the suspected carcinogen, have been shown to be strongly predictive of cancer production in the whole animal.[2]

If cells in culture are exposed to ionizing radiation then for most effects produced it can be shown that such effects are modified by a number of factors. Of these clearly the most important is the *amount* of radiation energy absorbed by the cells. (This absorbed energy per unit mass is measured in rads (*Table 6.3*) and the term 'dose' henceforth used in this chapter implies the more correct term 'absorbed dose'.)

**Table 6.3.** Summary of radiation units

| Quantity | Name | Symbol | Units |
| --- | --- | --- | --- |
| Radiation exposure | röntgen | r | 1 electrostatic unit (esu) in 0·001293 g air |
| Radioactivity | curie | Ci | $3·7 \times 10^{10}$ disintegrations/s |
|  | becquerel* | Bq | 1·0 disintegrations/s |
| Absorbed dose | rad | rad | 100 erg/g (0·01 J/kg) |
|  | gray* | Gy | 1 J/kg (= 100 rad) |
| Dose equivalent | rem | rem | rad $\times Q$[†] |
|  | sievert* | Sv | Gy $\times Q$ (= 100 rem) |

*SI units.
[†]*See Table 6.6*

Another important variable lies in the *quality* of the radiation. Since biological effects arise from the ionization or excitation of the irradiated molecules it follows that the greatest effects will be produced in the areas of most frequent ionization. This ionization depends on the amount of energy released by the radiation per unit length of its path. This quantity is termed the linear energy transfer or LET.

The LET of any type of ionizing radiation depends in a complicated way on the mass, energy and charge it possesses. Electromagnetic radiations such as X or gamma rays will have a low probability of interaction with the atoms of irradiated material and will release their energy over a relatively long path. Heavy particulate radiations will release their energy over a short track. Electromagnetic radiations are therefore termed *low* LET radiations, whereas the latter (alpha particles, neutrons etc.) are *high* LET radiations.

When LET values are high then within a given biological target area there will be many ionization events with a high probability of harmful effect even at relatively low doses. Conversely with low LET radiation these events will be isolated so that effects are low and molecular repair may be possible, although this separation will be less significant at higher doses.

When the effects of high and low LET radiation are contrasted experimentally on cell cultures then for most end-points a clear difference emerges, as illustrated in *Fig. 6.2*. Curve A shows that the incidence of effect with high LET radiation is uniformly proportional to dose down to the lowest dose measurable. Curve B shows that at lower doses the effect per unit dose of low LET radiation is less than at higher doses.

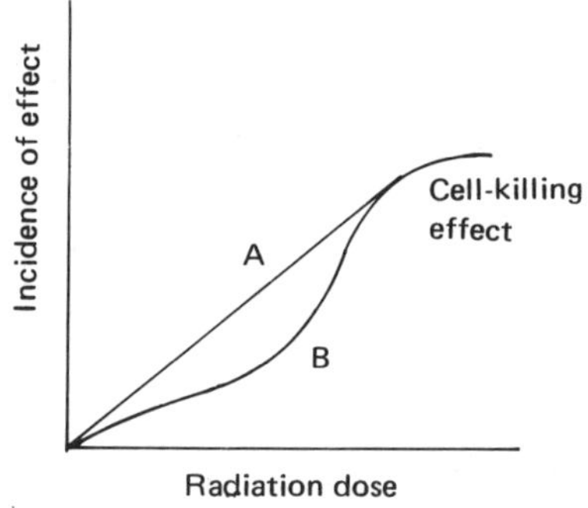


*Fig. 6.2*. Difference in effect of variation in LET.

Other important factors relative to both radiation protection and radiotherapy can be demonstrated with cell studies. The effect of radiation can be shown to be markedly less for a given radiation dose if that same total dose is given as a series of smaller doses *(fractionation)* or over a longer period of time *(protraction)*. This is found only in the case of low LET radiation and in both instances is attributed to repair processes in the cell. Certain chemicals have been shown to increase or reduce the sensitivity of cells to radiation. The most dramatic example of this chemical interaction is the presence or absence of oxygen. Oxygen-deficient cells are to some extent protected from low LET radiation. This effect is much less apparent with high LET radiation and there is consequent interest in the therapeutic benefits to be gained by the use of high LET radiation such as neutrons for the destruction of those cancers that are deficient in oxygen supply.

Cellular radiobiology has also been useful in confirming the increased sensitivity of cells in the process of division. This phenomenon seems to depend more on the length of the mitotic phase rather than its frequency, but has obvious implications for the possible effects of radiation on the fetus or in children.

The variations with dose, dose rate and LET found in cell studies cannot be applied with confidence at the exposure levels associated with occupational irradiation. Furthermore, effects produced in cells in artificial conditions and isolated from other body controls should not be regarded as necessarily representative of effects produced in tissues in the whole animal. There are hormonal and immunological regulatory mechanisms which are absent in cell cultures. Damage to a cell will not inevitably lead to mutation and cancer. Watson has described gross chromosomal damage in the skin cells of eight patients who had received radiotherapy years previously, without subsequent cancer development, illustrating that carcinogenic transformation is a rare event.[3]

## Animal studies

In addition to experimentally induced cancers, chromosome aberrations and hereditary effects have been studied following the irradiation of animals. For hereditary effects, even with low LET radiation the dose–effect curve seems to be linear, but for other end-points the differences between high and low LET radiation remain with regard to linearity, dose and dose rate.

## Human experience

Pope's dictum that 'The proper study of mankind is man' applies as much in the field of radiation exposure as elsewhere. As stated earlier much human experience was derived from fortuitous radiation exposure in the first half of this century and many of the more immediately apparent effects of high doses of radiation are well known. The more insidious effects of lower doses have been recognized only within the past 25 years.

It is now customary to describe the effects of radiation as being *somatic* if they occur in the individual who is irradiated or *hereditary* if they occur in his or her progeny. Effects produced by irradiation of the fetus are somatic effects whether they result in malformations or in malignancies in post-natal life. In considering somatic effects these are thought of as being produced in a *non-stochastic* or a *stochastic* fashion. The term 'stochastic' applies to events which occur with random probability.

## Non-stochastic effects

Non-stochastic effects, which are produced by relatively high doses of radiation, will occur in all or most irradiated subjects according to the level of dose. They are strongly dose-dependent and for a given effect there will be a threshold of dose below which this effect will not appear. For example, above 50 rad delivered at a high dose rate to the testis there will be depression of the spermatozoa population. Above 100 rad of high-dose-rate irradiation of most of the body there will be a depopulation of the circulating blood cells, more marked at higher doses. Other non-stochastic effects include loss of hair, inflammation and ulceration of the skin, loss of the lining of the gastrointestinal tract and so on. Death will follow high-dose-rate whole-body irradiation of around 600 rad if medical intervention is not possible and will certainly occur at doses of 1000 rad even with treatment.

These effects are attributable in the main to cell death or loss of proliferative ability in sensitive tissues. Although most evident at high dose rates there may be non-stochastic effects produced by smaller but repeated radiation doses which ultimately will also affect the function of an organ. Fibrosis and chronic ulceration are longer term effects, well recognized as following such doses to, say, lung or skin. Such effects, then, can be generally regarded as totally avoidable if these doses are not sustained.

### Stochastic effects

It has become evident that there is a *chance* that any irradiated individual may develop certain effects attributable to the radiation which will only appear at long time intervals after the irradiation event. This evidence has been derived from the study over many years of large groups of persons who were known to have received significant doses of radiation. Since such effects are chance and cannot be predicted for any one individual, they have been termed 'stochastic'. The most notable of these effects are carcinogenesis and mutagenesis, and the possibility of their development in an individual exposed at low doses of radiation is now regarded as the most important consideration in radiation protection. It is this consideration which forms the basis of the ICRP risk estimates and their consequent recommendations. It is appropriate therefore to review the most important human experience in this respect. The prerequisites for the validity of such information have been well set out in the UNSCEAR report of 1977.[4] They are:

A sufficiently long period of study (several decades) of the irradiated population to ensure development of all the effects which may occur.

Reliability of diagnosis and detection of such effects.

A sufficient number of effects to give statistical confidence when compared to those of a control population.

A valid control population.

Knowledge of the absorbed dose, dose rate or dose fractionation, and variations of these for different individuals.

A sufficiently uniform dose distribution in the whole body or tissue of interest.

A method of comparing the effect of radiation of different quality at different dose levels.

When these criteria have been fulfilled, another problem remains. Although there are many groups of persons in whom the late effects of radiation are undisputed, only a few studies can be regarded as quantitatively reliable. Even in these the dose–effect response can usually be estimated only for doses in the higher ranges. To gauge the effects of the doses of the few rads encountered in the radiation protection situation, certain assumptions have to be made. The first is that the incidence of the effect — for example the frequency of induction of cancer — is proportional to the dose received; the second is that every dose above zero has a probability of causing such an effect, regardless of the rate at which it is received. This concept of a linear, non-threshold dose–effect relationship is now central to the recommendations of the ICRP, but is still conjectural. The bulk of the evidence from cell and animal experimental data suggests that this concept will tend to overestimate the risks from low doses of radiation when derived from frequency of effect at high doses, especially in the case of the more commonly encountered low LET radiation.

## Epidemiology

Although there have been many heavily irradiated groups in whom malignancies have subsequently appeared, only a limited number fulfil the epidemiological criteria specified above. More detailed accounts of these are available in the United Nations report[4] and also in the 1980 report from the US National Academy of Sciences.[5] A point worth noting is that radiation-induced cancers are indistinguishable from those which occur 'naturally'. There are no esoteric tumours such as those which characterize asbestos or vinyl chloride exposure. The sources which provide the most reliable evidence are summarized in *Table 6.4*.

In addition to demonstrating cancer induction, these studies also show that some organs seem more radiosensitive than others, and that such radiosensitivity is often dependent on other factors such as age and sex. Leukaemia, for example, was more common in those irradiated over the age of 50; breast cancer induction seems greatest during reproductive life; cancer of the thyroid is more common in females than in males.

## Hereditary stochastic effects

No hereditary effects have been observed in human populations at any dose level, even in the progeny of the survivors of Hiroshima and

**Table 6.4.** Sources of information on risk of cancer from radiation

| Type of cancer | Reason for exposure | Mean risk estimate per $10^6$ per rem |
|---|---|---|
| Leukaemia | A bombs (Japan)<br>X-rays for ankylosing spondylitis<br>X-rays for artificial menopause | 20 |
| Bone | $^{226}$Ra intake in dial painters<br>$^{224}$Ra treatment<br>X-rays for ankylosing spondylitis | 5 |
| Breast | A bombs (Japan)<br>Fluoroscopy for artificial pneumothorax<br>X-ray treatment for mastitis | 25 |
| Lung | Hard rock mining (radon)<br>X-rays for spondylitis<br>A bombs (Japan) | 20 |
| Gastro-intestinal | A bombs (Japan)<br>X-rays for spondylitis | 10 |
| Thyroid | A bombs (Marshall Islands)<br>A bombs (Japan)<br>X-rays for thymus irradiation<br>X-rays for scalp fungal infection | 25 |

Nagasaki, in spite of popular belief to the contrary. The estimates of human radiation genetic effects therefore rely on data from animal experiments but are established with a reasonable degree of confidence (*Table 6.5*).

**Table 6.5.** Estimates of the human effect of radiation

| Somatic effects | *Fatal cancers per $10^6$ per rem* |
|---|---|
| ICRP (1977)[6] | 125 |
| Other estimates | |
|   UNSCEAR (1977)[4] | 100 |
|   US National Academy of<br>  Sciences (1980)[5] | 70–353 |

| Genetic effects | *Genetic disorders per $10^6$ live births per lifetime rem* | |
|---|---|---|
| | *1st* | *Subs.* |
| ICRP (1977)[6] | 20 | 80 |
| Other estimates | | |
|   UNSCEAR (1977)[4] | 20–63 | 185 |
|   US National Academy of<br>  Sciences (1980)[5] | 2–30 | 30–550 |

1st, First generation; Subs., all subsequent generations.

Such effects are no longer considered as important as the induction of malignancy. It is now believed that if control measures are adequate to guard against induction of malignancy then hereditary effects will be sufficiently minimized.

## Risk estimates

In estimating the risk from radiation the ICRP takes some account of the difference in effect between different types of radiation. This is done by using a 'quality factor', $Q$, to convert the absorbed dose in rads into a *dose equivalent* in *rems*. The recommended quality factors for the most common radiation are shown in *Table 6.6*, where it can be seen,

**Table 6.6.** Values of $Q$ (quality factor)

| | |
|---|---|
| X-rays, gamma rays and electrons | 1 |
| Neutrons, protons and singly charged particles of rest mass greater than one atomic mass unit of unknown energy | 10 |
| Alpha particles and multiply charged particles (and particles of unknown charge) of unknown energy | 20 |
| Thermal neutrons | 2·3 |

for example, that whereas 100 rad X-rays will be equivalent to 100 rem, it will require only 5 rad alpha particle irradiation to produce the same biological effect. At present new units are in the course of introduction to conform to the Système Internationale. The *gray* (Gy), the *becquerel* (Bq) and the *sievert* (Sv) will eventually supersede the rad, the curie and the rem. The relationship between these units is shown in *Table 6.3* above. Risk estimations incorporate the assumption that there is no dose threshold for stochastic effects and that these effects are linearly dependent on radiation dose to the lowest dose level (curve A in *Fig. 6.2*). They also average the risk of all ages and both sexes, and take account of the fact that both breast and thyroid cancers are not necessarily fatal. Risk estimations from different authorities are shown in *Table 6.5*. The ICRP estimate implies that 125 fatal cancers will occur for 1 000 000 persons receiving 1 rem whole body dose equivalent above natural background radiation levels or for 10 000 persons receiving 100 rem whole body dose equivalent.

## CURRENT ICRP RECOMMENDATIONS[6]

The recommendations for control of radiation dose over the years are summarized in *Table 6.7*.

In 1959, introducing the no-threshold concept, the ICRP stated that 'all doses should be kept as low as practicable and all unnecessary

**Table 6.7.** Recommended limits for worker exposure

| Year | Exposure or dose limit* |
|------|-------------------------|
| 1925 | 70 rem/year (0·1 erythema dose) |
| 1934 | 0·1 rem/day or 0·5 rem/week |
| 1949 | 0·3 rem/week or 15 rem/year |
| 1966 | 3 rem/13 weeks or 5 rem/year |

*Units normalized to rem.

exposure avoided'. In 1965 this was altered to a recommendation that 'any unnecessary exposures be avoided and that all doses be kept as low as is readily achievable, economic and social considerations being taken into account'. The latest recommendations of 1977 and 1980 finalize this philosophy by specifying three principles of radiation protection:

No practice shall be adopted unless its introduction produces net benefit (Justification).

All exposures shall be kept as low as reasonably achievable, economic and social factors being taken into account (Optimization).

The dose equivalent to individuals shall not exceed the limits recommended for the appropriate circumstances by the commission (Dose limitation).

With regard to the third of these principles the commission has gone further than ever before into societal considerations. Since every dose of radiation (additional to that already received from natural sources) apparently carries some risk there must be a decision on the level of risk which people find acceptable. The difficulties in deciding (on behalf of others) what that level might be are discussed in an earlier commission document,[7] but clearly these can be partly resolved by distinguishing between persons who choose to work in situations where the possibility of receiving radiation doses exists and the public at large who do not (although in many instances they may receive some benefit from the existence of such sources).

In the case of workers, the ICRP has set levels of exposure which correspond with the mortality rates for industries that are generally regarded as 'safe'. In such industries the average annual risk of death is of the order of $10^{-4}$, that is, with less than 100 deaths per million employed per year. Even within such industries the risk to any individual will vary with his task and the distribution of individual risks will turn about the mean with a few values being considerably higher. The commission has therefore taken account of the risk estimates derived from all the validated information available and has set limits of exposures to workers which will achieve a level of mortality risk comparable to industries having a high standard of safety (*Table 6.8*).

The ICRP feels that this is achieved by maintaining the dose limit

**Table 6.8.** Recommended annual dose equivalent limits

| Recommended limit | Application | Tissue or organ |
|---|---|---|
| 50 rem (0·5 Sv) | Workers | All tissue except lens of eye |
| 30 rem (0·15 Sv) | Workers | Lens of eye |
| 5 rem (50 mSv) | Workers | Uniform irradiation of whole body |
| 0·5 rem (5 mSv) | Individual members of the public | Whole body |
| 5 rem (50 mSv) | | Any one organ or tissue including skin and lens of eye |

for uniform whole body irradiation at 5 rem in a year. Although exposure at this level would lead to a risk greater than $10^{-4}$, experience has shown that the mean dose for radiation workers is consistently much lower than this and that the majority of workers have annual doses of less than 1 rem.

The dose limitation system is also based on the principle that the risk should be equal whether the body is irradiated uniformly or non-uniformly. If the latter, then a weighting factor can be used to adjust the risk of irradiation of only certain tissues. This takes account of the variation in radiosensitivity of various tissues and the consequent variation in risk (*Table 6.9*).

**Table 6.9.** Tissue weighting factors

| Tissue | Weighting factor ($W_T$) |
|---|---|
| Gonads | 0·25 |
| Breast | 0·15 |
| Red bone-marrow | 0·12 |
| Lung | 0·12 |
| Thyroid | 0·03 |
| Bone surfaces | 0·03 |
| Remainder | 0·30 |

The annual dose equivalent in any tissue can be multiplied by its weighting factor. The sum of such separate dose equivalents however should never be greater than the limit for uniform whole body irradiation of 5 rem.

Moreover to prevent the possible development of non-stochastic effects there is also a limit of 50 rem placed on irradiation of any single tissue. For example, the stochastic dose equivalent limit for the thyroid gland, obtained by multiplying the limit of 5 rem by its weighting factor 0·03, would give an implied dose equivalent of

170 rem. This would exceed the non-stochastic limit of 50 rem which must be the over-riding constraint.

A special non-stochastic limit of 30 rem is imposed for the lens of the eye to avoid cataract formation.

## Limits for the intake of radioactive materials[8]

Harmful irradiation can occur from radionuclides deposited or incorporated in body organs. The ICRP has set limits for such intakes in a new way. In order to do this, knowledge must be available for any radionuclide concerning its various chemical forms, its metabolic fate with regard to excretion, retention and distribution in the body, its own radiation characteristics and those of its daughter products. In addition such a situation is postulated to take place within a 'reference man' whose anatomical and physiological characteristics are standardized. Some radionuclides will persist for many years in the body and it is recommended that the dose equivalent be integrated over a working lifetime of 50 years and be termed the committed dose equivalent. In any working year this value in all organs of the body must be limited so that the resulting total risk of both cancer and hereditary disorder is no greater than that from uniform whole body irradiation of 5 rem.

## Pregnant women

Since the fetus seems to be especially sensitive to radiation, the ICRP considers the case of a female worker who becomes pregnant. The commission believes that if she is working under the recommended limits and at regular dose rates it is unlikely that the embryo will receive more than 0·5 rem during the first 2 months of life. This is sufficient to provide protection during the critical period of organ formation. It is unlikely that a pregnancy of more than 2 months' duration will go unrecognized. When pregnancy has been diagnosed arrangements should be made so that the woman can continue to work only where it is most unlikely that her annual exposure will exceed 1·5 rem, that is, three-tenths of the annual whole body dose equivalent limit.

## Classification of conditions of work

This decision in the case of a pregnant woman conforms to the ICRP recommendation regarding conditions of work. They suggest that these be classified as either A or B in terms of radiation work. In class A the conditions might be such that the annual exposures might exceed three-tenths of the dose equivalent limits. Workers in this situation must have medical supervision and individual dose monitoring. In working condition B, that is where doses are not likely to exceed 1·5 rem annually, such supervision is not necessary.

## Limits for the public

In setting these limits the ICRP has again made its own estimate of acceptable risk. On the basis of comparison with activities involving the public in risk, such as usage of public transport, the ICRP concludes that a level of acceptability of fatal risks to the public is much lower than for occupational risks and is of the order of one in a million per year. On the basis of radiation risk estimates this would correspond to a dose equivalent of 0·1 rem per year for life-long whole body exposure. The commission concludes that an annual limit of 0·5 rem for individual members of the public will result in average dose equivalents of less than 0·05 rem, and hence recommends this limit.

## CONCLUSION

The principles on which the ICRP bases its recommendations depend on a well-founded estimation of risks from radiation and a judgement on the acceptability of risks. The estimation of risk may require adjustment if further knowledge shows that these risks, particularly at low doses, are greater or lesser than they are presently thought to be. Such a change will only take place if appropriate scientific data emerge.

The question of the acceptability of risk, however, is not based on science, and it may be that it is in this area that most discussion and argument will take place in the future.

REFERENCES
1. Alper T. *Cellular Radiobiology*. London: Cambridge University Press, 1979.
2. Purchase IF, Longstaff E, Ashby J et al. Evaluation of six short term tests for detecting chemical carcinogens. *Nature* 1976; **264**:624–7.
3. Watson G. Chromosome changes in skin fibroblasts of radiotherapy patients. *Radiat. Res.* 1974; **84**:61–5.
4. Report of the Scientific Committee on the Effects of Atomic Radiation. New York: United Nations, 1977.
5. Report of the Committee on Biological Effects of Ionising Radiation. Washington DC: National Academy of Sciences, 1980.
6. IRCP. *Recommendations of the International Commission on Radiological Protection*. ICRP Publication 26. Oxford: Pergamon Press, 1977.
7. IRCP. *Problems of Developing an Index of Harm*. ICRP Publication 27. Oxford: Pergamon Press, 1977.
8. IRCP. *Limits of Intakes of Radionuclides by Workers*. ICRP Publication 30. Oxford: Pergamon Press, 1979.

# 7. NON-IONIZING RADIATION
*S. Kanagasabay*

Radiation may be defined as a flow or stream of atomic and subatomic particles and waves. We are constantly exposed to various kinds of radiation in the form of visible light, infrared and ultraviolet rays, radio waves, X-rays (from terrestrial and cosmic sources) and particulate cosmic rays. These are manifestations of energy transfer from one place to another. The results of energy transfer on a living organism may be described as the biological effects of radiation. These effects include normal life processes such as vision in animals and photosynthesis in plants, and the abnormal or injurious effects resulting from exposure to excessive amounts of commonly encountered radiations or to unusual types of radiation.

The two main types of radiation are electromagnetic waves and moving atomic particles. In the case of electromagnetic waves energy transfer is by means of oscillating electric and magnetic fields. Particulate radiation on the other hand depends on the atom, which consists of a nucleus of neutrons and protons surrounded by a cloud of electrons. In addition, the atom has other special particles. When separated, these particles, charged or uncharged, are capable of transferring their energy, wholly or in part, to any substance through which they pass. Streams of such particles constitute particle radiations.

In relation to biological effects, radiation can be categorized as ionizing and non-ionizing. The term 'ionizing radiation' includes all forms of radiation that directly or indirectly ionize the atoms with which they interact. These may be electromagnetic waves such as X-rays or they may be particle radiations such as beta particles. Non-ionizing radiation on the other hand is electromagnetic radiation, where the absorbed radiation energy is insufficient to ionize the atoms with which it interacts. This category includes radiofrequency, microwave, infrared, visible light and ultraviolet radiations. For example, the energy of ultraviolet radiation is absorbed by the planetary electrons of atoms; an electron jumps to a higher orbit, thereby bringing the atom into a state of excitation in which it is chemically reactive. The energy from X-rays is also absorbed by the planetary electrons of atoms but the amount absorbed is usually great enough to induce an electron or electrons to escape, thereby ionizing the atom.

It was the discovery of X-rays, the description of the characteristics of gamma radiation, as well as alpha and beta particles from naturally occurring isotopes, that led to the identification of ionizing radiation. Further work showed that X and gamma radiation were electromagnetic radiations differing from infrared, ultraviolet and visible light in wavelength or quantum energy with specific values for ionizing radiation ranging from 0·1 to 0·001 nanometre (nm). By virtue of this designation, electromagnetic radiation with wavelengths greater than 0·1 nm was regarded as non-ionizing radiation because the quantum energy of this radiation is much too low to produce the type of excitation necessary for ionization.

A proper understanding of non-ionizing radiation may be regarded as originating from James Clerk Maxwell's work in 1864. Based on the work of Faraday, Oersted and Ampère on electricity and magnetism, he deduced the existence of a wave with the property that its speed in free space bore a simple relationship to the dielectric constant and the magnetic permeability. By using known values for the dielectric constant and the magnetic permeability he was able to obtain a result identical to the previously known speed of light in a vacuum. He concluded that light was electromagnetic radiation. Further work established a connection between the various types of electromagnetic radiations and the spectrum shown in *Fig. 7.1* emerged. In this chapter the non-ionizing region of the electromagnetic spectrum relating to radiofrequency, microwave and laser radiation will be considered.

## RADIOFREQUENCY AND MICROWAVE RADIATION

The earliest record of the production of electromagnetic waves dates back to 1886, when the German physicist Heinrich Hertz tested Maxwell's theories and set up oscillating electric and magnetic fields by discharging an induction coil across a spark gap thus generating electromagnetic waves. He showed that these waves could be reflected and refracted in the same way as light. Since then scientists have developed generators capable of producing radiation in all regions of the electromagnetic spectrum, spanning wavelengths from hundreds of metres to one millimetre. In this review radiofrequency radiation will be regarded as that portion of the spectrum from frequencies of 30 kHz, wavelength $10^4$ metres, to 30 MHz, wavelength 10 m, and microwave radiation from frequencies of 30 MHz, wavelength 10 m, to 300 GHz, wavelength 1 mm. Boundary variations in frequency demarcation between radiofrequency and microwave radiation may be seen in various texts. These are of little significance for environmental control or assessment of biological effects. Additionally, the microwave region of the spectrum is divided into bands allocated for broad services as defined by the International Telecommunications Union. *Table 7.1*

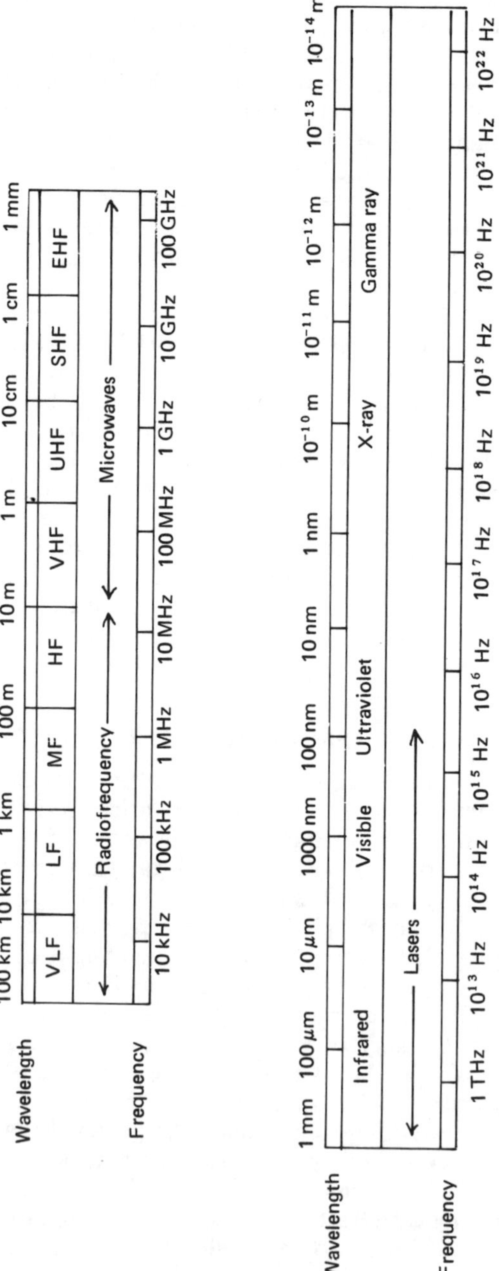


*Fig. 7.1.* The electromagnetic spectrum.

shows the band designations from the United States Military Standard MIL STD 463. Some of the old frequency designations are still used in the United Kingdom and care needs to be taken to avoid confusion in respect of previously assigned letters such as C, K, L etc. A number of frequencies have also been assigned for industrial, scientific and medical uses. Commonly used frequencies for these are 27·12 MHz, 896 MHz and 2·45 GHz. For example, radiofrequency drying and shortwave diathermy used 27·12 MHz, and 2·45 GHz is used for many heating purposes, such as microwave ovens and microwave diathermy.

**Table 7.1.** Frequency and designations

| Band designation | New frequency (GHz) | Old frequency (GHz) |
|---|---|---|
| A | 0·10– 0·25 | |
| B | 0·25– 0·50 | 3·90 – 6·20 |
| C | 0·50– 1·00 | |
| D | 1·00– 2·00 | |
| E | 2·00– 3·00 | |
| F | 3·00– 4·00 | |
| G | 4·00– 6·00 | |
| H | 6·00– 8·00 | |
| I | 8·00– 10·00 | |
| J | 10·00– 20·00 | |
| X | DISC | 5·70 – 10·90 |
| K | 20·00– 40·00 | 10·90 – 36·00 |
| Ka | DISC | 26·00 – 40·00 |
| Ku | DISC | 12·40 – 18·00 |
| L | 40·00– 60·00 | 0·34 – 1·55 |
| M | 60·00–100·00 | |
| P | DISC | 0·225– 0·39 |
| Q | DISC | 36·00 – 46·00 |
| S | DISC | 1·55 – 5·20 |
| V | DISC | 46·00 – 56·00 |
| W | DISC | 56·00 –100·00 |

Though frequencies between 10 MHz and 30 MHz have been used for induction heating and dielectric heating for processing of materials such as wood, plastics, paper, ceramics and textiles, radiofrequency radiation is used principally for communications. Microwave radiation, in addition to its use in communications, is used in surveillance and ranging systems and in a variety of heat treatment processes in industry. The rapid proliferation of radiofrequency and microwave systems over a short period of time has emphasized the need for an understanding of the potential hazards from exposure to radiation from these sources. It is therefore necessary to have a clear understanding of the physical characteristics of the radiation, the reasons underlying the current exposure standards for the protection of those at risk, and the methods of measurement for environmental control.

A summary of some of the applications of radiofrequency and micro-wave sources by Baranski and Czerski[1] is shown in *Table 7.2*.

## Physical characteristics

Hertz showed that charged particles such as electrons, when acce-lerated, set up electric and magnetic fields. Such electromagnetic radiation in its polarized form may be considered as sinusoidal electric and magnetic waves oscillating at right-angles to one another and to the direction of propagation. This is shown diagrammatically in *Fig. 7.2*. The fundamental correlation between the vector components of such a wave are based on Maxwell's deductions. The scalar components of the electric and magnetic field vectors are related to the intrinsic impedance $Z_0$, of a medium with negligible magnetic loss by the expression:

$$Z_0 = \frac{E}{H} = \left[ \mu \left( \varepsilon - \frac{j\sigma}{\omega} \right) \right]^{-\frac{1}{2}}$$

where $E$ and $H$ are the electric and magnetic fields respectively, $\mu$ the permeability, $\varepsilon$ the dielectric constant, $\sigma$ the conductivity of the medium, $\omega$ the angular velocity and $j = \sqrt{-1}$. In a non-conductive dielectric ($Z_0 = \sqrt{\mu/\varepsilon}$), and the velocity of wave propagation in a vacuum and to a good approximation in air is equal to the velocity of light.

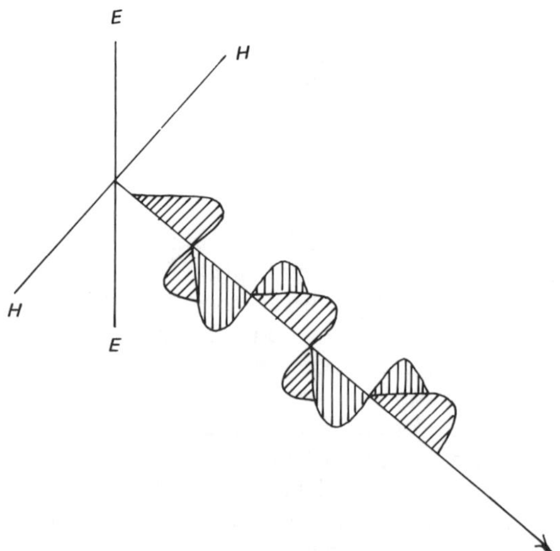


*Fig. 7.2*. An electromagnetic wave. The electric field $E$ oscillates at right-angles to the magnetic field $H$ and both are at right-angles to the direction of propagation.

**Table 7.2.** Selected examples of typical uses of equipment generating radiofrequency and microwave radiation

| Frequency | Use | Occupational exposure | Examples of potential incidental exposure (general population hazards) |
|---|---|---|---|
| Below 3 MHz | Metallurgy: welding, melting, tempering etc. Broadcasting, radio-communications, radionavigation | Various factory workers, e.g. furniture veneering operators, drug and food sterilizers, car industry workers | Factory executive personnel, watchmen, guards |
| 3–30 MHz | Many industries, e.g. motor, wood, chemical, food, heating, drying, welding, gluing, polymerization, sterilization of dielectrics, agriculture, food processing, medicine, radioastronomy, broadcasting | Electronic engineers, technicians, air-crewmen, missile launchers, radar mechanics and operators, microwave oven operators, and maintenance workers | Airport and seaport personnel of various professions, inhabitants of areas in the vicinity of highpower radar installations, broadcasting stations and TV transmitters |
| 30–300 MHz | Many industries as above: medicine, broadcasting, TV, air traffic control, radar, radio-navigation | Scientists, physicists, microwave development workers | |
| 300–3000 MHz | TV, radar (troposcatter and meteorological), microwave point to point, telecommunication, telemetry, medicine, microwave ovens, food industry | Microwave testers, microwave diathermy operators and maintenance workers, medical personnel | Housewives and children (microwave ovens in private homes) |
| 3–30 GHz | Altimeters, air and shipborne radar, navigation, satellite communications, microwave point to point | Broadcasting, transmitter and TV personnel | |
| 30–300 GHz | Radioastronomy, radiometeorology, space research, nuclear physics and radio spectroscopy | Marine and coastguard personnel, sailors, fishermen, persons professionally present on board ships | |

Source: Baranski and Czerski.[1]

The frequency of oscillation of the electric or magnetic field ($f_0 = \omega/2\pi$) is related to the wavelength ($\lambda$) and to the wave velocity ($v$) by the relationship $\lambda = f_0 v$. The wavelength thus depends on the velocity of propagation which in turn is a function of the parameters, namely the dielectric constant, the permeability and the conductivity of the medium. The relationship between the wavelength $\lambda$ of an electromagnetic wave and the corresponding free space wavelength $\lambda_0$ is:

$$\lambda_0 = \lambda \left[ \frac{\mu}{2} \left\{ \sqrt{\varepsilon^2 + \left(\frac{4\pi\sigma}{\omega}\right)^2} + \varepsilon \right\} \right]^{1/2}$$

Thus the free space wavelength may be significantly reduced when the wave is propagated in a medium having large values for the dielectric constant, the permeability and the conductivity, relative to air; depending always on the electric and magnetic variables. Examples of the wavelength (in centimetres) for some biological tissues and air at different frequencies are shown in *Table 7.3*.

**Table 7.3.** Wavelength in various biological tissues and air for different frequencies

| | Frequency (MHz) | | | | |
|---|---|---|---|---|---|
| | 400 | 1000 | 3000 | 10 000 | 24 000 |
| Air | 75 | 30 | 10 | 3·33 | 1·67 |
| Skin | 10·12 | 4·41 | 1·49 | 0·51 | 0·25 |
| Fat | 30·90 | 12·42 | 3·79 | 1·45 | 0·68 |
| Muscle | 9·41 | 3·09 | 2·12 | 0·62 | — |
| Bone | 32·19 | 12·63 | 3·97 | 1·25 | 0·21 |

For a time-varying electromagnetic wave in free space the energy through a unit area per unit time is referred to as the Poynting vector. The Poynting vector is the vector product of the electric and magnetic vectors. Since $E$ and $H$ are at right-angles to one another and to the direction of propagation, the Poynting vector is also the direction of propagation by virtue of its relationship to the intensity of the wave. Taking the time average of the Poynting vector and using the definition of the intrinsic impedance of the medium, the intensity of energy flow per unit area or power density, in the direction of propagation in free space, is directly proportional to the square of the electric field strength and inversely proportional to the intrinsic impedance. Because the intrinsic impedance and the wavelength are functions of the frequency, the energy density or power density is also dependent on the frequency and so it can be assumed that radiofrequency or microwave radiation absorption in a medium will vary with the radiation frequency.

The simple proportionality between the magnitudes of $E$ and $H$ is true only in free space or in the far field, which is the region far enough from the radiating source to eliminate any interaction between the source and the wave. In this region the amplitude of the $E$ or $H$ field is inversely proportional to the distance from the source; the energy density or power density is thus inversely proportional to the square of the distance in the far field and measurement of either $E$ or $H$ values would be adequate. In the near or inductive field, on the other hand, $E$ and $H$ are not represented by a simple proportionality and both components of the field must therefore be measured to obtain the power or energy density. However, because the near field is not a radiative field, the more meaningful measure is the energy density. The distance from the source at which the near and far fields interact is usually inversely proportional to the wavelength of the radiation. Because of the small wavelengths, microwave measurements are feasible in the far field and measurement of the electric field will suffice to obtain the power density in watts per metre squared ($W/m^2$). The longer wavelengths of radiofrequency radiation make it difficult to obtain such measurements even in the far field, and the field intensity in volts per metre ($V/m$) is therefore used for radiofrequency sources.

In certain applications of microwave and radiofrequency radiation, for example radar, the waves are pulse-modulated instead of being operated in the continuous wave mode; *Fig. 7.3* is a schematic representation of pulse modulation. The average power $P_{av}$ is related to the peak or maximum power $P_{max}$, the pulse duration $\delta$ (s), and the pulse repetition period $r$ ($s^{-1}$) by the relationship:

$$P_{av} = P_{max}\delta r^{-1}$$

Thus for a given set of variables, a microwave generator may be operated at a relatively low average power even though the peak power is extremely high.[2]

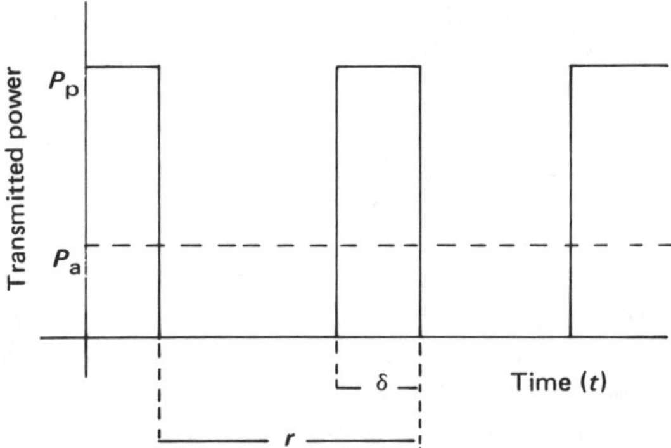


*Fig. 7.3.* Relationship between peak power ($P_p$) and average power ($P_a$) for a pulse modulated microwave generator with a pulse repetition period ($r$) and pulse duration ($\delta$). (After Cleary.[2])

Since Hertz first demonstrated electromagnetic radiation, technological advances have resulted in exploitation of the non-ionizing region of the electromagnetic spectrum. Systems have proliferated and power outputs have increased by several orders of magnitude and man-made non-ionizing radiation has been added to the natural background radiation. The sources contributing to the natural background radiation include lightning discharges (producing radiation from 100 MHz to 10 GHz), high frequency electromagnetic radiation from cosmic noise, radio galaxies and solar radiation. However, the net background radiation from all these sources is lower than the current occupational health exposure standard of 10 mW/cm² by a factor of about $10^9$. If man-made radiation is included, the factor is of the order of $10^6$, thus leaving a substantial margin of safety in respect of the current exposure standard.

## Biological effects and exposure standards

These will be considered under one heading because the biological effects chosen will determine the exposure standards. In broad terms the biological effects are designated 'thermal' or 'non-thermal'. Thermal effects may be regarded as those effects which result from heating of biological material and which can be reproduced by conventional heating. Non-thermal effects on the other hand are those effects which result from the interaction of the electromagnetic field with biological material and which cannot be reproduced by conventional heating techniques. In western countries the emphasis has been largely on the study of thermal effects. In the Soviet Union and other eastern European countries, however, the emphasis has been on non-thermal effects associated with the central nervous system, the cardiovascular system and the endocrine system. These include data on a variety of psychological and physiological effects of a reversible nature resulting from long term exposure to electromagnetic fields of low power density. Though studies in western European countries do not corroborate these findings, there is disquiet that the simplistic approach of gross thermal damage may not take account of the overall effects. This fundamental difference in approach is the basic reason for the great disparity in standards between the two groups.

In terms of electromagnetic radiation, dielectrics are materials which are not perfect, and the amplitude of an electromagnetic wave is attenuated when it passes through such materials with the resultant energy dissipated as heat in the material; these materials are called lossy dielectrics and include biological material. Electromagnetic radiation when incident on an object made of lossy dielectric is reflected, transmitted through and around the object, or absorbed by it. Absorption in biological material results in heating which is due both to ionic conduction and vibration of the dipole molecules of water and protein.[3] The degree of absorption is dependent on the electrical properties of the material, namely the dielectric constant and the electrical conductivity. Between 500 MHz (wavelength 60 cm) and 30 GHz (wavelength 1 cm) radiation is readily absorbed inside the human body.[4] The depth of penetration decreases rapidly with increase in frequency for a given set of parameters. For example, the depth of penetration in muscle at a frequency of 900 MHz is twice that which is achieved at 2500 MHz.[5] Additionally, tissue with a low water content such as fat is penetrated to a greater extent than muscle which has a high water content. This would account for the penetration of fatty tissue with relatively little loss of energy which is then dissipated in the deeper tissues. Because of this selective absorption, whole body radiation normally results in different parts inside the body being heated by different amounts. However, because of the circulation and other compensatory mechanisms this heating effect is rapidly distributed over the whole body; the only significant effect being a rise in whole body temperature. It has been estimated that for a power density of 100 W/m$^2$ (10 mW/cm$^2$) wholly absorbed by the body, a rise of body temperature would be less than 1°C under normal conditions.[6] It is on the basis of this work that a number of countries including the UK and the USA have adopted 10 mW/cm$^2$ as the maximum safe power density.

**Thermal effects**

In assessing the thermal effects of whole body radiation by microwaves, extensive investigations have led to the conclusion that exposure to power density of 100 mW/cm$^2$ for several minutes or hours, depending on the animal species, can result in pathological manifestations of a thermal nature with a rise in temperature which is dependent on the thermoregulatory process and active compensation by the animal. The studies were carried out on rabbits, rats and dogs exposed to 2800 MHz, 1240 MHz and 200 MHz radiation. At power densities greater than 100 mW/cm$^2$ the thermal response in dogs was seen to comprise three phases — initial heating, followed by thermal equilibrium and finally thermal breakdown resulting in death. The breakdown of thermal equilibrium was associated with a rapid rise in rectal temperature and acute distress followed by death. Several factors which are seen to influence temperature rise are summarized by Michaelson,[7] and these include the following:

Animal species.
Specific area of body exposed.
Skin thickness.
Amount of subcutaneous fat.

Covering of exposed body areas and body mass.
Position in electromagnetic field.
Frequency or wavelength of radiation.
Intensity or field strength.
Duration of exposure.
Mode of radiation — pulsed or continuous wave.
Ambient environmental conditions.
Condition of exposed subject.

Previously it was pointed out that radiation between frequencies of 500 MHz and 30 GHz is readily absorbed inside the human body, and that the depth of penetration decreases rapidly with increasing frequency. At the higher frequencies the site of interaction is the skin and the major biological effects result from the skin surface acting as a reflector or absorber with heating effects. At frequencies below 100 MHz (wavelength 300 cm) the body is penetrated with very little loss of energy. A summary by McRee on the sites of interaction and major effects is shown in *Table 7.4.*[8] Though actual power densities are not stated it must be assumed that these effects are at levels very much below the estimated lethal power densities. At these higher power densities there is an inverse relationship between elapsed time till death following exposure. Also, the most rapid lethal outcome in rats is following exposure to a radiation frequency of around 3300 MHz at high power density and lethality is more rapid and higher with pulsed rather than with continuous wave radiation. In practical terms microwave ovens at 2450 MHz come within this critical wavelength band, exploiting the features of good penetration and efficient absorption.

**Table 7.4.** Thermal biological effects of microwaves

| Frequency (MHz) | Wavelength (cm) | Site of major effects | Major biological effects |
| --- | --- | --- | --- |
| >10 000 | 3 | Skin | Skin surface acts as a reflector or absorber with heating effects |
| 10 000 | 3 | Skin | Skin heating with sensation of warmth |
| 10 000–3300 | 3–10 | Top layers of skin, lens of eye | Lens of eye and testicles particularly susceptible |
| 10 000–1000 | 3–30 | Lens of eye | Critical wavelength band for eye cataracts and testicular damage |
| 1200–150 | 25–200 | Internal body organs | Damage to internal organs from overheating |
| < 150 | above 200 | | Body is transparent to waves above 200 cm |

Source: McRee.[8]

The most susceptible organ of the body to thermal effects at microwave frequencies is the testicle because of its sensitivity to thermal stress and its poor ability to dissipate heat. These characteristics together with the possibility of genetic damage led to many investigations into microwave-induced testicular damage. A thorough and systematic study was carried out by Ely and his associates.[9] Based on these animal experiments and the studies of earlier workers, the maximum *normal* human testicular temperature was set at 35·6°C. The experimental animal was an anaesthetized dog, and a criterion of 'any demonstrable damage' was used to assess the critical power densities and temperatures. From these findings threshold values for temperature and exposure times were set at 37°C for 5 days, 38°C for 1 hour and 38·2°C for 1 minute. The power densities used were lower than 100 mW/cm². The histological changes were similar to those caused by other forms of heating and resembled changes associated with nutritional effects, toxins, trauma and certain diseases. Because of the patchy distribution and small number of severely damaged testicular tubules seen 5 days after initial assessment, the assumption was made that permanent damage would be unlikely. However, this was qualified by the caution that weight change, spermatozoa counts and fertility studies would need to be carried out to validate such an assumption. In man rectal temperatures of 40·5°C have been associated with a transient decrease in spermatozoa. Based on the data of Ely et al.,[9] the minimum power density required to maintain a testicular temperature of 37°C is 5 mW/cm²; at this level minimal histological changes have been noted.

In earlier studies the effects of infrared and microwave radiation at 2500 MHz on the testes of anaesthetized rats were compared. Though power densities were not measured, microwave exposure of 5–15 minutes showed evidence of testicular degeneration at intra-testicular temperatures of 31–35°C, while no histological change was seen in testes maintained at 38°C for 10 minutes with infrared radiation. The testicular changes following microwave radiation were associated with testicular temperatures lower than the *normal* range of testicular changes quoted for the rat. Though the histological picture appeared to be similar in both microwave and infrared radiation, the results seem to indicate that some mechanism other than simple heating might be involved in the case of microwaves. Very little is known about possible mutagenic effects, although there has been reference to possible association with Down's syndrome and paternal exposure to microwave radiation.[10] Chromosome aberrations in plant cells have been reported at a radiation frequency of 27 MHz.[11] In studies on adult male Chinese hamsters exposed to radiation at 2450 MHz, Leach[12] reported chromosomal aberrations in bone-marrow cells. Rugh and McManaway[13] compared the effects of exposures to

2450 MHz radiation with those due to ionizing radiation in respect of teratogenicity in developing rodent embryos. It was concluded that teratogenesis follows both ionizing and non-ionizing radiation at comparable gestation stages. An interesting feature is that there is essentially no thermal stress with ionizing radiation and no ionization with microwave heat energies. These studies have been criticized on the grounds that the systems were subject to a thermal stress and the experiments have not been independently validated. Despite these criticisms, the observations are worthy of note.

The other organ which has been studied in some depth is the eye, where microwave radiation has been shown to produce lens opacity in some experimental animals. Microwaves have also been shown to cause lens opacities in man. An advantage of the study of changes in the lens is that a definite defect can be precisely identified by modern techniques. It would be appropriate to consider briefly the normal growth of the lens of the eye and to relate it to abnormal or externally induced changes. A schematic representation of the cross-section of a dog lens is shown in *Fig. 7.4*. The lens grows throughout life by cell

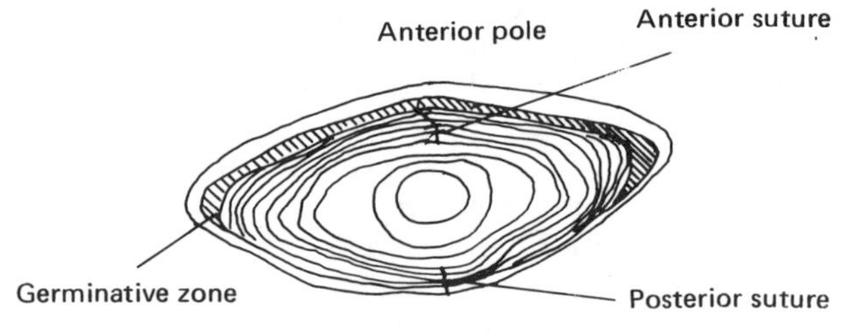


*Fig. 7.4.* Cross-section of a dog lens.

division within the single layer of cells at the anterior surface. The cell divisions take place mainly in the germinative zones and the new cells elongate to form new lens fibres. These fibres grow and curve round older and deeper fibres, aligning themselves at the anterior and posterior poles. Thus the junctions of the fibres at the poles form a short suture line. As the lens grows, the older cells lose their nuclei and get compressed as new fibres are added from the surface. As long as growth proceeds like this the lens remains transparent. When there is interference to this pattern of growth the fibres are deformed and lose their transparency. Areas of loss of transparency are called cataracts.

Cataracts are known to follow exposure to ionizing radiation. These occur at the posterior pole in the region of the suture within months after exposure. Microwave-induced opacities may occur at either pole depending on the method of exposure and appear in a much shorter time after exposure. Eight 30-minute exposures of the eye of a dog once daily to 2450 MHz radiation at a power density of 460 mW/cm² produced an observable anterior cataract within 6 days of the last exposure. Many studies using different animal species have attempted to quantify the microwave radiation parameters, which include frequency, threshold power density, duration of exposure, frequency of exposure and mode of radiation (CW or pulsed) in relation to the formation of lenticular opacities. Much of the available information is based on the work of Carpenter and his colleagues[14-16] and Birenbaum.[17] Two basic systems were used:

1. A closed waveguide system, in which the eye of the animal is placed at the end of the waveguide. The power absorbed by the eye is determined by measurements of the forward and reflected power.

2. A free field system, in which the animal is placed in an anechoic chamber and phantom calorimetric measurements are made at the position of the eye.

The two systems do not yield similar results, in that the waveguide system produces an anterior cataract and the anechoic method produces a posterior cataract. The frequency of radiation did not influence the position of the cataract, and there was no difference in the ability of continuous wave or pulsed radiation to produce a cataract. This implies that average power and not peak power is important in the production of cataracts. The average power causing an observable loss of transparency in at least 50 per cent of the animals was defined as the *threshold value*. Time and power density thresholds at 2450 MHz causing lens opacities in the rabbit eye by a single exposure[15] are shown in *Fig. 7.5*.

It was also noted from single or repeated exposure of the rabbit eye that opacities were seen when the temperature was increased by 4°C. In the case of repeated exposures a cumulative effect was postulated. This cumulative effect must not be confused with cumulative injury from exposure to ionizing radiation where a certain portion of the injury is beyond repair and these are designated 'residual radiation injuries'; such residual portions are additive. In the case of microwave-induced lens opacity, evidence in animals shows that when lens opacity is produced, a threshold power density of greater than 100 mW/cm² for a duration of over one hour must be available. The minimum power density below which lens opacities have not occurred is 80 mW/cm². Most studies indicate that a critical intraocular temperature ranging from 45 to 55°C must be attained for opacities to develop. For a cumulative rise of temperature the time interval between exposures should be less than the time required for the tissue to return to its normal temperature.

The earliest recorded case of microwave cataract recorded in man was reported by Hirsch.[18] This followed exposure from a microwave generator at a frequency of 1500–3000 MHz with an estimated power

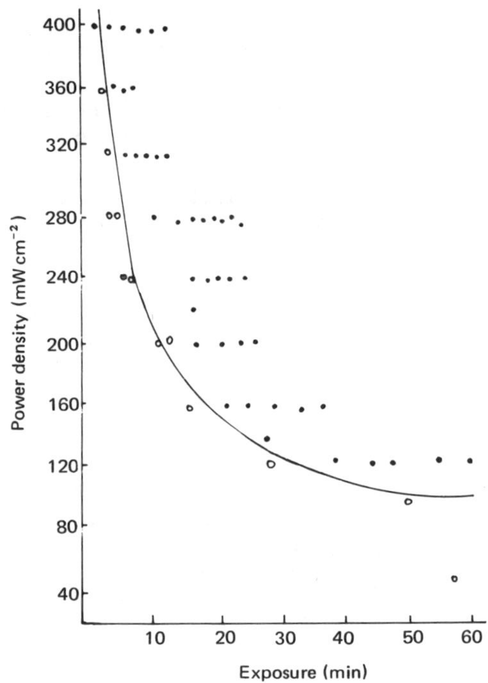



*Fig.* 7.5. Time and power density thresholds for the induction of lens opacities in the rabbit eye by single dose irradiation at 2·45 GHz: Carpenter and Van-Ummersen.[15]

density of 100 mW/cm² for one year. Since then several other cases have been reported as microwave-induced cataract.[19,20]

## Non-thermal effects

These have been studied in relation to high power density exposures and low power density exposures. At high power densities it is not easy to isolate the non-thermal from the dominant thermal effects. At low power densities non-thermal effects appear to result from a direct inter-action between the electromagnetic field and the biological material. The distinction between thermal and non-thermal effects is often therefore made on the basis of power density. In 1973 the following ex-ternal field intensity ranges were adopted internationally.[21]

| *Effect* | *Range* |
| --- | --- |
| Thermal, heating | 10 mW/cm² |
| Thermal, non-heating | 1–10 mW/cm² |
| Non-thermal | 1 mW/cm² |

Studies on low level non-thermal effects from Russian and Eastern European literature were first noted in the early 1960s. As mentioned earlier, these related to the central nervous system, the cardiovascular system and the endocrine system. After initial disbelief followed by a period of scepticism, scientists in the West displayed a more rational approach to these data and indeed some investigations into the effects of exposure to low power microwave radiation have been carried out. A summary of the reproducible effects from different countries by Baranski and Czerski[1] is shown in *Table 7.5*.

In the Russian and Eastern European data the bulk of significant clinical findings relates to CNS disorders. These, together with cardiovascular and endocrine data, are shown in *Tables 7.6–7.8*.

Closely related to neurological findings are various biochemical effects. These include increased levels of fasting blood sugar, depressed creatinine excretion and depressed serum levels of pyruvate and lactate.[23] These were seen in exposures in the centimetre band. In another study of persons exposed to 30–800 MHz radiation, changes included increased levels of serum protein, cholesterol, 17-ketosteroids and $\beta$-lipoproteins; elevated gamma globulin and total protein were noted[24] at exposures in the range 640 kHz to 300 MHz. It is suggested that these effects reflect imbalances, possibly related to the CNS, and may be significant since hormonal imbalances are linked to oncogenesis in tissues dependent on hormonal balance for their maintenance. The ability of electromagnetic fields to interact with cellular membranes may explain the findings of mitotic and chromosomal abnormalities following long term exposure to radiofrequency and microwave radiation. The frequencies and intensities at which mitotic and chromosomal abnormalities have been noted are shown in *Table 7.9*.

**Exposure standards**

Microwave and radiofrequency radiation at certain frequencies, power levels and exposure durations have been shown to produce biological effects. There was a need to set limits on the amount of exposure to such radiations which personnel can accept with safety. The development of these standards has been on the basis of gross thermal damage in Western European countries and on the criterion of reversible specific functional changes in Eastern European countries. This has led to widely differing standards, as can be seen from *Table 7.10*.

The present standard was developed in 1966 as an exposure safety standard by the American National Standards Institute. This standard was reaffirmed in 1973 and, at 10 mW/cm², is deemed to be at least a factor of 10 below the thresholds of damage by thermal effects. The Russian standard, which is three orders of magnitude lower at 0·01 mW/cm², is based on other biological effects. Many scientists doubt

**Table 7.5.** Some reproducible effects of microwaves

| Effect | Power density (mW/cm²) | Research | | | |
|---|---|---|---|---|---|
| | | USA | USSR | Czech. and Poland | France |
| Altered conditioned and unconditioned reflexes in dogs, rats and mice | 0·2–10+ | + | + | + | |
| Behavioural changes in rats and mice | 0·2–10+ | + | + | + | |
| EEG changes in rabbits and humans | < 10 | + | + | + | |
| Microwave 'hearing' in humans | < 10 | + | | | |
| Altered calcium binding in neuronal membranes | < 10 (8–16 Hz) | + | | | |
| Pathological changes in nerve tissue and brain | < 10 | + | + | | |
| Increased sensitivity to drugs | < 10 | + | + | + | + |

Source: Baranski and Czerski.[1]

**Table 7.6.** Occurrence of some symptoms in humans exposed occupationally to high frequency electromagnetic fields (750 kHz–200 MHz)

| | Length of employment | | | |
| | 1–6 years* | | 7–16 years† | |
| Symptoms | % | No. | % | No. |
| --- | --- | --- | --- | --- |
| Headache | 20·5 | 15 | 32·9 | 24 |
| Disturbance of sleep | 13·7 | 10 | 23·3 | 17 |
| Fatigue | 12·3 | 9 | 17·8 | 13 |
| General weakness | 7·0 | 5 | 12·3 | 9 |
| Disturbance of memory | 5·5 | 4 | 8·2 | 6 |
| Lowering of sexual potency | 5·5 | 4 | 8·2 | 6 |
| Drop in body weight | 2·7 | 2 | 12·3 | 9 |
| Disturbance of equilibration | 5·5 | 4 | 11·0 | 8 |
| Neurological symptoms | 0·0 | 0 | 15·1 | 11 |
| Changes in ECG | 17·8 | 13 | 28·8 | 21 |

After data by Minecki as quoted by McRee.[8]
*Average 4·3; 73 persons.
†Average 9·6; 73 persons.

**Table 7.7.** Subjective effects on persons working in RF fields

| | |
| --- | --- |
| Headaches | Feelings of fear |
| Eyestrain | Nervous tension |
| Fatigue | Mental depression |
| Dizziness | Memory impairment |
| Disturbed sleep at night | Pulling sensation in the scalp and brow |
| Sleepiness in daytime | Loss of hair |
| Moodiness | Pain in muscles and heart region |
| Irritability | Breathing difficulties |
| Unsociability | Increased perspiration of extremities |
| Hypochondriac reactions | Difficulty with sex life |

After data by Marha et al. as quoted by McRee.[8]

**Table 7.8.** Clinical manifestations of chronic occupational exposure to microwave radiation in 525 workers

*Symptomatology*
    Bradycardia
    Disruption of the endocrine–humoral process
    Hypotension
    Intensification of the activity of thyroid gland
    Exhausting influences on the central nervous system
    Decrease in sensitivity to smell
    Increase in histamine content of the blood

*Subjective complaints*
    Increased fatigability
    Periodic or constant headaches
    Extreme irritability
    Sleepiness during work

After data by Letavet and Gordon.[22]

**Table 7.9.** Effects of RF/MW radiation on mitosis and chromosomal structure

| System | Frequency | Power | Mitotic anomalies | Chromosomal anomalies | Ref. |
|---|---|---|---|---|---|
| Garlic root tips | 27 MHz (pulsed) | 1000–1500 V/cm | + | + | Heller and Teixeira-Pinto[11] |
| Stimulated lymphocytes, erythro-blasts (guinea-pigs, rabbits) | 2950 MHz (pulsed) | 7–20 mW/cm² | + | + | Stodolnik-Baranska[25] |
| Human lymphocytes stimulated with phytohaemagglutinin | | | | | |
| Chinese hamster cells, human amnion cells | 2450 MHz | 20–50 mW/cm² | | + | Chen et al.[26] |
| Slime mould *(Physarum polycephalum)* | 45, 60, 70 Hz | 2·0 G, 0·7 V/m | + | | Goodman et al.[27] |
| Chinese hamsters (femur bone-marrow) | 2450 MHz | < 200 mW/cm² | + | + | Leach[12] |
| Cultured human lymphocytes, Chinese hamster cells | 15, 19, 21, 25 MHz (pulsed) | 30 000 V/cm | | + | Mickey[28] |

**Table 7.10.** Personnel exposure standards for microwaves

| Maximum permissible power density (mW/cm²) | Frequency (MHz) | Country or agency | Specifications |
|---|---|---|---|
| 10 | 10–100 000 | USA* | |
| | | ANSI | 1 mW h/cm², 24 h |
| | | NIOSH | 8 h work day |
| | 100–100 000 | ACGIH | 10 mW/cm² TLV 8 h<br>10–25 mW/cm², 10 min/h<br>25 mW/cm² ceiling value |
| | 300–300 000 | Army/Air Force | 10–55 mW/cm²<br>min = 6000/(mW/cm²)² |
| 1 | 300–300 000 | Poland | 0·2 mW/cm²–10 mW/cm²<br>(8 h–11·5 s) (SF)†<br>1·0 mW/cm²–10 mW/cm²<br>(8 h–4·8 min) (NSF) |
| | | USSR‡ | 15–20 min/day |
| 0·1 | | Poland | 0·2 mW/cm², 8 h (SF)<br>24 h (NSF) |
| | | USSR | 2–3 h/day |
| 0·025 | | Czecho-slovakia | 8 h (CW) |
| 0·01 | | Poland | 24 h (SF) |
| | | USSR | 8 h |
| | | Czecho-slovakia | 8 h (pulsed) |

*Also with slight modification in Canada, United Kingdom, German Federal Republic, Netherlands, France, Sweden.

†SF = stationary field (h = 32/W/m²); NSF = non-stationary field (h = 800/W/m²).

‡MPE × 10 for exposure to movable beam or antenna.

Source: Michaelson.[29]

the validity of the very low level set in the Russian standard, but it must be accepted that the Russian scientists have placed more emphasis and research effort into the study of the effects on the brain and central nervous system. Their safe level of exposure is based on these effects and it is felt that these should not be dismissed till there is sufficient proof that they are not detrimental to the health of exposed persons. It is interesting to note that in 1977 the Canadian Department of Health and Welfare proposed a reduction in the exposure standard from 10 to 1 mW/cm². Additionally, there are proposals for a review of the present levels from the American National Standards Institute and the National Institute for Occupational Safety and Health, US

Department of Health, Education and Welfare. These, together with the Canadian proposals and the present Western European standard, are shown in *Fig. 7.6*. The proposals include standards for occupationally exposed and non-occupationally exposed persons and the levels vary also with the frequency of radiation.

In applying the present standard of 10 mW/cm², it should be considered in relation to the equivalent free space electric and magnetic field strengths, which are approximately 200 V/m and 0·5 A/m respectively. For modulated fields, power density and squares of field, strengths are averaged over any 6-minute period. Though the standard of 10 mW/cm² is deemed a safe value based on an average dose for a period of 6 minutes, it does not specify an upper limit of permissible short duration exposure. Because of this it was felt that the duration of exposure was important and that higher levels could be tolerated for shorter periods. A guide was therefore developed for short duration exposure of personnel for the frequency range 300–300 000 MHz based on the following equation: $T_\mathrm{p} = 6000/W^2$, where $T_\mathrm{p}$ is the permissible time of exposure in minutes during any one hour period and $W$ is the power density in mW/cm² in the area concerned. This exposure level is applicable to military personnel in the UK and the USA.

Finally, there is a product emission standard for microwave ovens. Currently this standard specifies an upper limit of 1 mW/cm² at 5 cm when manufactured and an allowance for degradation in use to a limit of 5 mW/cm² at 5 cm, both measurements being from the external

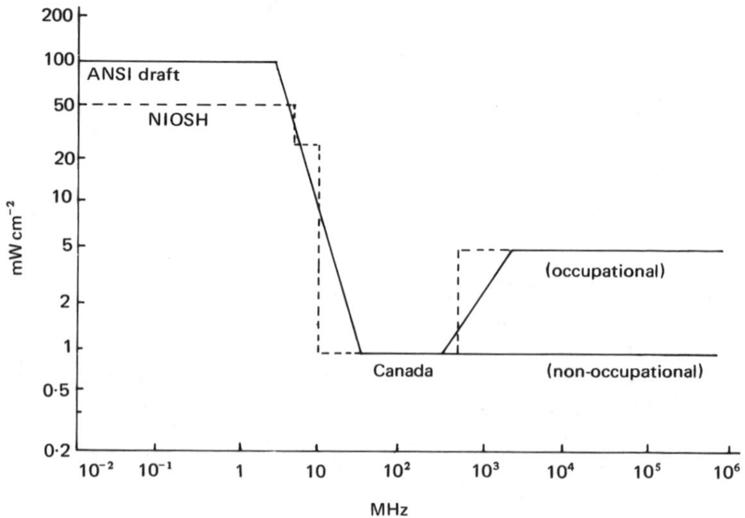


*Fig. 7.6.* Proposed ANSI and NIOSH standards together with Canadian standard.

surface. It is emphasized that this is an emission standard and should not be confused with the exposure standard detailed earlier.

## Measurement

Microwave or radiofrequency radiation may be emitted from radiating devices such as microwave horns and various types of antennas, or as a result of leakage from such equipment, and from microwave and radiofrequency processing units such as microwave ovens. In the latter group no radiation should normally be found outside the device, and measurement therefore is to verify the integrity of the safety mechanisms. Additionally, periodic measurement should be carried out to ensure that any leakage in use is within the limits of the emission standard for the product. Radiating devices such as radar and communications equipment have performance data from which theoretical determination of the power density distribution can be made. Measurement in this group, therefore, is to establish the validity of theoretical data, under actual operating conditions.

The measuring instruments are described as electromagnetic radiation survey meters or electromagnetic leakage monitors as appropriate. Essentially they consist of small dipoles with diode or thermocouple-type transducers which convert microwave or radiofrequency energy into electrical signals. A typical meter consists of a sensor with two or more orthogonal electric dipole elements. Each element ends in a thermocouple or microwave diode connected to a voltmeter which gives the power density in $mW/cm^2$. The instruments cover the frequency range up to around 40 GHz and would normally be adequate for most of the applications currently in use. In certain civil and military applications very much higher frequencies are used and commercially available measuring devices are inadequate. In these circumstances it is possible to evaluate the power density using a microwave power meter.[30]

In summary, any programme to assess the potential hazard in relation to current exposure limits should take account of:

(i) An inventory of all radiation sources;
(ii) Evaluation of hazard areas based on power output and mode of operation;
(iii) Planning and execution of a survey of areas calculated as hazardous;
(iv) Correlation of measured levels with calculated values;
(v) Designation of personnel at risk.

It is important that personnel required to carry out such surveys are trained and experienced in the use of monitoring equipment in accordance with the specifications provided by the manufacturer.

**Medical supervision**

Microwave workers have not been supervised medically as a special risk group. Any special examination has been for a suspected or actual overexposure. In these cases the medical examination is usually in respect of any damage to the lens. The recent proposal from the Commission of European Communities for a directive for the protection of workers and the general public from the harmful effects of microwave radiation requires that all microwave workers should be provided with occupational health supervision. The details have not been spelt out and it is likely that interpretation will be left to individual nations; also the definition of 'microwave worker' has not been detailed. However, in relation to the current Western standard of 10 mW/cm$^2$, and taking account of the extremely low level of background radiation, it would seem appropriate to consider persons working in areas where levels of up to 10 mW/cm$^2$ are present as 'microwave workers'. Additionally, it is hoped that the paucity of data relating to microwave workers and the possibility that there may be low level effects will influence occupational health physicians to carry out pre-placement, periodic and post-employment medical examinations, together with details of exposure patterns and type of radiation. Data from such records will be of immense value in future epidemiological studies.

*Visual display units (VDUs)*

There has been widespread concern in recent years about the potential health hazards associated with visual display units based on cathode ray tubes. It has been known for many years that these devices could emit soft X-rays under certain conditions. Though there was concern about the possible health implications, the operating and electronic conditions under which such emission could take place were not regarded as posing a significant problem to operators of such equipment, or to the general public. However, with increasing use of computers (linked intelligent terminals and microprocessors), both at home and at work, there has been speculation regarding exposure to radiations covering the whole electromagnetic spectrum. It is common knowledge that most electrical equipment generates low levels of electromagnetic radiation over a broad spectrum. Hitherto this has been regarded as an interference problem, detected only by the more sensitive receivers. Because users of VDUs are spending more time with such equipment, the possible effects of long term exposure to these radiations compounded by a plethora of anecdotal incidents, have resulted in many surveys being carried out to quantify any radiation.

In the United Kingdom, the Health and Safety Executive made arrangements for the National Radiological Protection Board to carry out a comprehensive survey, in various regions of the electromagnetic spectrum, of all known types of VDU currently manufactured or marketed. More than sixty firms were inspected and measurements were made on more than two hundred types of VDU. The conclusions are that the measured radiation from correctly operating VDUs is less than the exposure limits for the various spectral bands laid down in many national and international standards, and the radiations therefore do not pose a hazard to the operators either in the long or short term.[31] A similar survey carried out in the United States of America on a number of VDUs used by the Bell system found that the measured levels from correctly operating VDUs were far less than the most restrictive permissible exposure levels stipulated by any agency or government of any nation.[32] Even accepting the fears of Eastern European nations of low levels of exposure causing CNS, cardio-vascular and endocrine disturbances, the conclusion must be that on the present evidence there is no hazard to the operators from VDUs which are operating and being operated correctly.

## LASER RADIATION

Lasers are devices which produce a unique kind of radiation in the form of an intense beam of light of a very pure single colour. The term 'laser' is an acronym derived from Light Amplification by Stimulated Emission of Radiation. The laser operates in that part of the electromagnetic spectrum which includes a little of the ultraviolet, the visible and the infrared regions. It extends roughly from $0.2\ \mu m$ to $340\ \mu m$. Lasers may be considered to fulfil one of the oldest dreams of technology, that of producing a beam of light intense enough to vaporize the hardest and most heat-resistant materials. They have been used to drill holes in diamonds, to 'weld' the retina of the eye when detachment has occurred and for microsurgery on single cells. In industry lasers have a host of applications, which include shock hardening, glazing, drilling, deep penetration welding, aligning, transformation hardening and so on.

The existence of stimulated emission was recognized by Einstein in 1917, but not until the 1950s were ways found to use it in devices. United States physicist Charles Townes used the principle of stimulated emission in the microwave region of the electromagnetic spectrum to produce the maser, the acronym derived from Microwave Amplification by Stimulated Emission of Radiation. Shortly after this Townes and A. L. Shawlow demonstrated that it was possible to extend this principle to the optical region, and called the resulting device a laser.

## Physical characteristics

The phenomenon of emission and absorption needs explanation. In the simplest atom, the hydrogen atom, which consists of a nucleus and a single electron, the orbit of the electron is confined to a particular range. The energy of the atom is dependent on which orbit the electron occupies. If the energy of the atom is reduced by the electron occupying a different orbit then a photon is emitted. The energy of this photon ($E_\mathrm{p}$) is equal to the difference in energy of the atom before and after its transition, and the photon itself is a wave which follows the Einstein relationship

$$E_\mathrm{p} = h\nu \tag{1}$$

where $E_\mathrm{p}$ is the energy of the photon, $h$ is Planck's constant and $\nu$ the radiation frequency (Planck's constant $= 6 \cdot 626176 \times 10^{-34}$). Similarly, the emission spectra of more complicated atoms may be explained on this basis. The simplest atom, having only two energy levels, is shown schematically in *Fig. 7.7.*

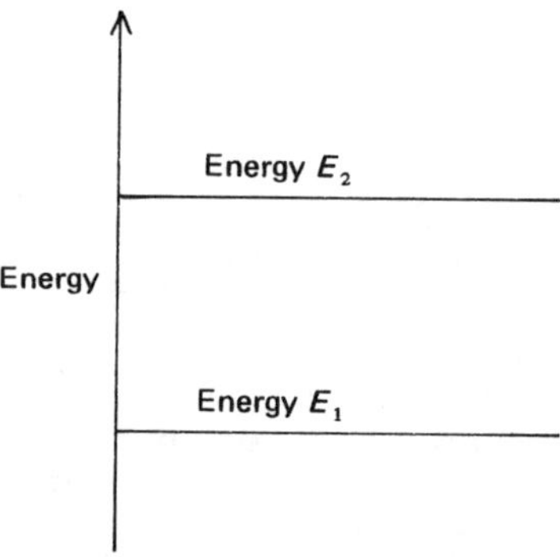


*Fig. 7.7.* Energy system with two levels.

If an atom in the upper level moves to the lower level the energy can be emitted as radiation whose frequency $\nu$ is given by

$$\nu = \frac{E_2 - E_1}{h} \tag{2}$$

where $E_1$ and $E_2$ are the two energy levels. Conversely, if the atom moves from the lower energy level to the upper energy level, radiation of a frequency determined by equation (2) must be absorbed. Einstein showed that emission can occur in two ways:
1. By the random change of an atom from the upper level to the lower level, called spontaneous emission; the converse is also valid and is called spontaneous absorption (*Fig. 7.8*).

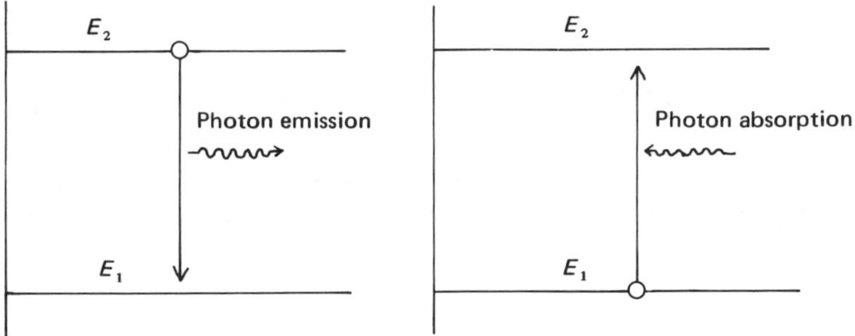

*Fig. 7.8.* Spontaneous emission and absorption.

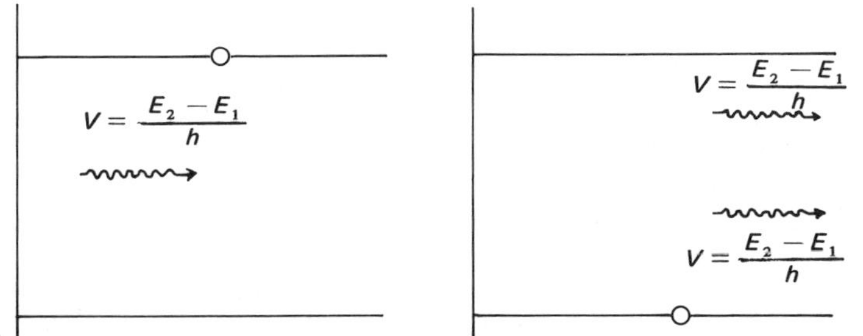

*Fig. 7.9.* Stimulated emission.

2. By the interaction of a photon having an energy level equal to
   the difference in the two energy levels causing an atom to change
   from the upper level to the lower level with the creation of a
   second photon, described as stimulated emission (*Fig. 7.9*).

The earliest application of stimulated emission was in the ammonia
maser and this led to the development of the laser on similar principles. It was found that the ammonia molecule could take on two
energy states depending on the position of the hydrogen atom. It was
argued that in a tank of ammonia, if sufficient atoms could be made
to exist in the upper energy level, then the interaction by a photon of
energy equal to the difference between the upper and lower states
could trigger off a chain reaction where stimulated emission would
dominate over absorption, resulting in an amplified wave (*Fig. 7.10*).

In a two-level energy system containing an equal number of atoms
in each level, a photon of energy equal to the difference in energy
levels could cause either stimulated emission or absorption with equal
probability. Thus for the dominant process to be emission, the system must contain a large number of atoms in the upper level. The
process of increasing the number of atoms in the upper level is called

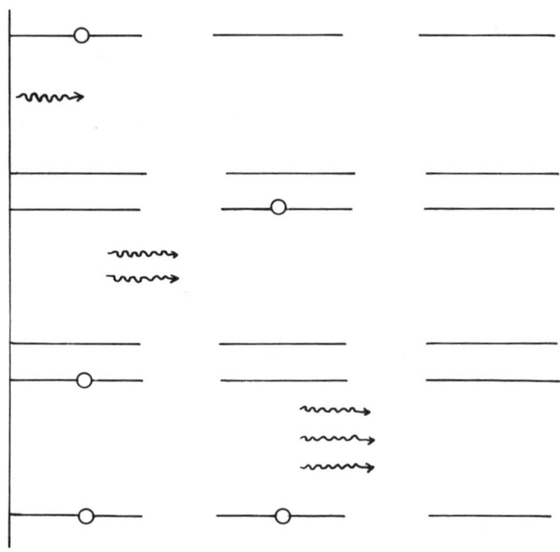

*Fig. 7.10.* Amplification.

population inversion. This is a necessary prerequisite for stimulated emission. In the ammonia maser the two-level system was achieved by physically removing the molecules in the lower state. This is not generally possible and a three-level system was envisaged where the population of each level would increase from the top down to the bottom level. Then by pumping atoms from the lower level to the upper level using photons of the correct frequency both stimulated emission and absorption are equally possible and a stable state would be achieved between the levels. Under these conditions if the spacing of the middle level is carefully chosen the population here can be made to exceed that in the middle level and laser action will take place. An example of the three-level system is seen in the ruby laser, which has a ruby crystal as the lasing medium (ruby crystals are made of aluminium oxide with a small proportion of chromium which gives the characteristic pink colour). Amplification is achieved by parallel ends to the ruby, and these are polished so that one end is totally reflecting and the other partially reflecting (about 10 per cent transmitting), to ensure laser output from this end. The chromium ions in the ruby are energized to the upper level by a xenon flash tube. From the upper level there is a transition to the middle level where there is no emission but a release of energy to the crystal and consequent heating. The atoms remain in the middle stage for a time before returning to the lower or ground level with spontaneous emission. Because of this metastable state the middle level builds up while the ground level is

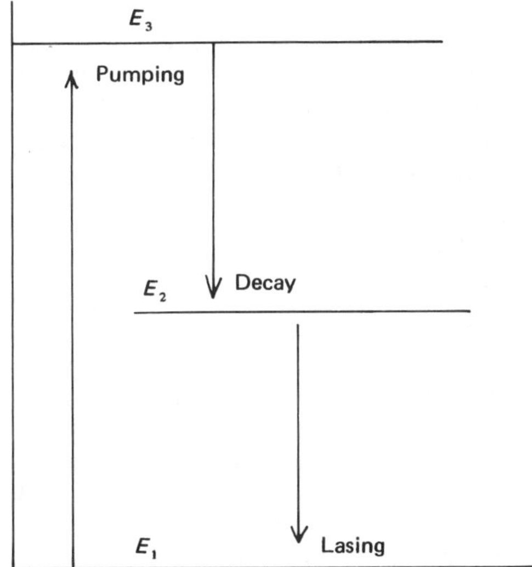



*Fig. 7.11.* The three-level system.

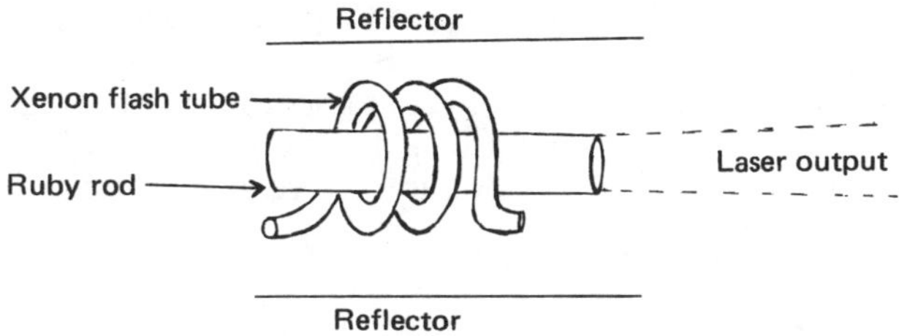



*Fig. 7.12.* The ruby laser.

depleted resulting in population inversion. *Fig. 7.11* shows the three-level system and *Fig. 7.12* shows a diagrammatic representation of the ruby laser.

Based on this principle many materials are used for lasers. The ruby is an example of an optically pumped solid laser. Other types include:

*Gas lasers:* Atoms in a gas discharge can be made to produce light, as in a neon sign. A particular energy level will cause an exceptionally large number of atoms to accumulate and, by the use of parallel mirrors, one of which is partially reflecting, lasing will take place. The beam produced is nearly a straight line and such devices are used for aligning in the construction industry.

*Gas dynamic lasers:* The principle is that when a hot gas is allowed to cool the number of molecules in the lower energy state decreases more rapidly and falls below the number in the higher state with resulting laser action. High outputs can be obtained from these lasers.

*Semi-conductor lasers:* These consist of a flat junction of two pieces of semi-conductor material each of which has been treated with a different type of impurity. When a large current is passed through such a system lasing action takes place.

*Dye lasers:* Certain organic dyes are capable of fluorescing, that is radiating light of a different colour. Though the excited state of their atoms lasts only a fraction of a second and the light emitted is not concentrated in a narrow band, many such dyes have been made to exhibit laser action, with the advantage that they can be tuned to a wide range of frequencies.

A laser may be made to produce short intense pulses by incorporating a shutter between the amplifying column and the end mirror. This prevents laser action so long as the shutter is closed. If conditions are otherwise correct for laser action and the shutter is suddenly opened, the stored energy is released as a giant pulse in a fraction of a second, having power which may be as high as several thousands of kilowatts. This is termed Q-switching. The Q-switch may be a mechanical shutter or more commonly a liquid or solid optical shutter. Normally a laser operates in several modes, and at several different frequencies. By synchronizing these modes, a process called mode-locking, even shorter and more powerful pulses may be generated. These are used in scientific investigations and for puncturing holes so rapidly that the surrounding material is not affected. Thus lasers can be operated in different temporal modes — continuous wave, pulsed, pulse-repetitive, Q-switched and mode-locked.

## Laser hazards

In assessing a laser for the protection of personnel the hazards fall into five categories: optical radiation hazards to the skin, optical radiation hazards to the eye, chemical hazards, electrical hazards and miscellaneous hazards.

### Skin hazards

The primary effect of a laser beam on the skin is one of surface burning, and the extent of this is dependent on the incident energy per unit area. Thresholds of injury to the skin and eye are comparable in the far infrared and ultraviolet regions of the spectrum. In the visible and near infrared regions of the spectrum the levels of radiation required to produce skin injury are very high, a minimum of several watts per square centimetre. Skin injury is therefore unlikely with low and medium power lasers under normal operating conditions.

The skin is particularly sensitive in the actinic ultraviolet region

(200–320 nm). It is well documented that radiation in this spectral region causes sunburn, and the exposure levels required for this are very low — of the order of microwatts per square centimetre. Radiation in this region has also been regarded as the cause of certain types of skin cancer. It should also be noted that the sensitivity of the skin is increased by photosensitizing chemicals and by previous exposure to specific wavelengths, usually in the visible and near ultraviolet regions of the spectrum. Some drugs and certain diseases such as lupus erythematosus, herpes simplex and xeroderma pigmentosa are also known to cause photosensitization.

### Eye hazards

Eye damage is potentially the most dangerous effect of laser radiation. The hazard to the eye is increased in visible and near infrared laser radiation because of the focusing action of the lens. This has the effect of concentrating the laser beam by several orders of magnitude. All lasers operating in the visible and infrared region of the spectrum, except for the very small lasers, should be regarded as being hazardous to the eye at wavelengths between 400 and 1400 nm. In the ultraviolet and far infrared regions, however, the energy is absorbed in the anterior portion of the eye, mainly in the cornea and the lens. The absorption properties of the human eye for electromagnetic radiation in this spectral region are shown in *Fig. 7.13*.

It would now be appropriate to look at the biological effects in the structures of the eye where laser radiation is absorbed. The structures in the anterior portion of the eye that are at risk are the cornea, conjunctiva, aqueous humour, iris and the lens. The cornea and the lens are the more important components of the optical system of the eye.

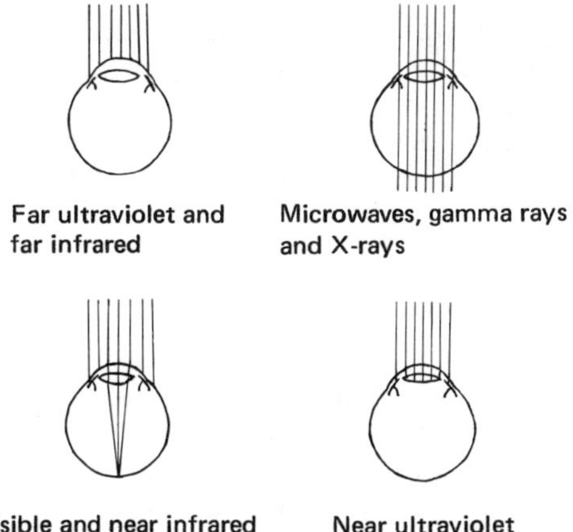



*Fig. 7.13*. Absorption properties of the human eye for electromagnetic radiation.

Any loss of transparency in this system is serious. The corneal epithelium is completely renewed almost every two days. The far ultraviolet and the far infrared radiations are almost wholly absorbed in the cornea. (These are the very short wavelengths in the ultraviolet and the long wavelengths in the infrared region.) Damage to the corneal epithelium is usually temporary and recovery is the rule. This is comparable to the situation in 'welder's flash'. However, if the deeper tissues are affected scarring and permanent damage will result. In the near infrared and near ultraviolet region the lens absorbs a great portion of the radiation. Though cataracts as seen with ionizing radiation, microwave radiation or in glass blowers have not been reported, the mechanics of the insult are not dissimilar and delayed effects cannot be discounted entirely. In the case of the iris it is difficult to envisage an injury which would not have already affected the other structures to a greater extent. The aqueous humour is not a cellular structure and as such similar damage would not be expected, though injury of a photochemical nature affecting metabolites is a possibility.

The most important and vulnerable structure in the eye is the retina. Visible and near infrared radiation is sharply focused on the retina. The area of importance is the fovea, which accounts for central vision, and is in the centre of the macula. This central area is approximately 3 mm in diameter in man. Whilst a retinal injury in the periphery may go unnoticed, an injury in the central part will result in a blind spot or scotoma. This would occur if one were looking along the laser beam towards the source, either direct or by specular reflection. The extent of the injury would depend on the degree of accommodation, the beam divergence and the power of the laser source. The power of the laser source together with the duration of exposure would determine the total energy dissipated at the point of focus. The damage to the retina, if in the fovea, will result in functional loss which until recently was thought to be permanent. It has been postulated by some workers that in the boundary area between no injury and permanent injury partial repair can take place. This theory has yet to be validated, so any functional or ophthalmologically identified defect resulting from laser radiation must be regarded as permanent.

### Chemical hazards

The chemical hazards are greatest in laser research laboratories where highly toxic, explosive and flammable materials are used. There is also hazard from fume and metal oxide during laser welding, drilling or cutting. With high power lasers, vaporization of the target or backstop could produce hazardous airborne contaminants. These could include asbestos, cadmium oxide fume, vanadium fume, copper fume and iron oxide fume. Additionally, the lasing medium itself could present a hazard, which could include carbon monoxide, hydrogen fluoride, fluorine and carbon dioxide.

*Electrical hazards*

Power supplies to most lasers have the potential to cause serious if not fatal electric shocks. The high power lasers are even more hazardous, sometimes requiring extremely high voltages.

*Miscellaneous hazards*

The risk of explosions of flash tube systems used to energize the lasing medium is present in certain lasers. There is also the potential hazard from soft X-rays from high voltage vacuum tubes used in laser power supplies. Noise can be a hazard, the sources being high capacitance condensers, and there is noise also from the laser beam itself hitting targets and backstops; impulse noise levels can exceed 120 dB(A).

**Protection of personnel**

Guidance on the protection of personnel against the hazards of laser radiation is contained in the British Standard BS 4803: 1972; the revised edition of this standard was issued in October 1982. Lasers are put into four categories according to their hazard rating together with the protective measures to be adopted. Each category should be considered in its own right with respect to exposure limits. It is important that any protective equipment, in this case goggles, should be compatible with the wavelength of the particular laser. For example, a carbon dioxide laser would require goggles with lenses which attenuate at 10·6 $\mu$m. In terms of measurement, parameters which can be considered are radiant power, radiant energy, irradiance and radiant exposure. Measurements are usually carried out to assess the laser output in order to classify the laser. This is normally carried out by the manufacturer. Routine monitoring is seldom necessary. The principle underlying safety is that a laser beam, either direct or reflected specularly, and with adequate energy, is hazardous. If a person looked along that beam towards the source, injury would result. Observance of the safety precautions for the particular laser type together with medical surveillance will minimize the potential hazard from lasers.

**Medical surveillance**

There has been much controversy as to the need to carry out medical surveillance — and if so, what sort of medical surveillance. In principle there are three medical examinations to be considered from the point of view of the occupational health physician. These are pre-placement, periodic and final medical examination. There is a large measure of agreement on the requirement for the pre-placement medical examination. This serves to establish a baseline against which any injury can be measured. There is also a large measure of agreement on the final medical examination or examination at cessation of work with lasers. It is generally accepted that the primary purpose of this examination is medico-legal. It is the periodic medical examination that is not straightforward because no chronic health problems have yet been associated with lasers. Moreover, any such examination can only detect the more serious injuries. If a person suffers an overexposure which results in a retinal burn and there is associated pain and loss of function the physician will be consulted, and a lesion in the retina will be identified by ophthalmoscopy. However, if the employee is not aware of the overexposure because of the absence of any symptom, and no obvious loss of function, it is not likely that the lesion would be identified by routine ophthalmoscopy, but a defect in the visual field may be identified by the use of the Amsler grid. Because of the certainty that routine non-invasive ophthalmoscopy would be negative and the probability that testing of visual fields could be negative, some physicians are not convinced of the need for periodic medical examinations. Testing of visual fields annually to ensure no obvious functional deficit using the Amsler grid is simple, effective and not time-consuming. It is my personal view that annual medical examinations using the Amsler grid is an effective method of medical surveillance of personnel working with lasers.


REFERENCES
1. Baranski S and Czerski P. *Biological Effects of Microwaves.* Stroudsberg, Pa: Dowden Hutchinson & Ross, 1976.
2. Cleary SF. Biological effects of microwave and radiofrequency radiation. *CRC Crit. Rev. Environ. Control* 1970; 1:257–306.
3. Johnson CC and Guy AW. Non-ionizing electromagnetic wave effects. *Proc. IEEE* **60**:692.
4. Shinn DH. Avoidance of radiation hazards from microwave antennas. *The Marconi Review* 1976; **39**:201.
5. Michaelson SM. Effects of exposure to microwaves: problems and perspectives. *Environ. Hlth Persp.* 1974; **8**:133–56.
6. Schwan KP and Li K. Hazards due to total body radiation by radar. *Proc. Inst. Radio Eng.* 1956; **44**:1572.
7. Michaelson SM. Thermal effects of single and repeated exposures to microwaves — a review. In: Czerski P, Ostrowski K, Shore ML et al. (ed.) *Biologic Effects and Health Hazards of Microwave Radiation.* Proceedings of an International Symposium. Warsaw: Polish Medical Publishers, 1974.
8. McRee DI. Environmental aspects of microwave radiation. *Environ. Hlth Persp.* 1972; **2**:41–53.
9. Ely TS, Goldman DE and Hearon JF. Heating characteristics of laboratory animals exposed to ten centimeter microwaves. *IEEE Trans. Biomed. Eng.* 1964; **11**:123.

10. Sigler HP, Lilienfield AM, Cohen BH et al. Radiation exposure of parents of children with mongolism (Down's syndrome). *Johns Hopkins Hosp. Bull.* 1965; **117**:374–99.

11. Heller JH and Teixeira-Pinto AA. A new physical method of creating chromosomal aberrations. *Nature* 1959; **183**:905–6.

12. Leach WM. On the induction of chromosomal aberrations by 2450 MHz microwave radiation. First International Congress on Cell Biology. Boston: HEW MA, 1976.

13. Rugh R and McManaway M. Comparison of ionizing and microwave radiations with respect to their effects on the rodent embryo and fetus. *Teratology* 1976; **14**:251.

14. Carpenter RL. In: Cleary SF (ed.) *Biological Effects and Health Implications of Microwave Radiation.* Bureau of Radiological Health, US Dept of Health, Education and Welfare, Maryland, BRH/DBE 1969: 70–2.

15. Carpenter RL and Van-Ummersen CA. The action of microwave radiation on the eye. *J. Microwave Power* 1968; **3**:3.

16. Carpenter RL, Biddle DK and Van-Ummersen CA. Biological effects of microwave radiation with particular reference to the eye. In: *Proceedings of the Third International Conference on Medical Electronics,* London, 1960.

17. Birenbaum L. Effects of microwaves on the eye. *IEEE Trans. Biomed. Eng.* 1969; MBE-**16**:7.

18. Hirsch FG and Parker JT. Bilateral lenticular opacities occurring in a technician operating a microwave generator. *Arch. Ind. Hyg.* 1952; **6**:512.

19. Zaret MM and Snyder WZ. Cataracts and avionic radiations. *Br. J. Ophthalmol.* 1977; **61**:380.

20. Zaret MM. *Health Hazards of VDUs.* Proceedings of a Conference. The Husat Research Group, Loughborough University of Technology, 1980: 49–60.

21. Czerski P, Ostrowski K, Shore ML et al. (ed.) *Biologic Effects and Health Hazards of Microwave Radiation.* Proceedings of an International Symposium. Warsaw: Polish Medical Publishers, 1974.

22. Letavet AA and Gordon ZV. The biological action of ultrahigh frequencies. *Acad. Med. Sci. USSR* (translated by *US Joint Pub. Res. Serv.*) 1962: 12471.

23. Klimova-Deutschova E. Neurologic findings in persons exposed to microwaves. In: Czerski P et al. (ed.) *Biologic Effects and Health Hazards of Microwave Radiation.* Proceedings of an International Symposium. Warsaw: Polish Medical Publishers, 1974: 268–72.

24. Pazderova J, Pickova J and Bryndova V. Blood proteins in personnel of television and radio transmitting stations. In: Czerski P et al. (ed.) *Biologic Effects and Health Hazards of Microwave Radiation.* Proceedings of an International Symposium. Warsaw: Polish Medical Publishers, 1974: 281–8.

25. Stodolnik-Baranska W. The effects of microwaves on human lymphocyte cultures. In: Czerski P et al. (ed.) *Biologic Effects and Health Hazards of Microwave Radiation.* Proceedings of an International Symposium. Warsaw: Polish Medical Publishers, 1974: 189–95.

26. Chen KM, Samuel A and Hoopingarner R. Chromosomal aberrations of living cells induced by microwave radiation. *Environ. Lett.* 1974; **6**:37–46.

27. Goodman EM, Greenbaum B and Marron MT. Effects of extremely low frequency electromagnetic fields on *Physarum polycephalum. Radiat. Res.* 1976; **66**:531–40.

28. Mickey GH. Electromagnetism and its effect on the organism. *NY State J. Med.* 1963; **63**:1935–43.

29. Michaelson SM. Protection guides and standards for microwave exposure. AGARD Lecture Series No. 78: Radiation Hazards 1975: 12-1–12-6.

30. Kanagasabay S. Non-ionizing radiation. In: Waldron HA and Harrington JM (ed.) *Occupational Hygiene.* Oxford: Blackwell Scientific, 1980: 257–69.

31. Cox EA. *Health Hazards of VDUs*. Proceedings of a Conference. The Husat Research Group, Loughborough University of Technology, 1980: 27–38.
32. Weiss MM and Peterson RC. Electromagnetic radiation emitted from video computer terminals. *Am. Ind. Hyg. Assoc. J.* 1979; **40**:300–9.

# 8. NOISE AND OCCUPATIONAL DEAFNESS
*Ian Acton*

## INTRODUCTION

Noise and occupational deafness are not new subjects. An Old Testament reference to blacksmiths experiencing ringing in their ears has been interpreted by some as a reference to tinnitus due to their noise exposure. Occupational deafness was accurately diagnosed by Dr Fosbroke[1] writing in *The Lancet* in 1831. In spite of this, the condition was accepted as inevitable and part of the job. The Chief Inspector of Factories and Workshops wrote in his Annual Report for 1934, 'Most industrial noises are inevitable and cannot be eliminated at the source. Only in comparatively few cases do the workers appear conscious of any inconvenience sufficient to justify the wearing of ear protectors...'.

Although compensation for occupational deafness has been payable in most American States for some time, limited Federal legislation requiring ear protection was not implemented until 1969. A Code of Practice was published by the United Kingdom government early in 1972. Attitudes have changed radically in the past ten years or so, and deafness is said to be the most common occupational disease of the present time. Occupational deafness has become a fashionable diagnosis for hearing loss in anyone who claims to have worked in noise. Occupational deafness often occurs mixed with hearing loss due to other causes, and diagnosis may not be easy. However, there are conditions which must be fulfilled before a conclusive diagnosis can be made, and it may be as well to examine some of these conditions at the outset.

### The 4 kHz dip

A dip in the audiogram at around 4 kHz is merely an indication of hearing damage in the cochlea, of which noise is only one, albeit a common, cause. A loss of hearing at 4 kHz or thereabouts is not, in itself, an absolute diagnostic indication of noise damage. Other causes of cochlear damage must be excluded, and some common examples are discussed later. Further clinical tests such as bone-conduction audiometry, acoustic impedance measurements and vestibular tests

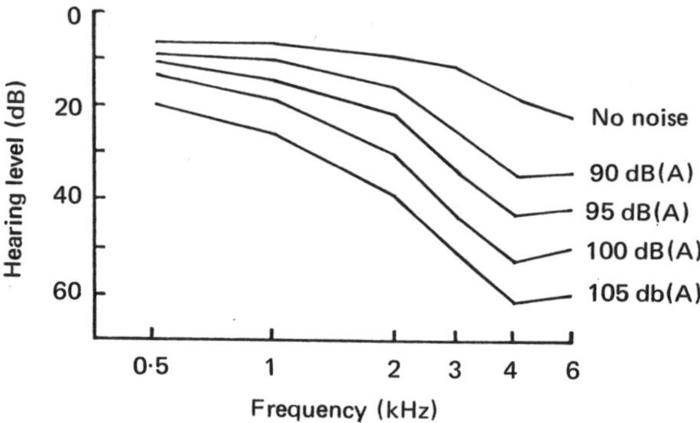


*Fig. 8.1.* Average hearing losses at age 60 years after 40 years' exposure to noise levels indicated. *After* Robinson and Shipton.[2]

may be necessary, but finally a diagnosis of noise-induced hearing loss is reached by a process of elimination.

It is a common sense conclusion that the more serious noise exposures will cause greater noise damage. The extent of the hearing loss in an individual case must be reasonable for the noise exposure sustained before a diagnosis is made. The proportion of a normal population of any age likely to have sustained a given hearing loss as a result of a specified noise exposure may be estimated on a statistical basis from tables published by the National Physical Laboratory.[2] These tables are based on noise immission level, which is a measure of the total noise energy received by a person. It is a function of the average noise level and duration of exposure. Partial noise immission levels from several different periods of employment or other noise exposures may be summated by the normal decibel addition process. The general form of the audiogram due to steady-state noise exposure, derived from the National Physical Laboratory data, is illustrated in *Fig. 8.1.*

Impulse noise situations were specifically excluded from the data on which the National Physical Laboratory tables were based. True impulse noise, especially due to small-arms fire, often produces a steeper audiogram with a greater loss at 4 kHz than would be expected from steady-state noise exposure. Most impulse noise situations in industry consist, in fact, of impulsive components superimposed upon a steady-state background which is potentially hazardous itself, for example drop-forging. True impulse noise situations are not common in industry. Average audiograms of three groups of persons exposed to true industrial impulse noise have been compared with those derived from the National Physical Laboratory data in *Fig. 8.2.*

Allowance should always be made for personal variations when assessing audiograms, but causes other than noise exposure should be

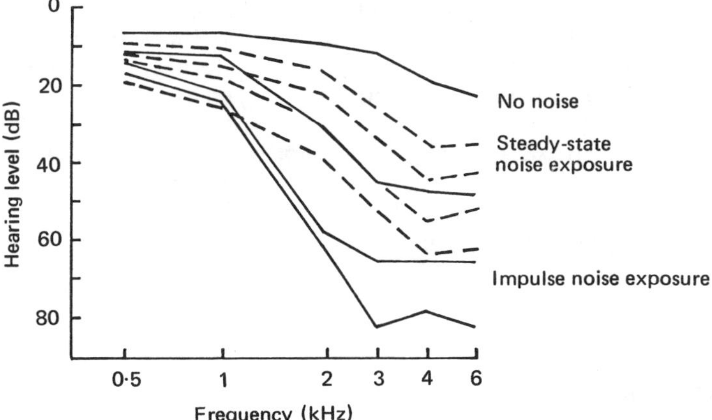


*Fig. 8.2.* Typical hearing loss due to true impulse noise compared with loss due to steady-state noise exposure.

suspected for those which are substantially flatter than the examples given in *Fig. 8.1* or show asymmetry. The most common forms of exaggerated hearing loss produce a flatter audiogram and will be discussed further. The acoustically reverberant conditions prevailing in most of industry cause both ears to sustain more or less the same noise exposure, and hence similar hearing losses. Asymmetrical hearing losses can usually be attributed to firearms if noise-induced, or to disease. The occupationally induced part of the total hearing loss will not exceed the better ear audiogram unless one ear has been fortuitously protected, for example by a plug of wax.

There are other causes of cochlear damage besides noise. The list of alleged ototoxic drugs grows steadily longer, but quinine, streptomycin and other -mycin derivatives, and aspirin in the doses used to treat rheumatic fever are commonly occurring examples. The administration of these drugs is now more carefully monitored than say 10 or 20 years ago, and hearing damage has become less common. The subject of ototoxic drugs has been reviewed by Ballantyne[3] and Hamernik and Henderson.[4] Exposure to industrial poisons, such as benzene, carbon disulphide, carbon monoxide, aniline dyes[5] and heavier alcohols[6] can cause cochlear damage. A report from Mexico[6] of hearing damage due to N-butyl alcohol has been quoted very widely, but normal safety procedures should prevent exposure to potentially ototoxic concentrations of such substances. Menière's disorder and other vestibular disturbances may also produce a hearing loss maximal at 4 kHz, but the audiogram is usually flatter than with noise damage and there should be other clinical signs.

Certain conditions and agents may also predispose towards hypersensitivity to noise exposure. These include the ototoxic drugs, although most patients are absent from work during drug therapy of this type. Unfortunately the sensitivity may persist after cessation of treatment, particularly with the -mycins. Post-stapedectomy operative cases seem at especial risk, and should not be exposed to noise. Some authorities would also include Menière's disorder and certain congenital hearing losses, and sufferers from Menière's disorder often become intolerant of high noise levels. The subject has also been reviewed by Hamernik and Henderson.[4] The hypersensitivity cannot be quantified, but any hearing loss could be in excess of that expected in a normal population.

## Tinnitus and vertigo

Tinnitus or noise in the ears may be present immediately following noise exposure and may become permanent. It does not usually arise later. The tinnitus due to noise exposure is typically high pitched in nature, and is often described as whistling, ringing or like escaping steam or air. It may be masked by everyday background noise, and is often not noticed until the evening or bed-time. Low pitched tinnitus, especially if pulsed, more often accompanies disease of the ear or vestibular system. Tolerance of tinnitus seems to be related more to personality than to the frequency or level.

Only very extreme noise exposures can cause vertigo. Temporary vertigo has been reported after exposure to the noise of unsilenced jet engines.[7-9] Temporary or even permanent vertigo can follow exposure to heavy gunfire or explosions. Vertigo definitely cannot be related to ordinary industrial noise exposure.

The triad of symptoms involving hearing loss, tinnitus and vertigo is often called Menière's syndrome. It is not connected with occupation. Nevertheless, the possibility should always be considered that noise-induced hearing loss can coexist with this or hearing loss due to other causes. Apportionment of the hearing loss to the various causes is difficult, but the noise-induced component likely in the average person can be estimated using the National Physical Laboratory data.

## Presbyacusis

Presbyacusis, which is the natural hearing loss due to advancing age, is more or less additive to noise-induced hearing loss. Unlike noise-induced hearing loss it does not cause a dip in the audiogram but increases steadily towards higher frequencies. There have been numerous surveys of presbyacusis and the data obtained by Hinchcliffe[10] during the 1950s from rural populations in Wales and Scotland are usually quoted in the United Kingdom. In spite of extensive precautions to exclude subjects showing ear disease or having noise exposure, males had significantly worse hearing than females at frequencies between 2 and 8 kHz. These are the frequencies damaged by noise exposure. The scope for finding non-noise-exposed populations has diminished during recent years. However, subsequent surveys have confirmed Hinchcliffe's figures.

Robinson and Sutton[11] have recently completed a numerical analysis of data taken from available literature on presbyacusis. Percentile tables of hearing loss against age calculated from this analysis have been published by the National Physical Laboratory.[12] The formulae and tables form the basis of a new draft International Standard.[13] Separate tables are given for males and females, with males again showing the worse hearing.

American Standards for normal hearing show significantly worse hearing than the British and international equivalents. The reason for the difference probably lies in the fact that the Americans tested the hearing of 'all comers' at State fairs, without selection to exclude non-normals. In consequence, these data are probably more representative of the population as a whole, though they do not form a good basis for standardization. The word 'sociacusis' has been coined to describe that part of the hearing loss due to causes other than occupational noise and age. More recently Ward[14] has proposed the further term 'nosoacusis' for hearing loss produced by causes other than noise, leaving 'sociacusis' purely for the effects of non-occupational noise exposure.

The results of a survey of the hearing of the Mabaan tribe from the Sudan[15,16] attracted the attention of the media and the data are still sometimes quoted because the average thresholds of young males and females were better than the American Standard. In fact, their hearing was worse than the British and International Standards and deteriorated at a faster rate with age than the presbyacusis data reviewed above. Although the tribe are not noise-exposed, there are a number of differences between them and Western civilizations which may more than offset this. These include difficulty in assessing age due to lack of documentation, genetic differences, dietary differences and deficiencies, incidence of disease and environmental differences.

## NOISE MEASUREMENT

The first stage in any hearing conservation programme is a noise survey. In the present context it is important to assess the average noise levels experienced by personnel. This requirement should not be confused with the procedures adopted in taking measurements for noise control or test purposes, where a series of fixed points is usually specified.

Most codes of practice or similar documents recommend that precision grade measuring instruments should be used, although the Health and Safety Executive Code of Practice[17] allows the use of industrial grade meters 'provided that allowance is made for the reduced accuracy of these instruments'. This would seem to imply comparison of the results with a criterion of 87 dB(A) instead of 90 dB(A) when an industrial grade instrument is used. The Code of Practice is written in terms of the equal energy principle; this means that every 3 dB(A) increment in sound level corresponds to a halving or doubling of exposure time. The economic consequences of a 3 dB(A) error could far outweigh the savings to be gained by purchasing an instrument with the poorer specification because of the reduction in permitted exposure times.

The various standards for sound level meters have recently been consolidated into a single international recommendation,[18] and equivalent British Standard.[19] Two new grades of instrument have been introduced, and the grades are now as follows:

Type 0:    laboratory grade meters.
Type 1:    precision grade meters.
Type 2:    industrial grade meters.
Type 3:    survey grade meters.

Caution should be exercised when specification or advertisements claim compliance with type 3 only, as the tolerances for survey grade instruments allow considerable inaccuracies in taking measurements.

The latest generation of noise average meters can cope with fluctuating and even impulse noises. The relevant International Standard[20] requires the use of equivalent continuous sound level calculated on the equal energy principle, and this system is widely accepted throughout the world. An exception is America, where the Occupational Health and Safety Act requires the use of increments of 5 dB(A) for every doubling or halving of exposure time, with a ceiling of 115 dB(A). Instruments designed to conform strictly to the requirements of this Act will indicate an overload at 115 dB(A), and may show a low average reading where the noise contains significant components in excess, for example drop-forge noise. It may be said in favour of the 5 dB(A) system that it recognizes the recovery periods in intermittent noise patterns, and the 3 dB(A) rule will be over-conservative in these circumstances.

Advances in microelectronics have meant that averaging meters can be made small enough to be worn on the body. These are known as personal noise dosimeters or exposimeters. They can be made to conform to the most stringent electronic standards, but the effect of attaching the microphone to the user's body may modify the sound field to such an extent that the acoustic specification is no longer met. The result of a laboratory test[21] is reproduced in *Fig. 8.3,* which shows

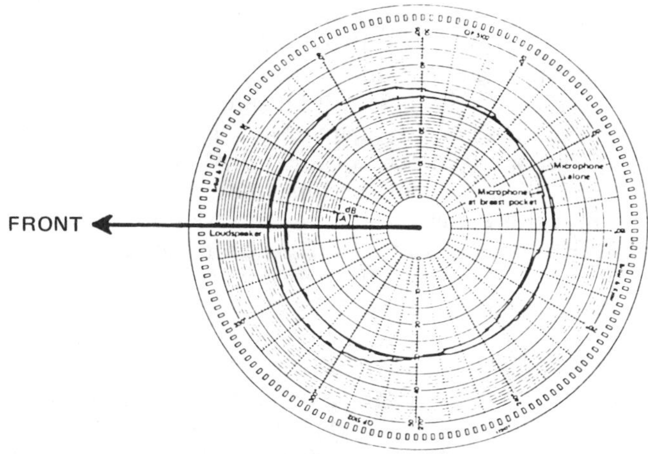



*Fig. 8.3.* Directional effects of wearing a microphone in breast pocket in semi-reverberant conditions. *After* Christensen and Hemingway.[21]

the modification to the sound field around a microphone worn on the right lapel. The level of frontally incident sound is enhanced by reflections off the body, whereas the microphone is shielded from sound coming from behind. In practice the reading may be enhanced by up to 5 dB when the sound field is directional, for example in a drop-forge,[22] whilst in a reverberant sound field the difference may prove insignificant.

## Exposure criteria: 90 or 85 dB(A)?

It is impractical to set limits which would protect everyone from any noxious agent. For noise, the level would perhaps have to be as low as 70 dB(A). A criterion of 90 dB(A) has been adopted almost universally as an economic compromise. Only Sweden has a limit of 85 dB(A) among those countries which enforce their legislation, although lower limits have been under discussion in the United Kingdom, the United States and certain other countries for some time. It should be noted that all instances of enforced legislation only require the provision and use of ear protection, and not the reduction of noise beyond taking 'reasonably practicable' measures.

The British Occupational Hygiene Society[23] accepted an 'impairment of hearing acuity' in not more than 1 per cent of a population exposed to an average noise level of 90 dB(A) for a working lifetime. The more recent British Standard[24] predicts that 4 per cent of a similarly exposed population will reach its definition of 'hearing handicap', excluding the effects of presbyacusis. Reducing the level to

85 dB(A) would protect 99 instead of 96 per cent of the working population from incurring 'hearing handicap'.

It is relevant to examine some of the consequences of a change of exposure criteria from 90 to 85 dB(A). Informed estimates[25] suggest that over one million people employed in manufacturing industry are exposed to noise levels in excess of 90 dB(A) for some part of their working day. The figure could become one and a half million if construction, agriculture and the service industries are included. If the criterion were to be lowered to 85 dB(A), then the number of persons over the limit would be approximately doubled. The relative proportions are roughly similar for the United States and Canada, and probably other manufacturing countries also. An estimate based on data available from the oil industry indicates that it would cost about 2·8 times as much to reduce noise levels to 85 dB(A) instead of to 90 dB(A). The alternative of providing and enforcing the use of ear protection would involve something more than a doubling of expenditure, as the cost of education and enforcement increase disproportionately when the hazard is less obvious. These sums of money must be considered in relation to an additional 3 per cent of the population protected from hearing handicap.


## EAR PROTECTION

### Attenuation

A British Standard[26] method of test of the attenuation of ear protectors at threshold was published in 1974 which required the use of omni-directional sound. This was a departure from previous test methods employing directional sound, and was intended to simulate industrial conditions more realistically. Measurements carried out on the same protectors to both British and then American Standard test procedures indicated that there was generally little difference between the results obtained by the two methods.[27] Where statistically significant differences had occurred, the British test method gave lower and presumably more realistic test results.

Another innovation in the British Standard was a requirement that statistical information by way of standard deviations and inter-quartile ranges of the results should be given in addition to the mean. This gives a measure of the fit of the protector on various people, and enables the 'assumed protection' to be calculated according to the requirements of the Health and Safety Executive Code of Practice. This is defined as the mean less one standard deviation of the results, and indicates the level of protection afforded to 83 per cent of an exposed population. The other 17 per cent will receive less than the 'assumed protection'.

The Code of Practice suggests that the attenuation of ear protection should be compared with an octave band analysis of the noise, and the differences summated to give an overall dB(A) exposure level to the user. In practice, the attenuation in the octave bands centred on 500 Hz and 1 kHz is often critical when making comparisons with common types of industrial noise. In 1977 Martin[27] published the acoustic attenuation characteristics of 26 ear protectors tested according to the British Standard procedure. The results were revealing; two types of ear plug gave a negative 'assumed protection', and a number of other protectors less than 10 dB protection at 500 Hz. Since then, most reputable manufacturers have had their products tested according to the Standard, and the purchase of ear protection for which attenuation data are not available should not be contemplated. The Health and Safety Executive are currently preparing a more up-to-date collection of test data.

Other countries have followed the British lead, and have published or are preparing similar national test standards. Work is progressing at international level also. Considerable effort has been devoted to evolving a single figure specification for the attenuation of ear protectors, especially in North America, but agreement has not yet been reached.

More recent work by Martin[28] using microphones introduced into cadaver ears showed that all types of ear protection behaved in a linear manner in steady-state noise levels up to 125 dB. It is valid, therefore, to apply attenuation figures obtained by threshold shift methods such as the British Standard test procedure to industrial situations. Protectors embodying mechanical devices which purported to operate in an amplitude-sensitive manner do not perform in the way advertised. They offer virtually no protection to industrial noise although they are often popular with work people because they allow conversation to pass unimpeded.

Ear plugs incorporating small holes have been found to offer increasing attenuation to high level, short duration, impulsive noise such as that produced by gunfire.[28,29] Shaw[30] suggested that the impedance of the orifice increases due to non-linearity, thereby reducing the rate of pressure rise in the ear canal. Such effects are not observed below peak pressure levels of about 110 dB, and a reasonable degree of attenuation is not achieved until much higher levels. Clearly these devices have no place in industry.

An important factor determining the actual protection afforded is the proportion of time for which ear protection is worn. No device, no matter how good, can provide more than 3 dB reduction in the equivalent continuous sound level if it is only worn for half the time because 3 dB represents a halving of the noise energy. This concept was developed by Else[31] into the set of curves reproduced in *Fig. 8.4.*

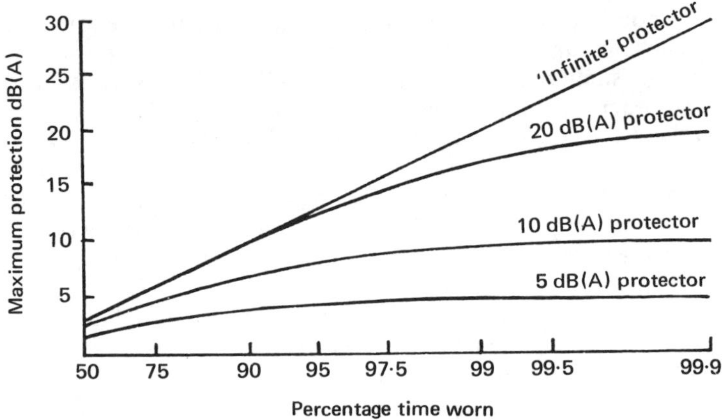


*Fig. 8.4.* Maximum protection as a function of percentage time worn for different hearing protectors. *After* Else.[31]

It therefore follows that ear protection must be worn for a very high proportion of the time to be effective.

So far, only the results of laboratory tests have been considered. The spread of the results indicated by the standard deviation of the attenuation suggests that the fit of the protector on a person is an important factor. Canadian experience[32] of testing the attenuation of protectors as worn in industry confirmed that the user fits the protector less carefully than the person conducting the attenuation test. Berger[33] suggested that the 'assumed protection' obtained from laboratory tests was an approximate rule of thumb indication of the average protection achieved in practice, although there is no scientific foundation for this relationship.

Riko,[34] also in Canada, analysed the reasons for the failure of ear protection to achieve the 'advertised' attenuation in practice. Some reasons were obvious, such as a craze for 'personalizing' ear muffs by drilling holes in the shells or damage to the seals. Personally moulded semi-insert protectors showed great promise in laboratory tests, but in practice gave poor results and deteriorated rapidly due to lack of care at the time of moulding. Different results were obtained in two situations where expanding-foam ear plugs were used. The variation was traced to differences in the extent of instruction given on the packets. The above reasons may all be summarized in one word: education. This is the most important single factor in achieving the maximum potential from the use of personal ear protection.

### Comfort, durability and usage

Professor A. Glorig is reputed to have once said 'the only effective ear protector is the one that gets worn' and this statement remains true today. The attenuation of the better types of ear protection approaches the theoretical maximum imposed by bone conduction of sound directly to the cochlea. Attention is now turning, quite rightly, to producing ear protection which is more acceptable to the user.

The British Occupational Hygiene Society[35] recently published a design guide for ear protectors. This was concerned primarily with such factors as materials, cleaning, durability, physical dimensions, comfort, safety and the like rather than attenuation. Acton, Lee and Smith[36] recognized that pressure on ear muff seals was more important than total force in determining the comfort of ear muffs. Protectors should not be damaged by common industrial liquids, or by dropping from ear-height. Colour may even make a contribution to safety by enabling other persons to see the protector more easily and to appreciate the limitations of hearing and movement which the user may have. Ear muffs fixed to the hard shell of safety helmets seemingly violate the requirements of British and International Standards which require 25 mm free movement of the shell or attachments with respect to the head. A British Standards committee is considering the user aspects of ear protection, and will publish a Standard for ear muffs as a first step. Canada and West Germany have similar standards also.

Users of ear protection often complain that the quality of speech communication is degraded, and they express fears that they might miss a vital signal such as an alarm. Early experiments by Kryter[37] showed that the understanding of tape-recorded speech was not degraded by the use of ear plugs, and was actually improved when the volume was reduced by the plugs to levels which did not cause distortion within the ear. This accords with simple theory which predicts that the speech and the noise are both reduced by a similar amount, leaving the same signal to noise ratio. However, this finding does not agree with common experience.

Wilkins and his colleagues[38] have shown that voice level has an important effect in speech communication when using ear protection. A natural reflex action, sometimes known as the 'Lombard' effect, causes the voice to be raised against increased noise levels. The ratio of vocal effort to noise level is not one to one, but with choice of words and a familiar subject matter there may be considerable compensation. If the noise level reaching the ear is attenuated, for example by ear protection, then the voice level is not raised accordingly and communication is degraded. This is a function of the speaker and not the listener using protection. It is therefore most important to instruct users of ear protection to maintain their vocal output.

There was further degradation of the speech signal when the listener had some loss of hearing. Physically produced sounds, such as warning signals, alarms and the like, were heard just as well when using ear protection as without.

There is some evidence[39,40] that subjects who wear ear muffs for the first time suffer some loss of their ability to localize the direction of sound sources. It has been suggested that the effect may be less when ear plugs are worn because the external ear (pinna) is not covered, although most directional localization takes place within the brain by a cross-correlation process utilizing time delays due to distance between the ears. Nevertheless, experiments indicate that subjects do not adapt to the adverse effects of the muff.[41]

The effect that education can play in the fitting and hence the protection achieved from ear protection has been mentioned already. Education by any means available in industry can also play an important part in the initial acceptance and usage of protection, and is more effective than blatant propaganda. Persuasion is another technique. Ear muffs embodying communication receivers were described by Acton and Childs[42] and are available from several manufacturers. Trials with females engaged on a repetitive task showed that increased usage of the muffs coincided with relaying a music programme through the receivers.

Certain constraints must be placed on the design of such systems. The equivalent continuous sound level of the relayed signal must not exceed 90 dB(A) or whatever hearing damage risk criterion is accepted locally. The use of high quality ear muffs in the first instance maximizes the dynamic range available for reproduction, but may still give only 10–15 dB. Amplitude compression of the transmitted signal makes the most of the available range in these circumstances. Restriction of the signal to one ear only, though not popular with the users, limits masking of external signals to about 3 dB. This is an important consideration in maintaining safety where it may be necessary to perceive warnings, alarms or other external signals.

One final point concerning the usage of ear plugs, which has been emphasized many times before, still needs repeating. The modern types of glass-down and expanding-foam-plastic ear plugs do not need fitting for size, and there is a temptation to make them available 'ad lib' to all comers. A medical examination of the ear canal and ear drum is essential before any type of ear plugs is used. Perforations of the ear drum or other middle ear trouble, skin conditions of the ear canal, and a predisposition to dermatitis or eczema are all contra-indications to the use of ear plugs.


## MONITORING AUDIOMETRY

Numerous arguments have been advanced both for and against the introduction of monitoring audiometry in industrial hearing conservation programmes. Ideally, audiometry should not be necessary if all other hearing conservation measures are followed. Because of human fallibility, people inevitably take chances. Failure to use ear protection is common; figures available from the Factory Inspectorate in 1977 showed that overall only 11 per cent of persons exposed to noise levels in excess of 90 dB(A) were found to be wearing protection during spot checks. Furthermore, currently accepted hearing damage risk criteria are a compromise, and are only intended to protect a certain percentage of an exposed population.

An argument often advanced by management against audiometry is that it would attract the attention of workpeople to their hearing losses and thence precipitate claims for compensation. Such claims must be accepted as inevitable sooner or later, and they cannot be suppressed indefinitely. It is more economic to deal with any claim without delay, and more correct ethically to institute hearing conservation measures forthwith.

One of the biggest advantages of audiometry is that it provides a personal contact on a one-to-one basis between those people who are sustaining hearing damage and medical staff. The evidence that they are losing their hearing is available in graphic form, and the opportunity may be taken to demonstrate the proper fitting of ear protection at the same time. All companies which have achieved a high utilization of personal ear protection incorporate audiometry in their hearing conservation programmes, and usage figures as high as 90 per cent have been claimed.

At the time of writing there is no statutory obligation to test hearing in the United Kingdom although it seems that a limited requirement may be introduced in forthcoming noise regulations. Audiometry should be viewed in the same light as many of the other screening tests which are practised routinely in occupational medicine. It must be regarded as good practice where personnel are exposed to potentially hazardous noise levels and have to rely on ear protection.

An audiometric programme cannot be undertaken lightly. The physical requirements for the test environment, equipment and calibration, and the training of staff both to undertake the tests and to interpret results have been dealt with adequately elsewhere.[43,44] Two important considerations are often omitted from such texts: when and where to start with tests, and how to detect and deal with persons who attempt to exaggerate their hearing losses.

There must be a strong temptation to commence testing with older men, and there will be certain satisfaction to be gained from discovering and recording substantial noise-induced hearing losses; but these are cases who have already suffered injury. Prevention of hearing loss should be the primary objective of audiometry. A more beneficial, more logical and less daunting introduction to an audiometric programme, and one that brings less industrial relations problems, is to commence testing persons to be employed in noise for the first time. These will usually be new starters. Besides providing a base-line audiogram, this will screen out some persons particularly susceptible to noise exposure and will protect the employer from spurious claims from persons who already have a hearing loss.

A gradual build-up to include in the programme all persons at risk is provided by extending the tests as follows:

1. Re-testing at regular intervals any person who gave an abnormal initial audiogram, whether or not he is exposed to noise. Medical treatment may be indicated in such cases.

2. Carrying out audiometry on any person who has been employed for less than 2 years in a noise-hazardous situation, then after an interval of 1 year, and thereafter at 2- or 3-year intervals. The first audiogram should ideally be carried out at the beginning of a working day and preferably after a weekend or shift break away from noise, otherwise the results are likely to include temporary losses due to noise exposure. A practical compromise for all re-tests is to ask the person to wear good quality ear muffs during all noise exposure in the preceding 24 hours. If no significant deterioration[44] relative to the initial audiogram is detected, the test result can be accepted. If an apparent deterioration is found, that person should be tested again after a rest period as above.

3. For presently noise-exposed long service personnel re-testing should be done at 2- or 3-year intervals. It may be possible with experience to relax the intervals at which certain long service employees are tested once stable audiometric patterns have been established, providing that their noise exposure does not increase. These will probably be persons who use ear protection conscientiously, or who have 'tough' ears. There is no way of distinguishing 'tough' from 'tender' ears prior to noise exposure.

Outright feigning of occupational hearing loss is uncommon,[45] but exaggeration of audiometric responses occurs more frequently. Outbreaks of exaggeration often occur in geographical areas or within specific industries, and are often associated with claims situations. The apparatus necessary to measure hearing objectively is unlikely to be available within industry, but experience can play a large part in indicating when exaggeration has occurred. Inconsistency in audiometric responses may occur, even over quite short time intervals. Many audiometers incorporate a switch which gives a 10 dB or 20 dB instantaneous step in the signal, and this should be matched by the subject's responses. Audiometric results should agree with the tester's impression of hearing ability without lip reading. The response to a quietly spoken egocentric word such as the subject's name, or a psychologically sensitive phrase such as a reminder to adjust clothing, will often provide confirmation of a suspicion of exaggeration.

The commoner forms of exaggeration all lead to a greater enhancement of audiometric thresholds at lower frequencies than at high. The subject usually either attempts to respond at a constant loudness level of the stimulus across all frequencies, or he adds a common subjective increment to all his responses. Noise-induced hearing loss is accompanied by loudness recruitment. At levels above his hearing threshold where damage has occurred, a person experiences a greater subjective increase in loudness than the true change in sound level. Sounds which are loud enough will be heard at their true level in spite of the loss of hearing. As an example, an increase in sound level of 5 dB above threshold at 4 kHz may sound like 20 dB. By the time the level is 20 or 30 dB above threshold, sounds may be heard more or less at their correct level. An attempt at exaggerating by, say, 20 dB may only lead to an increase in the recorded audiogram of 5 dB at 4 kHz. At lower frequencies, where noise damage has not occurred, the audiogram will be inflated by the full amount of the exaggeration. The result is a flatter audiogram than predicted by the National Physical Laboratory data. A typical example is shown in *Fig. 8.5.*

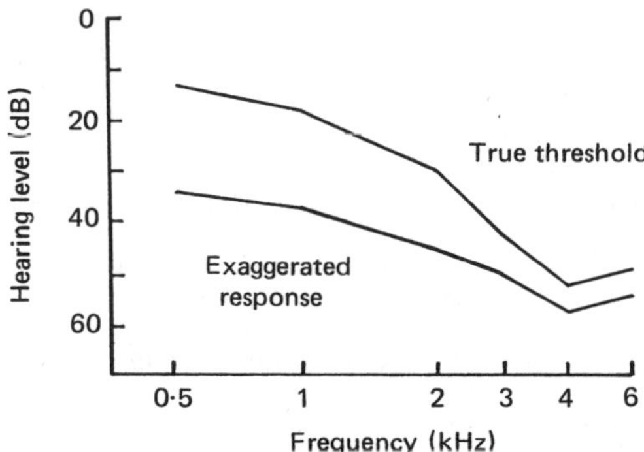


*Fig. 8.5.* Example of an exaggerated audiogram.

A flatter audiometric configuration is also typical of certain pathological conditions, and care should be taken to exclude the possibility of pathology or an equipment malfunction before concluding that exaggeration has occurred. Where noise-induced hearing loss is accepted but there are differing audiometric responses, the better hearing level will always be more representative of the true state of hearing. When the full cooperation of the subject is not obtained, it may be necessary sometimes to build up a 'composite best' audiogram from tests conducted on two or more occasions.

REFERENCES
1. Fosbroke J. Practical observations on the pathology and treatment of deafness. *Lancet* 1831; pp. 645–8.

2. Robinson DW and Shipton MS. *Tables for the Estimation of Noise-induced Hearing Loss,* 2nd ed. NPL Acoustics Report Ac61. Teddington: National Physical Laboratory, 1977.
3. Ballantyne J. Ototoxic drugs. In: Hinchcliffe R and Harrison D (ed.) *Scientific Foundations of Otolaryngology.* London: Heinemann, 1976.
4. Hamernik RP and Henderson D. Potentiation of noise by other ototraumatic agents. In: Henderson D, Hamernik RP, Dosanjh DS et al. (ed.) *Effects of Noise on Hearing.* New York: Raven Press, 1976.
5. Lenhardt E. Die Berufsschäden des Ohres. *Arch. Ohren. Nasen. Kehlkopfhlk.* 1965; **185**:11–242.
6. Velazques J, Escobar R and Almaraz A. Audiologic impairment due to N-butyl alcohol exposition. *Proceedings of the 16th International Congress on Occupational Health,* Tokyo, 21–28 September 1969.
7. Allen CH, Frings H and Rudnick I. Some biological effects of intense high-frequency airborne sound. *J. Acoust. Soc. Am.* 1948; **20**:62–5.
8. Dickson EDD and Watson NP. A clinical survey into the effects of turbo-jet engine noise on service personnel. *J. Laryngol. Otol.* 1949; **63**:276–85.
9. Dickson EDD and Chadwick DL. Observations on disturbances of equilibrium and other symptoms induced by jet engine noise. *J. Laryngol. Otol.* 1951; **65**:154–65.
10. Hinchcliffe R. The threshold of hearing as a function of age. *Acustica* 1959; **9**:303–8.
11. Robinson DW and Sutton GJ. *A Comparative Analysis of Data on the Relation of Pure-tone Audiometric Thresholds to Age.* NPL Acoustics Report Ac84. Teddington: National Physical Laboratory, 1978.
12. Shipton MS. *Tables Relating Pure-tone Audiometric Threshold to Age.* NPL Acoustics Report Ac94. Teddington: National Physical Laboratory, 1979.
13. International Organization for Standardization. *Normal Threshold of Hearing by Air Conduction as a Function of Age and Sex:* Draft Proposal DP 7029. Geneva: International Organization for Standardization, 1979.
14. Ward WD. Acoustic trauma and noise-induced hearing loss. In: Tower DB (ed.) *The Nervous System,* vol. 3. *Human Communication and Its Disorders.* New York: Raven Press, 1975.
15. Rosen S, Bergman M, Plester D et al. Presbyacusis study of a relatively noise-free population in the Sudan. *Ann. Otol. Rhinol. Laryngol.* 1962; **71**:727–43.
16. Rosen S, Plester D, El-Mofty A et al. High frequency audiometry in presbyacusis: a comparative study of the Mabaan tribe in the Sudan with urban populations. *Arch. Otolaryngol.* 1964; **79**:18–32.
17. Health and Safety Executive. *Code of Practice for Reducing the Exposure of Employed Persons to Noise.* London: HMSO, 1972.
18. International Electro-technical Commission. *Publication 651: Sound Level Meters.* Geneva: International Electro-technical Commission, 1979.
19. British Standard 5969: 1981. *Sound Level Meters.* London: British Standards Institution, 1981.
20. International Organization for Standardization. *Recommendation 1999: Assessment of Occupational Noise Exposure for Hearing Conservation Purposes.* Geneva: International Organization for Standardization, 1971.
21. Christensen LS and Hemingway JR. Sources of error in noise dose measurements. *Bruel and Kjaer Technical Review* 1973; No. 3, 3.
22. Kershaw JG. Environmental survey: noise measurement and results. In: Medical Research Council Project Report, Universities of Dundee and Salford. *Noise Levels and Hearing Thresholds in the Drop Forging Industry.* GKN Forgings Limited and GKN Group Technological Centre, 1976.
23. British Occupational Hygiene Society. Hygiene standard for wide-band noise. *Ann. Occup. Hyg.* 1971; **14**:57–64.

24. British Standard 5330: 1976. *Method of Test for Estimating the Risk of Hearing Handicap due to Noise Exposure.* London: British Standards Institution, 1976.
25. Dove AR. Regulation of Occupational Noise Exposure in Great Britain. Paper presented to Institution of Mechanical Engineers meeting on noise in industry. London, 1–2 February 1977.
26. British Standard 5108: 1974. *Method of Measurement of the Attenuation of Hearing Protectors at Threshold.* London: British Standards Institution, 1974.
27. Martin AM. The acoustic attenuation characteristics of 26 hearing protectors evaluated following the British Standard procedure. *Ann. Occup. Hyg.* 1977; **20**:229–46.
28. Martin AM. Dependence of acoustic attenuation of hearing protectors on incident sound level. *Br. J. Ind. Med.* 1979; **36**:1–14.
29. Forrest MR and Coles RRA. Problems of communication and ear protection in the Royal Marines. *J. R. Nav. Med. Serv.* 1970; **56**:162–9.
30. Shaw EAG. Hearing protector attenuation: a perspective view. *Appl. Acoust.* 1979; **12**:139–57.
31. Else D. A note on the protection afforded by hearing protectors — implications of the energy principle. *Ann. Occup. Hyg.* 1973; **16**:81–3.
32. Abel SM, Alberti PW and Riko K. User fitting of protectors: attenuation results. In: Alberti PW (ed.) *Personal Hearing Protection in Industry.* New York: Raven Press, 1982.
33. Berger EH. Laboratory estimates of the real world performance of E.A.R. hearing protectors. In: Alberti PW (ed.) *Personal Hearing Protection in Industry.* New York: Raven Press, 1982.
34. Riko K. How ear protectors fail: a practical guide. In: Alberti PW (ed.) *Personal Hearing Protection in Industry.* New York: Raven Press, 1982.
35. British Occupational Hygiene Society. Guide for the design of hearing protectors for general industrial use. *Ann. Occup. Hyg.* 1979; **22**:203–11.
36. Acton WI, Lee GL and Smith DJ. Effect of head band forces and pressure on comfort of ear muffs. *Ann. Occup. Hyg.* 1976; **19**:357–61.
37. Kryter KD. Effects of ear protective devices on the intelligibility of speech in noise. *J. Acoust. Soc. Am.* 1946; **18**:413–17.
38. Wilkins PA and Martin AM. The effects of hearing protection on the perception of warning sounds. In: Alberti PW (ed.) *Personal Hearing Protection in Industry.* New York: Raven Press, 1982.
39. Atherley GRC and Noble WG. Effect of ear-defenders (ear muffs) on the localization of sound. *Br. J. Ind. Med.* 1970; **27**:260–5.
40. Atherley GRC and Else D. Effect of ear-muffs on the localization of sound under reverberant conditions. *Proc. R. Soc. Med.* 1971; **64**:203–5.
41. Russell G. Limits to behavioral compensation for auditory localization in earmuff listening conditions. *J. Acoust. Soc. Am.* 1977; **61**:219–20.
42. Acton WI and Childs J. Background music as an incentive to wearing ear muffs. *Protection* **11(10)**:16–18.
43. Bryan ME and Tempest W. *Industrial Audiometry,* 2nd ed. St Annes: Bryan & Tempest, 1980.
44. Health and Safety Executive. *Audiometry in Industry.* London: HMSO, 1978.
45. Coles RRA and Priede VM. Non-organic overlay in noise-induced hearing loss. *Proc. R. Soc. Med.* 1971; **64**:194–9.

# 9. VIBRATION
*William Taylor*


## INTRODUCTION

In industrial processes vibration is frequently associated with noise, both emanating from the same source. The cause and effect relationship is more firmly established for noise in so far as measurement of the stimulus, assessment of the resulting medical effects and establishment of exposure criteria are concerned. In the case of vibration, the relationship between exposure to vibration and the resulting health hazards to exposed subjects required to establish safe standards has not yet been ascertained. Vibration is a relatively new physical agent compared with noise, where 50 years of noise investigation in industry are reflected in current legislation and, in the majority of countries, compensation.

Occupational exposure to vibration may arise in a variety of ways, often reaching the subject at intensities disturbing to comfort, efficiency, health and safety through different routes of transmission. In the case of *whole body vibration,* it may be transmitted to the worker through a contacting or supporting structure which is itself vibrating — the deck of a ship, the seat or floor of a vehicle (for example a tank), or a whole workplace shaken by machinery. In some processes whole body vibration is intentionally generated to compact or grade such materials as coal, iron ore or concrete. In many instances the route of entry into the human body is through the hands, wrists and arms of the subject — so-called *segmental vibration.* There are many examples in industry from early history, such as the use of pneumatic drills in France around 1840 for tunnel making. By 1890 drills were extensively used for mining. Air-powered tools are in general use in road making (jack-hammers), in foundries (pneumatic hammers for chipping and grinding) and in rock drilling (jack-leg drills). The use of electrically driven rotary tools, introduced around the 1930s for grinding and polishing, was followed by chain-saws in the 1950s for felling, de-branching and cross-cutting in forestry operations. These saws were powered by small capacity (30–80 cc) internal combustion engines giving rise to vibration on the saw handles emanating from reciprocating forces arising from the engine and from the bar and chain. The new forestry techniques were adopted world wide from 1955 onwards.

When considering the effects of vibration on man, it is both practical and convenient to treat whole body and hand-transmitted
segmental vibration in separate categories. This broad classification
follows the approach already adopted in vibration research, in
engineering and in setting regulatory standards. By international
agreement separate advisory standards have been drafted governing
whole body and segmental hand–arm systems. The purpose of this
chapter is to acquaint the reader with the present state of knowledge
and to interpret the Draft Regulations[1-3] and the general principles involved in the recognition, evaluation and control of workers exposed
to vibration.

## THE MEASUREMENT OF VIBRATION

In order to define vibration several units are in common use. These
are: *frequency, amplitude, direction of application with respect to an
anatomical axis, duration of vibration exposure* and *whether the
stimulus is continuous or interrupted.* The international standard unit
of vibration frequency is, as in acoustics, the Hertz (Hz), defined as
one cycle of operation per second and formerly presented in the form
c/s. The second unit of vibration is amplitude or intensity or severity
of the motion at any given frequency. Assuming that the vibration is a
simple harmonic oscillation (usually a rare condition), a *velocity* of
vibrational motion and an *acceleration* due to the restoring force are
both proportional to the amplitude (or displacement), the vibrational
velocity rising proportionally with frequency and the acceleration
increasing with the square of frequency. At low frequencies around
1 Hz, found with large ships or on oil exploration platforms rolling at
sea, an amplitude of several metres is required to generate appreciable
levels of acceleration. Higher up the frequency scale, however, around
the 30–60 Hz frequencies associated with hand-held power tools, high
accelerations may be generated by small displacements of fractions of
a millimetre. With modern electronic vibration measuring instrumentation, vibration is expressed in terms of acceleration, defined as
metres per second per second ($m/s^2$), over a known frequency band
such as ⅓ octave or octave. Transducers (or accelerometers) are commercially available which generate an electrical output directly proportional to acceleration. For practical purposes it is convenient to
express acceleration in terms of 'g', the standard gravity acceleration
at the earth's surface. Thus 1 g is equivalent to $9·81 \, m/S^2$. Vibration
spectra are rarely simple and so when describing a complex vibration
pattern, it is common practice to compute an average or root mean
square (rms) value recorded over a finite sample. The rms value of a
varying quantity (in case of pure harmonic motion) is 0·707 times
the maximum peak value. Whilst rms measurements are sufficient

for the evaluation of high frequency vibration, such an intensity-averaging technique must be applied cautiously if varying phase relationships are present, for example complex harmonic motion in the range below 1 Hz found in motion sickness.[4]

The direction of the vibrating forces applied to the human body determines the sensory and psycho-physiological reactions to the motion and are different when the subject vibrates from side to side ($x,y$ axis) or in the vertical mode ($z$ axis). The responses to vibration depend to a marked degree on *resonance,* where some anatomical organs (or part of) are set into an oscillatory motion (amplified or resonant modes) greater than that of related structures. The characteristic frequencies and magnification factors (mechanical amplification at the resonant frequencies) will be determined by the *mass,* the *elasticity* and the *damping* within the human body.

## WHOLE BODY VIBRATION

The principal resonance of the seated, recumbent or standing human body vibrated in the $z$ axis (cephalo-caudally) occurs in the region of 5 Hz with substantial amplification in the band 4–8 Hz. Accordingly, current recommended limits of human exposure to whole body vibration expressed in terms of acceleration as a function of frequency (1–80 Hz) are set at their lowest in the band 4–8 Hz.[1] In contrast, when a person is vibrated in the $x$ axis (anteroposterior) or $y$ axis (lateral), as in a swaying train, the principal mode of resonant vibration occurs between 1 and 2 Hz.

Whole body vibration in the range 1–20 Hz, predominantly vertical, is found in many types of commercial and earth-working vehicles, such as mobile heavy tractor equipment either driven or ridden by the operator. The physiological response is found in the cardiopulmonary system with increase in heart and respiratory rates, cardiac output and oxygen uptake. Such changes have been attributed to raised metabolic activity associated with increased activity of the skeletal musculature provoked by vibration. In general these changes reflect a non-specific response to mechanical vibration as a stressor. If the vibration is intense (up to 3 $g$) then gross mechanical interference with the haemodynamics of central and peripheral blood flow may result. Severe whole body vibration in rough riding vehicles could therefore disturb cerebral blood flow in drivers,[5] which in turn could result in a decreased work performance. Investigators in the Soviet Union[6] in particular maintain that chronic exposure to whole body vibration causes generalized debilitating effects mediated by the central nervous system together with cerebral oedema and degenerative changes in nerve fibres. In subjects exposed to long term whole body vibration, various disorders of the spine and internal systems have been described,[7,8] associated with farm tractors, haulage vehicles and mobile heavy equipment. The aetiology has been difficult to establish since, apart from vibration, many surveys have indicated poor ergonomic design factors, inefficient work methods, adverse climatic conditions, fatigue and non-specific stress. Therefore it must be emphasized that the extent to which chronic exposure to moderate levels of whole body vibration can injure normal subjects, as opposed to aggravating pre-existing weaknesses, is at the present state of knowledge an open question. Nevertheless, occupational health physicians should be aware[9,10] that long distance bus and heavy vehicle drivers may be more prone to chronic disorders affecting the gastrointestinal, respiratory, venous and muscular systems, in particular spinal ailments. In this research area it has not proved possible to distinguish vibration as a principal aetiological factor from other variables such as poor posture due to bad seating, dietary habits, long and irregular hours and the general stress of driving on crowded motorways. In the Soviet Union, however, and in other Eastern European countries many investigators have taken the view that continuous whole body vibration in factories, in cranes, in compactors in construction industry heavy equipment and in ships' engine rooms can produce debilitating stress and malaise mediated centrally by neuroendocrinological mechanisms. At somewhat higher frequencies than that experienced in rough terrain sites, these higher occupationally produced frequencies result in various syndromes termed vegetative — vascular, polyneuritic, asthenic, vestibular — all resulting in a reduction in working capacity. Experimental studies, however, have yet to establish a link between the physical agents in the working environment and the signs and symptoms reported, particularly if adaptation is taken into account over a reasonable time base of 6 months to a year. It is well established in other occupational health fields that the initial (temporary) signs and symptoms of stress disappear on prolonged exposure. With reference to female workers in industry exposed to whole body vibration, comparative data relating to sex differences are lacking. Consequently current standards on human exposure to vibration apply equally to both sexes.

When the occupational health physician is confronted with hazards associated with whole body vibration and an irreducible amount of vibration is transmitted to an employee in his vehicle seat or work place, he should advise management to minimize the duration of vibration exposure; to allow adequate rest periods; to inspect ergonomic design of seats, display and controls; to supply mechanical devices to restrain body oscillation; and to set up selection procedures pre-employment to exclude subjects medically unfit for exposure to vibration or motion, such as a history of low back pain. With the exception of the well-established remedies for motion sickness at frequencies below 1 Hz, no medication has yet been found to increase whole body tolerance above 1 Hz.

Recently, attention has been focused on the possibility that whole body vibration may act additively with other physical agents such as noise, and with chemical agents such as heavy metal toxicity and ionizing radiation. This approach examines the overall industrial environment and its adverse effect on the workers' physiological state mediated through the central and autonomic nervous systems. Such observations are based on animal experiments. Data establishing connections between physical and chemical agents and whole body vibration are, as yet, purely speculative and firm data for these noxious agents hazardous to occupational health are not available. Meantime physicians, through the vibration guidelines laid down in document ISO 2631: 1974,[1] must support vibration reduction programmes to protect their workers. To minimize the workers' exposure, especially in the biodynamic response range 1–20 Hz, is the physician's task, and therefore the following principles are important: (*a*) isolation of vibration sources; (*b*) location of recipient with respect to pathways of transmission; (*c*) isolation of recipient; and (*d*) reduction of flanking transmission.


## SEGMENTAL VIBRATION

Hand-held vibratory tools are universally used in industry and in a variety of occupations such as shipbuilding, mining, foundries, assembly plants and forestry. The vibratory energy when transmitted to the hands and arms produces, as early manifestations, numbness or tingling (or both) in the fingers and hands. With increasing vibration exposure time, the operators notice that the fingertips go white (blanching) when the fingers touch a cold object or are exposed to a low environmental temperature. The 'white finger' attacks increase in number and severity with further vibration exposure. The area of blanching increases, extending from the fingertips to the mid-phalangeal joint and then to the root of the finger. The digits on both hands are involved. In all cases the precipitating cause of these attacks is exposure to cold. Except in very severe cases the thumbs are never affected. Loriga[11] in 1911 was the first to describe 'vascular spasm' in the hands of rock drillers using pneumatic tools. The classic description of this condition was given in 1918 by Hamilton,[12] who studied miners using jack-hammers in the limestone quarries of Indiana. By the 1950s it was evident from the work of Agate[13] that 'vibration-induced white finger' (VWF) or 'Raynaud's phenomenon of occupational origin' was on the increase in most countries and that VWF prevalences of the order of 60–70 per cent in vibration-exposed populations were common. In 1960, however, Pecora[14] conducted a questionnaire survey in the USA of a group of occupational physicians and concluded that Raynaud's phenomenon of occupational origin could not be completely evaluated but that it may have become in the USA an uncommon occupational disease approaching extinction.

From 1964 onwards reports from all countries engaged in forestry operations indicated that power driven chain-saws were hazardous tools producing VWF prevalence of 40–60 per cent in Australia,[15] Japan,[16] Sweden,[17] Finland,[18] Norway,[19] Czechoslovakia,[20] New Zealand[21] and in the UK.[22–25]

The clinical evidence obtained from these and early surveys showed that systems other than the peripheral circulation of the hands could be involved. The following conditions have been described: (*a*) neurological degenerative changes in the fingers giving rise to loss of light touch, loss of finger co-ordination and therefore inability to perform fine work, loss of manipulation skills and loss of thermal appreciation; (*b*) muscular weakness, tendonitis, joint pains in fingers and wrists and stiffness of finger joints on waking, and with increasing vibration there is loss of grip force;[26] (*c*) skeletal changes in the wrists and elbow joints[27] resulting mainly from low frequency (less than 100 Hz) vibration transmitted through the hands to the elbows and as far as the shoulder joints giving rise to arthritic and peri-articular changes. Vascular areas of bone decalcification (terminal phalanges) and bone cysts (in wrist and phalanges) have been described.[27,28] A clear relationship between these skeletal changes and vibration has, however, not yet been definitely established. Carpal tunnel syndrome and Dupuytren's contracture involving the palmar fascia are often seen in association with vibrating hand tools but there is no firm evidence as yet reported of a causal connection. These disorders in systems other than hand blood circulation have led to the adoption of a new terminology more embracing than vibration-induced white finger, namely 'the vibration syndrome.'

## THE VIBRATION SYNDROME: ASSESSMENT OF DISABILITY

Difficulty has been experienced by occupational physicians both in quantifying the vibration stimulus and assessing the impairment, particularly in the area of compensation. The degree or severity of VWF signs and symptoms (clinical) and the extent to which other body systems are involved have been quantified by physicians in Japan by 'levels', in the UK by 'stages' and in Finland by a 'TVD (traumatic vascular disease) index'. In all methods the clinical assessment is made on the number of affected fingers that turn white, the number of VWF attacks, the interference with work and social activities, the interruption of sleep, the reduction in grip force, the amount of stress, the weather conditions necessary to provoke attacks and the interference with hobbies such as fishing, shooting, cycling and swimming. The system used in the UK is summarized in *Table 9.1*.

**Table 9.1.** Stage assessment for Raynaud's phenomenon

| Condition of digits | Work and social interference | Stage | |
|---|---|---|---|
| No blanching of digits | No complaints | 0 | |
| Intermittent tingling | No interference with activities | $0_T$ | No blanching |
| Intermittent numbness | No interference with activities | $0_N$ | |
| Both tingling and numbness | No interference with activities | $0_{TN}$ | |
| Blanching of one or more fingertips with or without tingling and numbness | No interference with activities | 1 | |
| Blanching of fingers beyond tips usually confined to winter | Slight interference with home and social activities. No interference at work | 2 | |
| Extensive blanching of digits. Frequent episodes summer as well as winter | Definite interference at work, at home and with social activities. Restriction of hobbies | 3 | Blanching |
| Extensive blanching. Most fingers: frequent episodes summer and winter | Occupation changed to avoid further vibration exposure because of severity of signs and symptoms | 4 | |

Complications are not used in this grading.

An occupational history is followed by a series of objective tests in an attempt to establish VWF outwith the subject's own history of the number and severity of his attacks and his subsequent disability. At present these objective tests are divided into two main groups as follows:

*Group 1:* Light touch (cotton wool), pain (pin prick) and temperature (hot probe). These are the standard established clinical tests attempting to establish sensory loss, pain and temperature appreciation loss in the digits. Sensory tests (two-point discrimination and depth sense aesthesiometry).

*Group 2:* Provocative cooling of the hands (for 1 minute at 0°C) and subsequently measuring recovery time by plethysmography.

No published data on the correlation of the above objective tests and the stages of VWF *(Table 9.1)* have yet enabled physicians to assess disability outwith the subject's own history. Preliminary evidence from American studies not yet published indicates that the depth sense and two-point sensory tests do not correlate with staging and that the provocative cooling test is promising but not by itself sufficient to establish VWF disability. In these cooling tests three variables require to be controlled: vascular tone, emotional state and core temperature. It has been found recently that time of day for the provocative cooling test must be standardized. The three clinical tests of light touch, pain and temperature when taken together give the highest correlation with staging (estimated at 70–80 per cent). In summary, a stage or level assessment can be made only by knowing a detailed occupational history with respect to vibration exposure (exposure times and tools used), a knowledge of the vibration characteristics of the vibratory tools used at the work place, previous vibration exposure and present VWF history with details of blanching attacks and their severity. The clinical tests used for evaluation are non-invasive, do not require elaborate instrumentation and are within the capabilities of an occupational health physician. It must be pointed out that to date no single objective test has been found which will alone assess VWF disability. Finally when confronted with a VWF case, all other causes of white finger attacks (such as primary Raynaud's disease) must be excluded.

## DIFFERENTIAL DIAGNOSIS OF VWF

It is not known whether vibration directly injures the peripheral nerves thereby causing numbness and subsequently sensory loss, or whether the paraesthesia of the hands is secondary to the vascular constriction of the blood vessels causing ischaemia in bones (osteoporosis) and in nerve-end organs. Whatever the aetiology, a detailed medical history is required to exclude all other conditions giving rise to impaired circulation of the digits and sensory changes in the hands. Of these, primary Raynaud's disease is the first to be excluded by establishing whether Raynaud's disease was present before the subject entered industry and before being exposed to vibration. There may be a strong constitutional or familial history of white fingers. The blanching is often symmetrical in distribution, may include cold feet, ears and nose and be present in childhood or early manhood. This exaggerated response to cold is difficult to distinguish from VWF without a detailed occupational history and without a summary of the vibration exposure. Collagen diseases often present initially with a history of white finger attacks triggered by cold and therefore connective tissue diseases such as lupus erythematosus, scleroderma and rheumatoid arthritis require exclusion. In a few cases there is interference with the main arteries leaving the thoracic cage, and so the costoclavicular and hyperabduction syndromes must be considered. Neurogenic disorders such as poliomyelitis and syringomyelia would confuse the VWF paraesthesia. Generalized atherosclerosis, a history of raised blood pressure and

Buerger's disease would require elimination when considering impaired circulation. Intoxication with nicotine and ergot will produce vasospasm and lower the temperature of the extremities. Simple lacerations or fractures of the digits, commonly found in the metal and stone cutting trades, often disturb the peripheral circulation. Finally blood dyscrasias such as cryoglobulinaemias and disturbance in fibrinogen levels require exclusion. Increases in $\beta$- and $\gamma$-globulins and antibodies (IgM and IgG) have been reported.[29] A summary of the conditions to be considered by the occupational health physician in this context is shown in *Table 9.2*.

It is advisable to take radiographs of the hands and wrists (bone vacuoles, previous fractures), the chest (cervical rib) and the cervical vertebrae. A clinical examination of the pulses is called for (including the Allen test*) to exclude abnormal configuration and distribution of the radial and ulnar arteries in the forearm.

**Table 9.2.** Differential diagnosis of Raynaud's phenomenon or vibration-induced white finger

| | |
|---|---|
| *Primary* | |
| 1. Raynaud's disease | Constitutional white finger |
| *Secondary* | |
| 1. Connective tissue disease | Scleroderma |
| | Systemic lupus erythematosus |
| | Rheumatoid arthritis |
| | Dermatomyositis |
| | Polyarteritis nodosa |
| | Mixed connective tissue disease |
| 2. Trauma | |
|    (i) Direct to extremities | Following injury, fracture or operation |
| | Of occupational origin (vibration) |
| | Frostbite and immersion syndrome |
|    (ii) To proximal vessels by compression | Thoracic outlet syndrome (cervical rib, scalenus anterior muscle) |
| | Costoclavicular and hyperabduction syndromes |
| 3. Occlusive vascular disease | Thromboangiitis obliterans |
| | Arteriosclerosis |
| | Embolism |
| | Thrombosis |
| 4. Dysglobulinaemia | Cold haemagglutination syndrome |
| | Cryoglobulinaemia |
| | Macroglobulinaemia |
| 5. Intoxication | Acro-osteolysis |
| | Ergot |
| | Nicotine |
| 6. Neurogenic | Poliomyelitis |
| | Syringomyelia |
| | Hemiplegia |

To complete the diagnosis the investigations will include a full blood count, erythrocyte sedimentation rate, blood lipids, lupus erythematosus cell and rheumatoid arthritis fixation tests, C-reactive protein, anti-streptolysin titre, serum uric acid, cryoglobulinaemias and urine analysis. If there is doubt about the deep and superficial palmar arches and their configuration and connections with the digital arteries, brachial arteriography is recommended.

## PREVENTION AND TREATMENT OF THE VIBRATION SYNDROME

At present there are no generally accepted International Standards for the protection of workers exposed to hand-transmitted vibration. A 1975 Draft British Standard[3] is in the process of revision. An ISO Sub-committee (1978)[2] has also prepared a guide for both the measurement techniques and for human exposure to hand-transmitted vibration. Proposed standard acceleration and velocity exposure limits are shown in ⅓ octave band analyses (*Table 9.3*) and in octave bands (*Table 9.4*), both extracted from the ISO 1978 proposals.

The vibration limits proposed are tentative and are based entirely on subjective laboratory data.[30] Owing to the lack of field study data correlating vibration parameters with VWF medical data, the introduction of guidelines for tool designers has been delayed. Whilst awaiting such surveys, vibration criteria are being presented as recommended limits enabling tool designers to make, in some areas, significant progress in minimizing vibrational energy entering the hands of the workers. Foremost in vibration reduction design have been the chain-saw designers. The introduction of anti-vibration (A/V) saws in

---

*The Allen test, first described in 1929, is a useful clinical test for the detection of occlusive lesions of the ulnar or radial arteries distal to the wrist. For doctors investigating VWF, and who are aware that major anatomical arterial variations may occur in both the upper and lower arm, the following normal findings are relevant.

1. The ulnar artery is the larger branch of the two terminal branches of the brachial artery and it is continued in the palm as the superficial volar arch.

2. The deep volar arch receives its blood supply primarily from the radial artery.

3. The branches of the superficial and deep volar arches anastomose in the palm but the digital arteries in the fingers arise mainly from the superficial arch.

To perform the Allen test the subject's radial artery is compressed at the wrist, the hand is raised whilst applying compression and a fist made several times to facilitate drainage of blood from the palm and digits. The hand is then lowered to heart level and the fist is opened whilst maintaining compression of the radial artery. Prompt flushing of the palm indicates that blood is entering the superficial volar arch and its digital branches by way of a patent ulnar artery. If the palm remains pale whilst compressing the radial artery it can be concluded that either the ulnar artery is occluded proximal to the wrist or that an anatomical anomaly exists in the arm arteries. To support this conclusion, removal of radial artery pressure should result in immediate marked flushing of the palm. Faint and delayed flushing to one or another digit will indicate local arterial occlusion. The Allen test should always precede brachial arteriography.

**Table 9.3.** Proposed international standard acceleration and velocity exposure limits for hand-transmitted vibration as a function of frequency*

| Frequency or centre frequency of ⅓ octave band (Hz) | Maximum rms intensity of hand-transmitted vibration in any direction (axis) | |
| --- | --- | --- |
| | Acceleration (m/s²) | Velocity (m/s) |
| 6·4 | 0·8 | 0·016 |
| 8 | 0·8 | 0·016 |
| 10 | 0·8 | 0·013 |
| 12·5 | 0·8 | 0·010 |
| 16 | 0·8 | 0·008 |
| 20 | 1 | 0·008 |
| 25 | 1·3 | 0·008 |
| 31·5 | 1·6 | 0·008 |
| 40 | 2 | 0·008 |
| 50 | 2·5 | 0·008 |
| 63 | 3·2 | 0·008 |
| 80 | 4 | 0·008 |
| 100 | 5 | 0·008 |
| 125 | 6·3 | 0·008 |
| 160 | 8 | 0·008 |
| 200 | 10 | 0·008 |
| 250 | 12·5 | 0·008 |
| 315 | 16 | 0·008 |
| 400 | 20 | 0·008 |
| 500 | 25 | 0·008 |
| 630 | 31·5 | 0·008 |
| 800 | 40 | 0·008 |
| 1000 | 50 | 0·008 |

*Proposed exposure limits for single frequency or narrow band vibration (third octave band analysis), where daily exposure is essentially uninterrupted for 4–8 hr.

**Table 9.4.** Proposed exposure limits for broad band vibration (octave band analysis)*

| Octave band centre frequency (Hz) | Maximum rms intensity of hand-transmitted vibration in any direction (axis) | |
| --- | --- | --- |
| | Acceleration (m/s²) | Velocity (m/s) |
| 8 | 1·4 | 0·027 |
| 16 | 1·4 | 0·014 |
| 31·5 | 2·7 | 0·014 |
| 63 | 5·4 | 0·014 |
| 125 | 10·7 | 0·014 |
| 250 | 21·3 | 0·014 |
| 500 | 42·5 | 0·014 |
| 1000 | 85 | 0·014 |

*Daily exposure uninterrupted for 4–8 hours. ISO Draft Proposal 5349.[2]

the early 1970s led to a significant reduction in VWF prevalence throughout the timber-producing countries, notably the UK[22-25] and Sweden.[31] It is a matter for regret that a similar improvement in pneumatic hammer design, albeit a more difficult engineering task, has not so far been achieved. Attempts to reduce pneumatic hammer vibration have resulted in a serious decrease in tool cutting efficiency and these new hammers have not been accepted in the work situation.

In many processes using pneumatic hammers, the possibility of automating chipping, scarfing and grinding processes is being actively considered by managements. Until this change is economically justified, the wearing of suitable gloves is recommended, particularly in hostile weather conditions (cold plus dampness). There are two principles operating: (*a*) the maintenance of peripheral temperature and (*b*) the absorption, it is hoped, of vibration in the damaging frequency range of 100–300 Hz. A recent design of chain-saw warms the saw handle to avoid hand temperature reduction. There may be, however, serious objections to the wearing of gloves, particularly if these are padded to absorb a proportion of the higher frequency vibration, because they reduce manipulative efficiency. As in other biological systems (for example noise and the ear), interrupted exposure to vibration is strongly recommended with, if the work schedule will allow, breaks of at least 10 minutes in every hour of continuous exposure. The coupling of the hand with the tool, that is the grip force, has been shown to be a major factor in producing VWF,[26] in that the tighter the grip on the work piece, the greater the transmission of vibration into the hand. The tighter grip also restricts the blood circulation. As tools become lighter, working techniques become even more important since a reduction in weight means more vibration entering the hands resulting from less damping within the tool. As the vibration exposure time rises in years, it is recommended that workers already afflicted (late stage 2 and stage 3 cases — advanced cases) should have an annual medical examination with VWF specifically in mind, in order to avoid cumulative vibration energy absorption which will lead to further VWF deterioration.

The evidence from recent surveys[32] indicates that the main factors involved in VWF stage progression (or deterioration) are (*a*) increasing vibration exposure time; (*b*) the intensity of vibration and its spectral content (hazardous frequency range 30–300 Hz); (*c*) whether the exposure is continuous or interrupted; and (*d*) the susceptibility of the subject. The so-called 'latent interval' in years, being defined as the time period between the first exposure to vibration and the appearance of the first white fingertip, is a guide to the energy content of the vibration: the shorter the latent interval the more hazardous the vibration. The latency may vary from 6 to 8 months in pedestal grinders to 5 to 6 years in chippers and grinders. Rapidly deteriorating cases of

VWF, where the latency is known to be short (weeks or months), should be removed from further vibration exposure. They are usually found to be the late stage 2 or stage 3 cases.

For subjects with the vibration syndrome no cure or treatment is presently available. Therapy is essentially palliative involving either blockade or severance of sympathetic nerve activity. Vasodilatation by direct action through chemotherapy has been unsuccessful and in some cases is dangerous from the point of view of undesirable side effects affecting balance. The failure of chemotherapy is due to the absence of a therapeutic drug which specifically acts on, and dilates, only the vessels of the digits. The broadest success has been with the sympathetic inhibitor reserpine. Recent reports[33] indicate some success with serial plasmapheresis although it is difficult to understand the mechanism of this treatment since involvement of plasma fibrinogen, cold haemagglutins or cryoglobulins has not been demonstrated conclusively in VWF. Some success has also been claimed for biofeedback therapy.[34,35]

Finally, some countries have not yet recognized the vibration syndrome as a disease of occupation. The Industrial Injuries Advisory Council of the UK Government has published reports in 1954, 1970 and 1975.[36-38] In all reports prescription for state compensation has been disallowed on the grounds of the absence of objective tests and the difficulty of assessing VWF impairment other than the subject's own description of his attacks. These decisions, when the majority of countries in Europe, Japan, Russia and the Eastern block have recognized the vibration syndrome as a compensatable disease, emphasize the importance of searching for and establishing objective tests for VWF. Recent attempts have been made to develop a multi-finger photocell plethysmography system for blood flow determinations,[39] and for the measurement of sensory loss in the fingertips.[40] Without such tests and their evaluation in surveys of exposed and non-exposed populations, it has not yet been possible to introduce legal measures to protect workers from the effects of segmental vibration.

*Note:* Since time of writing, the UK Industrial Injuries Advisory Council[41] has recommended that 'vibration white finger should be prescribed for industrial injury purposes'. Prescription should be limited at present to use of those tools in those processes that expert evidence has identified as presenting most risk of causing VWF (chain saws, grinding and other rotary tools, percussive hammers, drills and tools used in riveting, caulking, drilling, chipping, hammering, fettling and swaging).

REFERENCES

1. International Organization for Standardization. *Guide for the Evaluation of Human Exposure to Whole Body Vibration.* ISO 2631-1974. Geneva, 1974.

2. International Organization for Standardization. *Guide for the Measurement and the Evaluation of Human Exposure to Vibration Transmitted to the Hand:* ISO Draft Proposal 5349 [(Secr. Doc. 150/108/4) WG 3.20. (ISO/TC/108/SC4)]. Berlin, 1978.

3. British Standards Institution. *Guide to the Evaluation of Exposure of the Human Hand-arm System to Vibration.* Draft for Development: DD43. London, 1975.

4. Guignard JC and McCauley ME. Motion sickness incidence induced by complex periodic waveforms. Paper presented to the 21st Annual Meeting of the Human Factors Society, San Francisco, October 1977.

5. Edwards RG, McCutcheon EP and Knapp CF. Cardiovascular changes produced by brief whole-body vibration of animals. *J. Appl. Physiol.* 1972; **32**:386–90.

6. Melkumova AS and Russkikh VV. *Byul. Eksp. Biol. Med.* **9**:28–31. Cited by Guignard JC. In: *Patty's Industrial Hygiene and Toxicology,* vol 3. New York: Wiley, 1979:475.

7. Matthews J. Ride comfort for tractor operators, Part 1. Review of existing information. *J. Agric. Eng. Res.* 1964; **9**:3–31.

8. Rosegger R and Rosegger S. Health effects of tractor driving. *J. Agric. Eng. Res.* 1960; **5**:241–75.

9. Gruber GJ and Ziperman HH. *Relationship between Whole Body Vibration and Morbidity Patterns among Motor Coach Operators.* Publication 75-104, National Institute of Occupational Safety and Health. Cincinnati, Ohio: US Department of Health, Education and Welfare (NIOSH), 1974.

10. Gruber GJ. *Relationship between Whole Body Vibration and Morbidity Patterns among Interstate Truck Drivers.* Publication 77-167, National Institute of Occupational Safety and Health. Cincinnati, Ohio: US Department of Health, Education and Welfare (NIOSH), 1976.

11. Loriga G. Il labora con i martelli pneumatici. *Boll. Ispett. Lavora* 1911; **2**:35–60.

12. Hamilton A. A study of spastic anemia in the hands of stone-cutters; an effect of the air hammer on the hands of stone-cutters. *Bulletin 236, US Bureau of Labor Statistics* 1918; **19**:53–61.

13. Agate JN, Druett HA and Tombleson JBL. Raynaud's phenomenon in grinders of small metal castings. *Br. J. Ind. Med.* 1946; **3**:167–74.

14. Pecora LJ, Udel M and Christman RP. Survey of current status of Raynaud's phenomenon of occupational origin. *Am. Ind. Hyg. Assoc.* 1960; **21**:80–2.

15. Grounds MD. Raynaud's phenomenon in users of chain-saws. *Med. J. Aust.* 1964; **8**:270–2.

16. Miura T, Kimura K, Tominaga Y et al. On Raynaud's phenomenon of occupational origin due to vibrating tools. *Institute of Science of Labour (Japan)* 1966; No. 65.

17. Kylin B, Lidström I-M, Liljenberg B et al. I. Almänna hälsoundersökningen. II. Undersökning över vibrationsskador hos skogsarbetare. III. Undersökning över skogsarbetarens arbetsmiljo. Hälso-Och miljöundersökning Bland Skogsarbetare. *Al Rapport* 1968; No. 5.

18. Tiilila M. A preliminary study of the white finger syndrome in lumberjacks using thermographic and other diagnostic tests. *Work Environment — Health* 1970; **7**:85–7.

19. Hellström B, Stensvold I, Halvorsrud JR et al. Finger blood circulation in forest workers with Raynaud's phenomenon of occupational origin. *Int. Z. Angew. Physiol.* 1970; **29**:18–28.

20. Huzl F, Stolarik R, Mainerova J et al. Damage due to vibrations when felling timber by power saws. *Pràcov. Lek.* 1971; **23**:7–15.

21. Allingham PM and Firth RD. The vibration syndrome. *NZ Med. J.* 1972; **76**: 317–21, 486.

22. Taylor W, Pearson JCG, Kell RL et al. Vibration syndrome in Forestry Commission chain-saw operators. *Br. J. Ind. Med.* 1971; **28**:83–9.

23. Taylor W, Pelmear PL and Pearson JCG. Raynaud's phenomenon in forestry chain-saw operators. In: Taylor W. (ed.) *The Vibration Syndrome*. London: Academic Press, 1974; 121–40.
24. Taylor W and Pelmear PL (ed.) *Vibration White Finger in Industry*. London: Academic Press, 1975.
25. Taylor W, Pearson JCG and Keighley GD. Raynaud's phenomenon in chain-saw operators. In: Wasserman DE, Taylor W and Curry MG (ed.) *Proceedings of the Second International Hand-Arm Conference*. Publication 77-170, National Institute of Occupational Safety and Health. Cincinnati, Ohio: US Department of Health, Education and Welfare (NIOSH), 1977.
26. Färkkilä M, Pyykkö I, Korhonen O et al. Hand grip forces during chain saw operation and vibration white finger in lumberjacks. *Br. J. Ind. Med.* 1979; **36:** 336–41.
27. Agate JN. An outbreak of cases of Raynaud's phenomenon of occupational origin. *Br. J. Ind. Med.* 1949; **6:**144–63.
28. James PB and Galloway RW. Arteriography of the hand in men exposed to vibration. In: Taylor W (ed.) *Vibration White Finger in Industry*. London: Academic, 1975: 31–41.
29. Okada A, Yamashita T, Nagano C et al. Studies on the diagnosis and pathogenesis of Raynaud's phenomenon of occupational origin. *Br. J. Ind. Med.* 1971; **28:**353–7.
30. Miwa T. Evaluation methods for vibration effect, Part 3. Measurements of threshold and equal sensation contours on hand for vertical and horizontal sinusoidal vibrations. *Ind. Health (Japan)* 1967; **5:**182–205, 206–12.
31. Axelsson SA. Progress in solving the problem of hand–arm vibration for chain saw operators in Sweden, 1967 to date. In: Wasserman DE, Taylor W and Curry MG (ed.) *Proceedings of the Second International Hand–Arm Vibration Conference*. Publication 77-170, National Institute of Occupational Safety and Health. Cincinnati, Ohio: US Department of Health, Education and Welfare (NIOSH), 1977.
32. Wasserman D, Reynolds D, Behrens V et al. Vibration white finger disease in US workers using pneumatic chipping and grinding hand tools. Vol. II: Engineering testing. Publication 82-101. NIOSH, Cincinnati, Ohio: US Department of Health, Education and Welfare, 1981.
33. Talpos G, Horrocks M, White JM et al. Plasmapheresis in Raynaud's disease. *Lancet* 1978; **1:**416–17.
34. Jacobson AM, Silverberg E, Hackett T et al. Treatment of Raynaud's disease. Evaluation. *Psychiatr. Clin.* 1978; **3:**125–31.
35. Sedlacek K. Biofeedback for Raynaud's disease. *Psychosomatics* 1979; **8:**538–41.
36. Industrial Injuries Advisory Council. *Report on Raynaud's Phenomenon*. National Insurance (Industrial Injuries) Act 1946: Cmnd 9347. London: HMSO, 1954.
37. Industrial Injuries Advisory Council. *Interim Report on the Vibration Syndrome*. National Insurance (Industrial Injuries) Act 1965: Cmnd 4430. London: HMSO, 1970.
38. Industrial Injuries Advisory Council. *The Vibration Syndrome*. National Insurance (Industrial Injuries) Act 1965: Cmnd 6965. London: HMSO, 1975.
39. Wasserman D, Carlson W, Samueloff S et al. A versatile simultaneous multi-finger photo-cell plethysmography system for use in clinical and occupational medicine. *Medical Instrumentation 13* 1979; **4:**232–4.
40. Carlson W, Samueloff S, Taylor W et al. Instrumentation for measurement of sensory loss in the fingertips. *J. Occup. Med.* 1979; **4:**260–4.
41. Industrial Injuries Advisory Council. Vibration White Finger. Social Security Act 1975. Cmnd 8350. London: HMSO, 1981.

# 10. OCCUPATIONAL LUNG DISEASE
*A. J. Newman Taylor, C. A. C. Pickering
and I. I. Coutts*

This chapter is complementary to the chapter written by Dr K. H. Nickol which appeared in the first volume in this series, *Current Approaches to Occupational Medicine,*[1] in 1979. Dr Nickol painted on a broad canvas, and of necessity was unable to include some areas which since 1979 have assumed greater importance. We have chosen to focus on four different topics which we believe are now of particular interest. He called his chapter 'Some occupational respiratory disorders'. We could properly have entitled this 'Some other occupational respiratory disorders'.

## ASBESTOS-RELATED DISORDERS

Exposure to airborne asbestos fibre can cause intrapulmonary fibrosis, lung cancer, mesothelioma, pleural thickening, pleural plaques and possible pleural effusion. The first three of these conditions are particularly serious and this review will concentrate on them. It will try to outline the present state of knowledge about the relationships between asbestos exposure and the diseases which result from it rather than their clinical manifestations.

### Intrapulmonary fibrosis: asbestosis

Asbestosis causes respiratory disability and reduces life expectation. The Department of Employment and Productivity[2] reported on 430 cases certified by the Pneumoconiosis Medical Panels between 1956 and 1965 and concluded that male cases of asbestosis had a mortality two to three times that of the general male population. McVittie[3] found that 40 per cent of cases certified by the panels died of lung cancer or mesothelioma and a further 30 per cent died from other respiratory causes.

### *Dose–response relationships*

The first recorded case of asbestosis is probably that described by Montague Murray in 1907[4] to the Departmental Committee on

Compensation for Industrial Diseases, but it was not until the 1920s that a series of cases was reported associating pulmonary fibrosis with asbestos exposure.[5-7] These reports led to the investigations of Merewether and Price[8] who first showed a relationship between the duration and intensity of asbestos exposure and the development of asbestosis. Their report resulted in the Asbestos Industry Regulations (1931) which led to an improvement in working conditions but did not eliminate the risks associated with asbestos exposure; moreover certain types of asbestos work, notably the insulation trade, were not within the scope of the legislation.

The 1960s saw a resurgence of interest in the problems associated with asbestos exposure and the start of a series of attempts to quantify its effects in the hope that it might prove possible to define levels of exposure within which the health of those exposed was not at risk. The British Occupational Hygiene Society study of asbestos mill workers in Rochdale[9] was carried out to answer this problem and its results were used in framing the more comprehensive Asbestos Regulations of 1968. The study findings suggested that the risks of developing the earliest clinical signs of asbestosis were less than 1 per cent after an accumulated exposure to chrysotile of 100 fibre years per m³ (i.e. the equivalent of 2 fibres per m³ for 50 years).

The Canadian studies on chrysotile miners and millers[10] provide correlations between measurements of asbestos exposure and radiographic response. Despite powerful methodology they have consistently shown only weak correlations ($r < 0.3$) between measurements of exposure and radiographic evidence of asbestosis. These correlations are very much poorer than those obtained in similar studies of coal workers. This difference has been attributed to the fact that the radiographic appearances in coal workers pneumoconiosis are largely due to accumulated dust whereas in asbestosis the radiographic changes are the result of the biological reactions to inhaled asbestos fibres. Such poor correlations imply that much of the variation in the radiographic evidence of asbestosis is due to factors other than the dose of inhaled asbestos. This being so, a number of attempts have been made to identify other causes for the variability in response. So far these have looked for evidence of altered — usually increased — individual susceptibility on the basis of immunological differences and at the effects of cigarette smoking.


*Immunological differences*

Turner-Warwick and Parkes[11] found an increased prevalence of antinuclear antibody and rheumatoid factors in asbestosis but these changes are probably a consequence of pulmonary fibrosis. These changes do not allow the detection of susceptible individuals before disease develops. The possibility that histocompatibility (HLA) antigens might be linked to the development of asbestosis has been examined by a number of workers and these studies have been reviewed by Turner-Warwick.[12]. No strong or consistent associations have been found between asbestosis and HLA antigens and so tissue typing cannot at present provide a method of screening for susceptibility. Abnormalities of immunoglobulins[13] and lymphocyte function[13-15] have been shown in asbestosis but again it is unclear what is cause and what is effect. Long term studies are in progress which may clarify the position.

## Cigarette smoking

The effect of cigarette smoking on the development of fibrosis has been examined in a number of studies with conflicting results. Two studies from the United Kingdom[16,17] have shown that the incidence of asbestosis is increased in cigarette smokers, but the Canadian studies show no such effect.[18] Conflicting results have emerged from the United States where Weiss[19] found that cigarette smoking and asbestos exposure were at least additive in causing pulmonary fibrosis while Samet[20] found no evidence of synergism. Thus although the evidence suggests that factors other than asbestos exposure are important in the development of asbestosis, no other associated factor has yet been unequivocally demonstrated. So far attention has focused on immunological changes and cigarette smoking. Little attention has been given to the possibility that other environmental agents might interact with asbestos to cause fibrosis.

## Asbestos and lung cancer

In 1935 Lynch and Smith reported a case of lung cancer in a man with asbestosis.[21] Many similar reports followed but it was not until 1955 that Doll[22] demonstrated an approximately tenfold increase in the risk of lung cancer in asbestos workers. This increased mortality from lung cancer has been confirmed in the many subsequent mortality studies of asbestos-exposed cohorts. These studies have been summarized in *Asbestos:* Final Report of the Advisory Committee.[23] It is worth noting at this point that all the cases of lung cancer reported by Doll occurred in subjects with asbestosis. Later mortality studies generally provide no information as to whether or not the excess risk is confined to those with asbestosis. Cohort studies in which it has been possible to examine mortality in relation to cumulative exposure to asbestos have shown that the incidence of lung cancer increases with exposure and although it is difficult to determine the shape of the exposure–response relationship at low levels it is probably linear. All the commercially exploited asbestos minerals have been associated with an increased risk of lung cancer.

The relationship between asbestos exposure and the development of lung cancer is close and almost certainly is one of cause and effect but two aspects of the problem need further consideration. First, is the risk of lung cancer confined to those with asbestosis, and secondly, how do asbestos exposure and cigarette smoking interact?

### Are asbestosis and lung cancer independent risks of asbestos exposure?

This question cannot as yet be answered with certainty. The mortality from lung cancer in cases of asbestosis certified by the Pneumoconiosis Medical Panels is very high with around 40 per cent of cases ultimately developing a bronchial carcinoma. Of the major cohort studies only that of the Belfast insulation workers[24] has shown higher mortality ratios than occur in panel cases. The Belfast insulators were a very heavily exposed group, but it is of interest that in the first mortality study of this population no case of lung cancer was found in the absence of evidence of fibrosis. Similar findings emerged from the first mortality study of the Rochdale asbestos textile workers[25] where the excess mortality from lung cancer observed was less but again lung cancer only occurred in those with asbestosis. These studies which give some information about the presence of intrapulmonary fibrosis provide clear evidence that asbestosis and lung cancer often occur together, but they do not prove that the risk is confined to those with asbestosis.

Liddell and McDonald[18] have investigated the use of the chest radiograph as a predictor of death in Canadian chrysotile miners and millers. The analysis of the relationship between the excess mortality from lung cancer and the presence of asbestosis was difficult because of the interval between the chest radiograph and death but the authors believe that most, but not necessarily all, cases of lung cancer attributed to chrysotile exposure in mining and milling probably have small opacities before death.

Another approach to this problem has been to see whether there is an increased mortality from lung cancer in asbestos-exposed populations with little evidence of pulmonary fibrosis. Lumley[26] carried out a proportional mortality study of Devonport naval dockyard employees. He found an excess mortality due to mesothelioma but no excess of lung cancer deaths. Edge[27] did find a slight excess of lung cancer deaths in men with pleural plaques, but it was confined to the first 2 years of his study and almost certainly reflects the effects of selection. He was not able to define a cohort and many of the deaths occurred in men with pleural plaques detected at a hospital attendance. Lung cancer mortality was lower in cases detected by chest radiographs taken at work.

Is it possible that the excess cases of lung cancer among asbestos workers occur only in those with basal pulmonary fibrosis? Crypto-genic fibrosing alveolitis (CFA), a disease similar to asbestosis in many ways but not associated with asbestos exposure, is also com-plicated by an excess mortality from lung cancer.[28] Wagner[29] in a series of inhalation experiments in which rats were exposed to asbestos found that cancers only occurred in those with pulmonary fibrosis. A number of authors[30-32] have found that the usual upper lobe predominance of lung cancers is reversed in asbestosis, the tumours tending to arise in the lower lobes, the area usually affected by fibrosis.

If pulmonary fibrosis is a necessary precondition for the develop-ment of asbestos-related lung cancer then one might expect the incidence of lung cancer to rise with increasing fibrosis. This does not appear to happen in asbestosis[33] or in cryptogenic fibrosing alveolitis.[34] A close relationship between fibrosis and cancer might be expected to give rise to cancers of a particular histological type but again this is not the case in either asbestosis or CFA. There may be a slight increase in the proportion of adenocarcinoma[32] but all the com-mon histological types occur more frequently in asbestosis.

Thus the evidence suggests a close relationship between lung cancer and asbestosis and there is no unequivocal evidence of an excess mor-tality from lung cancer in the absence of asbestosis. The data on which to make a final statement on this question are not available.

### Asbestos exposure, smoking and lung cancer

The interaction of cigarette smoking and asbestos exposure in the causation of lung cancer has been a subject of some interest. It has been clear for some time that asbestos workers who smoke are at a very much increased risk of developing lung cancer,[35] but the nature of the interaction is still unclear. Are the two effects simply additive or are they multiplicative? And does asbestos cause lung cancer in non-smokers?

Two recent reports both suggest that the relative risk of lung cancer is increased in non-smoking asbestos workers. In the report from Canada[36] the relative risk of lung cancer actually appears to be higher in non-smokers while in Hammond's study from the United States[37] the relative risks are similar. In Hammond's study an attempt was made to verify the death certificate diagnosis and it appears that lung cancer may have been under-reported in the non-smokers. This study established a non-smoking, non-asbestos-exposed control group for comparative purposes. The ratio of observed deaths in the non-smoking non-asbestos-exposed group compared to expectation based on the mortality experience of the control group was 5·33. The evidence from both these studies is consistent in suggesting that asbestos exposure can act independently of cigarette smoking to cause lung cancer.

There are broadly two hypotheses which try to explain the interaction of cigarette smoking and asbestos exposure in the causation of lung cancer. The two factors may always operate independently and when they occur together the effects are summated. This is the additive hypothesis. The second hypothesis which has a number of variations is the multiplicative hypothesis. In its simplest form this states that when the two factors operate together their effects are multiplied. Variations of this model allow for the effects of cigarette smoking to be amplified in the presence of asbestos fibres.

Saracci[38] has reviewed the evidence for these alternative hypotheses and has concluded that on the available data the additive hypothesis is unlikely and that some form of multiplicative model is more likely. Enterline[39] believes that the recent data[37] on the risks of lung cancer in smoking and non-smoking asbestos workers are compatible with his view that asbestos amplifies the effects of other environmental carcinogens.

### Mesothelioma

Wagner[40] first noticed the association between mesothelioma and exposure to crocidolite (blue asbestos) in the asbestos mining area of the North West Cape Province in South Africa. It soon became apparent that this tumour occurred in many parts of the world where asbestos was used. In most work places exposure to a variety of asbestos fibres occurred and so it was not possible to know whether the risk of mesothelioma was confined to those working with crocidolite.

Anthophyllite, the least important asbestos mineral commercially, was mined in Finland, and although an excess of lung cancer was observed in the miners and millers no mesotheliomas have occurred.[41]

Among Canadian chrysotile miners and millers a small number of mesotheliomas have occurred[36] but it is uncertain whether they can be attributed solely to chrysotile exposure. Some Canadian chrysotile is contaminated with tremolite, an amphibole asbestos mineral which, although not commercially exploited, is under suspicion of causing mesothelioma.[42] Furthermore, during World War II some South African crocidolite was processed in the Canadian mills. If chrysotile does cause mesothelioma the risk appears to be small.

Amosite presents something of a paradox. On the one hand amosite mining in South Africa has rarely been associated with mesothelioma whereas its use in the United States has.[43] This discrepancy is almost certainly not due to failure to recognize the disease in amosite miners[44] nor does it seem likely that contamination by crocidolite of the amosite arriving in the United States is the explanation. Electron microscopy of the lungs of some of the mesothelioma cases from the

United States has revealed amosite fibres but in many cases crocidolite could not be detected. Possibly the further processing of amosite that occurs after leaving the mining area results in the production of fibres capable of causing disease. More information is required before the risks of amosite can be fully assessed.

The greatly increased risk of mesothelioma associated with crocidolite use has been confirmed in studies of two groups of workers employed in the manufacture of military gas masks during World War II.[45,46] These masks had an asbestos filter made from crocidolite and although the length of exposure was relatively short in both groups, the incidence of mesothelioma has been very high. Of the 56 deaths reported so far among 199 personnel involved in military gas mask manufacture in Canada, 9 have been due to mesothelioma. Pleural mesothelioma frequently occurs in the absence of intrapulmonary fibrosis and clinical observations suggest that the tumour can arise after much smaller exposure to asbestos than is usual in cases of asbestosis. Despite this, the exposure required to cause mesothelioma is appreciable and unlikely to occur outside an occupational setting. Whitwell[47] assessed the fibre content of the lungs in cases of mesothelioma, lung cancer without pneumoconiosis and controls by using the light microscope. The mesothelioma cases had higher counts than either the controls or the lung cancer cases. Cases with asbestosis had higher counts than mesothelioma cases.

Support for these observations comes from the experience of mesothelioma in Plymouth,[48] where exposure to crocidolite, amosite and chrysotile occurred among naval dockyard workers. One hundred and eight cases have occurred, mostly among laggers, boilermakers, painters, welders, burners and shipwrights who in the past were often heavily exposed to asbestos. Cases have also arisen in men whose only exposure has been to the environment within the dockyard but outside the ships and workshops where asbestos was handled. No cases have occurred so far in indoor office workers at the dockyard, and there is no evidence of increased risk in the general population of Plymouth outside the dockyard.

Irrespective of the length and intensity of exposure to asbestos there is a long latent period of at least 20 years, and often much longer, before mesothelioma develops. The incidence of the disease is currently rising and is likely to continue increasing for some years[49] until the improvements in environmental hygiene brought about during the 1960s and 1970s lead to the elimination of asbestos-related mesothelioma. The number of new cases of mesothelioma diagnosed each year now exceeds those of asbestosis.

Clinical features of the disease have been described by Elmes and Simpson.[50] No treatment has yet been shown to alter the course of the disease.

*Pathogenesis of mesothelioma*

The pathogenesis of mesothelioma has been investigated in a series of experiments where a variety of materials have been inoculated into the pleural space of rats.[51,52] They provide some grounds for thinking that the physical size and shape of fibres rather than their chemical composition is important in producing mesothelioma. It is worth noting, however, that silicon microspheres, barium sulphate and aluminium oxide, all non-fibrous substances, also produced small numbers of mesotheliomas when injected into the pleural cavity of rats. It is difficult to extrapolate from intrapleural exposures in rats to airborne exposures in man but if the development of mesothelioma is dependent on fibres of the appropriate size and shape, irrespective of chemical composition, reaching the pleura, then the problems might be anticipated from non-asbestiform fibres. So far there is no evidence of mesothelioma occurring among rock wool and fibre glass workers but there does appear to be a small epidemic of mesothelioma in part of Turkey[53,54] where no asbestos minerals occur, but where a fibrous zeolite has been found. If it is confirmed that non-asbestos fibrous materials can give rise to mesotheliomas, and if the risks can be assessed on the basis of the size and shape of the material, then careful consideration will be necessary before deciding to exploit such substances commercially.

## Conclusions

It is clear that asbestos exposure can give rise to a number of serious diseases but the pathogenetic mechanisms involved are still poorly understood. It is to be hoped that with improved environmental hygiene these diseases will disappear but the legacy of past asbestos exposure will be with us for some time yet.

## OCCUPATIONAL ASTHMA

Occupational asthma can be defined as variable airway narrowing causally related to exposure to airborne pollutants in the working environment. Both byssinosis and transient airway narrowing provoked by the inhalation of airway irritants such as chlorine gas are properly included in this definition. However, occupational asthma is usually restricted to those causes of variable airway narrowing which fulfil the clinical criteria of a hypersensitivity reaction to the specific cause. Asthma occurs in a proportion, usually a minority, of those exposed; it develops only after an initial symptom-free period of exposure, which can vary from weeks to years; and in those affected, it recurs on re-exposure to the causal agent at concentrations which do not cause reactions in others similarly exposed, and of which they were, themselves, previously tolerant.

## Causes

A large number of agents to which exposure occurs at work have been reported as causing asthma. Those of most importance are listed in *Table 10.1*, with some of the more common circumstances in which exposure to them occurs. It can be anticipated that with increasing recognition of the hazard of occupational asthma and the continuous introduction of new materials into industrial processes, the number of agents identified as causing occupational asthma will increase.

**Table 10.1.** Some causes of occupational asthma

| Causes | Exposures or uses |
| --- | --- |
| *Biological* | |
| *Bacillus subtilis* enzymes (alcalase etc.) | Enzyme detergent industry |
| Small mammal urine proteins (rats, mice etc.) | Laboratory animal workers |
| Grain, flour and contaminants | Food industry (millers, bakers etc.), farmers |
| Wood dusts (Western red cedar, Iroko etc.) | Wood mills, carpenters |
| Colophony (in soft solder flux) | Electronics industry |
| Antibiotics (penicillins etc.) | Pharmaceutical industry |
| *Chemical* | |
| Di-isocyanates (toluene (TDI), diphenylmethane (MDI), hexa-methylene (HDI) and naphthalene (NDI)) | Polyurethane foam manufacture<br>Painting industry<br>Paint and synthetic rubber |
| Epoxy resin curing agents (phthalic anhydride, trimellitic anhydride, triethylene tetramine etc.) | Surface coatings<br>Adhesives |
| Complex salts of platinum (particularly ammonium hexachlorplatinate) | Platinum refining |

## Prevalence

The relative importance of occupational exposures as causes of asthma is unknown. In Finland, where occupational asthma is a prescribed disease, 80 new cases were reported to the Finland Occupational Diseases Register in 1976.[55] We should know more about the importance of occupational causes of asthma in the United Kingdom when occupational asthma becomes a prescribed disease.

Studies of working populations exposed to specific causes of occupational asthma have shown that a relatively high proportion of those at risk are affected. In a prevalence study of an electronics factory where colophony was used as a soft solder flux, 22 per cent of 446 shopfloor employees had work-related respiratory symptoms, compared with only 6 per cent of those working in other parts of the factory.[56] Of those employees who had left during the 3½ years before this survey, a significantly greater proportion had worked on the shopfloor than in other parts of the factory.[57] This excess could largely be accounted for by those leaving with work-related respiratory symptoms. In a study of employees in an Australian factory manufacturing enzyme detergents, 49 of the 98 workers surveyed had work-related respiratory symptoms.[58] Of 144 employees of a pharmaceutical company working with laboratory animals, 28 had allergic reactions provoked by one or more of the small mammals (rats, mice, guinea-pigs and rabbits), of whom 13 had asthmatic symptoms.[59] In another survey of laboratory animal workers employed in a medical centre in the United States, 59 (15 per cent) of the 399 exposed had allergic symptoms provoked by animal exposure, of whom about a half, 30, had asthmatic symptoms.[60]

Few incidence studies of occupational asthma have been reported. Of 1642 employees in a factory manufacturing enzyme detergents, Juniper[61] found that during a 7-year period 53 (3·2 per cent) had been transferred from areas in the factory where enzyme exposure occurred because of respiratory symptoms provoked by enzyme dust.

Of 86 new employees in a platinum refinery in 1973 and 1974, 35 left with asthma due to sensitivity to ammonium hexachlorplatinate, the great majority within 18 months of first exposure.[62]

## Disease mechanisms

The relationship between asthma and the specific occupational exposure which is causing it can be validated by inhalation testing: exposure of the affected individual to the specific agent provokes an asthmatic reaction. The clinical features of this relationship suggest that those affected have developed a hypersensitivity reaction to the specific cause *(see above)*. Investigation into the mechanisms which cause occupational asthma has therefore been primarily directed towards the identification of underlying immunological reactions. Specific IgE antibody, usually demonstrated by the radio-allergosorbent test (RAST), has been found in the sera of those with asthma caused by several of the biological agents, including the *Bacillus subtilis* enzyme alcalase,[63] rat and mouse urine and serum proteins[64] and wheat and rye flour.[65]

Specific antibodies have also been identified in the sera of those with asthma caused by exposure to some low molecular weight chemicals. Specific IgE antibody to ammonium tetrachlorplatinate has been found in the sera of some platinum refinery workers with asthma.[66] In a study of the manufacturers of trimellitic anhydride (TMA), Zeiss[67] distinguished three clinically distinct respiratory reactions among those exposed, each with corresponding immunological differences. Some had developed rhinitis and asthma when exposed to TMA which occurred within minutes of exposure. These had specific IgE antibody to a TMA human serum albumin conjugate in their serum. In others, asthma associated with fever and constitutional upset ('TMA 'flu') occurred several hours after the onset of exposure. Specific IgG, but not IgE, antibody to the TMA conjugate was found in their serum. In those exposed to sufficient concentration, TMA provoked nasal discharge, cough with sputum and breathlessness ('irritant reaction'). Those who had had irritant, but not the other reactions, had neither IgE nor IgG antibody to the TMA conjugate in their serum. Following this report, Ahmad[68] found that TMA exposure could also cause haemoptysis, on occasions associated with shadowing on the chest radiograph, and haemolytic anaemia. Both those with this illness and those with 'TMA 'flu' had, in addition to specific IgG antibody in their serum to the TMA–HSA conjugate, IgG antibody to erythrocytes coated with TMA, and were also able to agglutinate TMA-coated erythrocytes.[69]

Inhaled TMA can provoke at least four clinically distinguishable reactions, which are also partly but not completely distinguishable immunologically. The mechanisms underlying asthmatic reactions provoked by toluene di-isocyanate (TDI) are less clearly defined. TDI has been found not only to stimulate IgE antibody production, but also to possess pharmacological activity *in vitro*. Butcher[70] found that about 15 per cent of a group of patients with occupational asthma due to TDI had IgE antibody to a conjugate of toluene mono-isocyanate (TMI). Investigating a different possible mechanism of action of TDI, Davies[71] found that TDI inhibited the *in vitro* stimulation of cyclic AMP by isoprenaline and prostaglandin E, in a dose-related fashion. He suggested that TDI could cause asthma by this $\beta$-adrenoreceptor activity, possibly in those who had pre-existing bronchial hyper-reactivity. This hypothesis does, however, fail to explain the latent interval which occurs between initial exposure to TDI and the development of asthma.

Good evidence has been found for immunological reactions, mediated by both IgE and IgG antibodies, to several of the causes of occupational asthma, both biological and chemical. However, there remain some agents such as colophony and formaldehyde to which neither specific antibodies nor sensitized lymphocytes have been identified to date. This may reflect the difficulties of preparing appropriate conjugates of these agents for *in vitro* testing, which only recently have been overcome with the complex platinum salts and epoxy resin curing agents such as trimellitic anhydride. Alternatively the chemical to which exposure occurs may not itself be antigenic, but the product of a chemical reaction derived from it. It must also be remembered that mechanisms other than immunological ones can cause asthma, and may indicate asthmatic reactions due to some occupational agents.

## Clinical features and diagnosis

In those with occupational asthma, avoidance of further exposure to its cause is often followed by complete remission of symptoms and restoration of previous lung function. Those who develop the disease are often recommended to change their job. Relocation within the same organization may be possible, but a change of job can mean the loss of current employment. Early recognition and accurate diagnosis of occupational asthma is therefore of great importance both to ensure that those who are affected avoid further exposure to its cause, as well as to prevent those without work-related asthma being advised to change their job unnecessarily. Both misdiagnosis and missing the diagnosis have serious implications. The characteristic symptoms of occupational asthma are episodic lower respiratory symptoms — breathlessness, cough, wheeze or chest tightness — temporally related to occupational exposures. Those affected may recognize that their symptoms develop during the working week, often increasing in severity as the week progresses, improve during periods of absence from work, such as at weekends or during holidays, and recur on return to work. Symptoms may develop either within minutes of exposure or only several hours after the onset of exposure, during the latter part of the working day or in the evening after returning home from work. On occasions symptoms may not be easily appreciated as due to asthma, or the relationship to occupational exposures may be unclear.

Symptoms can persist for several days or even weeks after exposure has ceased, and where little variation in symptoms occurs, they may be attributed to cigarette smoking or to bronchial infections. Of 21 employees in the electronics industry who had asthma demonstrated by inhalation testing to be provoked by soft solder flux fumes, 5 had had asthma before starting this work and 14 had never smoked. Nevertheless 19 were initially diagnosed as having 'bronchitis'. Those with occupational asthma may also find that in addition to specific occupational exposures, other non-specific stimuli such as exercise, upper respiratory tract infections, inhalation of cold air and emotional upset provoke asthmatic symptoms.

Confirmation of the diagnosis of occupational asthma may be obtained either by demonstrating changes in airway function related to occupational exposures, or by inhalation testing with the specific causal agent. Inhalation testing, particularly with occupational agents, is potentially hazardous and should be undertaken only by those experienced in the techniques. It is also very demanding on resources; those tested must be admitted to hospital during the period of testing for observation by medical staff which should continue for at least 24 hours after testing. Where appropriate other methods of investigation are employed.

Lung function measurements made in relation to periods of occupational exposure may demonstrate work-related airway narrowing, and skin tests, where available, can provide evidence of a specific immunological reaction to the causal agent. Such measurements of lung function made before and after a working shift may show that airway narrowing has developed during the working day, but such measurements are often unhelpful and can be misleading. At present the most satisfactory method of demonstrating work-related asthma is regular self-measurement of peak expiratory flow rates recorded at 2-hourly intervals from waking to sleeping, made over several weeks to include both periods at work and absences from work. Such records, however, can pose difficulties in interpretation and analysis. The patterns of change observed can show great variation between different individuals; the most important are the time taken for lung function to recover and the effect of cumulative exposures. Where lung function recovers in the interval between leaving work one day and returning on the following day, an equivalent deterioration occurs in lung function on each day with restoration of normal lung function during both days of a weekend. Where recovery takes longer than this, but is complete within 3 days, progressive deterioration in lung function occurs during the working week, with complete recovery requiring part or all of a weekend. If recovery takes more than 3 days, lung function does not return to normal by the end of a weekend away from work. Progressive deterioration both during each week and with each week may develop until a plateau is reached with little variation occurring within or between days. Recovery may not start for up to 10 days after exposure has ceased, and can continue for up to 3 months. In addition, marked diurnal variation in airway calibre may develop, with maximum narrowing occurring when away from work during the night or on first waking in the morning. Where this occurs the pattern of changes in lung function provoked by occupational exposures can vary considerably depending on the period of the day during which the individual is exposed. However, despite these problems the differences between periods at work and absence from work in mean daily peak flow measurements have been shown in those with occupational asthma due to colophony[72] and di-isocyanates[73] to be a sensitive and specific index of the disease.

A positive immediate 'weal and flare' skin prick test reaction is evidence of IgE antibody production to the specific allergen. Although such reactions have been found in those exposed to several of the agents which may cause occupational asthma — including the *B. subtilis* enzyme alcalase, wheat and rye flour, rat, mouse and guinea-pig urines and the complex platinum salt ammonium hexachlorplatinate — their value in diagnosis is limited by knowledge of their sensitivity and specificity in relation to disease in occupationally exposed populations. Mitchell and Gandevia[58] found that a positive immediate reaction to alcalase in enzyme detergent manufacturers lacked both sensitivity and specificity. In laboratory animal workers a positive reaction to specific animal extracts, particularly urine, appears a sensitive index of asthma, but not of rhinitis or urticaria.[59] In platinum workers a positive reaction to ammonium hexachlorplatinate is a more specific than sensitive index of disease.[62] The finding of a positive skin test to an occupational allergen in exposed individuals with respiratory symptoms cannot usually be regarded as sufficient evidence of occupational cause of asthma. Furthermore, it is only for a minority of the agents that cause occupational asthma that skin test reactions can be elicited.

There remain four main indications for inhalation testing:

1. Where the occupational agent suspected of causing asthma is not as yet recognized as a cause of occupational asthma.

2. Where an individual with work-related asthma is exposed to more than one recognized cause of occupational asthma in his work.

3. Where the asthmatic symptoms experienced at work are of such severity that it is considered unjustifiable for the individual to be further exposed in the work place.

4. Where after all other investigations (including work records of PEF measurements) have been completed genuine doubt remains about the diagnosis of occupational asthma.

Inhalation testing whose sole purpose is in support of a legal claim is, we believe, unjustified.

The aim of occupational type inhalation testing is to expose the individual to the possible cause of his asthma in circumstances which closely resemble the conditions of exposure at work. Exposure concentrations in inhalation tests should be based wherever possible upon knowledge of the exposures experienced at work. The physical conditions of exposure, such as the size of dust particles and the temperature to which materials are heated, should be repeated.

The different methods used in such inhalation tests depend primarily on the physical state of the different materials used. Inhalation of nebulized extracts in solution is used for exposure to soluble allergens, such as urine and serum proteins of laboratory animals. This, the traditional method of inhalation testing, is not applicable to the majority of occupational causes of asthma. Liquids which evaporate at room temperature, such as toluene di-isocyanate and formaldehyde, can be painted on to a flat surface. The testing is undertaken in an enclosed chamber, and the exposure concentration varied by using different concentrations of the material in solution, and measured with an appropriate monitor. Exposure to dusts such as drugs (for example, antibiotics) and complex platinum salts are achieved by the individual tipping the material diluted in dried lactose between two trays. Wood dust exposures may be obtained by sanding the appropriate wood with a power tool. The individual can be exposed to the soft solder flux colophony by soldering with Multicore solder at the same temperatures as at work.

Asthmatic reactions provoked by inhalation testing can be easily monitored with regular measurements of forced expiratory volume in 1 second ($FEV_1$) and forced vital capacity (FVC) or of peak expiratory flow rates (PEF). Any changes observed should be compared with the results obtained following an appropriate control exposure. Each test exposure should be made on a separate day.

At least four different patterns of asthmatic reaction provoked by inhalation testing may be distinguished on the basis of their time of onset in relation to the provoking exposure. 'Immediate' reactions develop within minutes and resolve spontaneously within 1–1½ hours. 'Late' reactions develop 1 or more hours after the exposure, most commonly after an interval of 3–4 hours, and may persist for up to 24–36 hours. A 'dual' reaction is a late reaction preceded by an immediate reaction. 'Recurrent nocturnal' reactions may develop following a single exposure of short duration; asthmatic reactions occur during the night time, waking the individual from sleep, with partial or complete recovery of lung function during the intervening day times. Because such reactions may persist or recur for several days in the absence of further exposure to the provoking agent, it is essential that an adequate interval is left between inhalation tests with different agents. This will prevent a reaction provoked by one agent being incorrectly attributed to the subsequent exposure to a different agent.

## Management of established cases

The basis of management of the established case is the necessity to avoid wherever possible further exposure to the causal agent. Where substitution of the causal agent is not practicable the most satisfactory solution is alternative employment within the same organization. Such relocation may not be possible, particularly where a relatively large number of employees have been affected, or for those whose work is very specialized, such as chemists working with di-isocyanates or pharmacologists undertaking experiments on small mammals, and also within small firms. In such situations, avoidance of exposure may only be achieved by leaving current employment. For social and financial reasons an affected individual, although made fully aware of the potential risks to his future health, may choose to remain in his present employment. Other measures, less satisfactory medically, have then to be adopted. At present these rely primarily on the use of personal protection, and where adopted their effectiveness should be monitored. Regular measurements of PEF can be made during periods of protected exposure and compared with measurements made during absences from exposure.

The long term effects on lung function of those with occupational asthma who are no longer exposed to the causal agent have to date been little studied. Clinical studies suggest that asthma may persist in a proportion of those who develop occupational asthma who previously were without respiratory symptoms. Chan-Yeung[74] found that of 38 patients with inhalation-test-proved occupational asthma, 8 continued to have attacks of asthma, not attributable to other causes, 6 or more months after their last exposures. In a case control analysis of 46 TDI-sensitive men who had avoided exposure for between 2 and 11 years, Adams[75] found a significant excess of respiratory symptoms in the TDI-sensitive group. Sufficient information on long term lung function will only be provided by longitudinal studies of populations exposed to the different causes of occupational asthma and followed both during and after their periods of exposure.

## Prevention

The problems of preventing the development of new cases of occupational asthma must be clearly distinguished from that of preventing reactions in those who are already hypersensitive to the specific agent. Asthmatic reactions in those with occupational asthma are elicited by exposure to low atmospheric concentrations of the specific agent. Exposure to 0·001 parts per million (ppm) of TDI (whose threshold limit value is currently 0·02 ppm) will provoke asthma in many of those sensitive to it. The exposures which cause occupational asthma are probably greater, although to date we have no knowledge of what is a 'sensitive dose'. In the absence of such information wholly satisfactory recommendations which will ensure prevention of disease in an exposed workforce cannot be provided. Current guidelines are based both on keeping exposures to a minimum and on medical surveillance. Where practicable a safe substitute should be used. Occupational asthma in printers caused by gum acacia used as an anti-set-off spray was eliminated following its substitution by dextrose, and alternatives to colophony as a soft solder flux are currently being investigated. Where a less hazardous substitute cannot be used, exposures to the causal agents should be rigorously controlled by the well-tried methods of industrial hygiene — enclosure of the process, local exhaust ventilation, limitation of the numbers exposed by segregation of the process and, least satisfactorily, by personal protection. Medical surveillance is undertaken both before and during employment. Before employment, those who are thought to be particularly at risk of developing occupational asthma to the specific agent should be identified, and where considered appropriate excluded from exposure. To date atopic individuals (those with one or more positive skin prick test reactions to common inhalant allergens) have been shown to have an increased risk of developing asthma due to *B. subtilis* enzymes,[61] complex platinum salts[62] and laboratory animals.[59] However, with each of these allergens, before adopting a general policy of excluding atopics it has to be remembered that asthma also occurs in non-atopics, and they constitute in excess of 30 per cent of the adult population. In addition, those with evidence of pre-existing cardiorespiratory disease should be identified, as the development of asthma in them is potentially more hazardous than for those with previously normal lung function.

Those entering employment where they will be exposed to a recognized cause of occupational asthma should be informed of the hazard, the measures to be taken to control exposures, and the nature of symptoms which may suggest an asthmatic reaction. Regular periodic surveillance allows early identification of those who have developed asthma, but probably of greater importance is immediate access to informed medical opinion for those in whom suggestive respiratory symptoms may have developed.


## BYSSINOSIS

### Introduction

Occupational respiratory disease in textile workers has been recognized since the early eighteenth century, when Ramazzini described respiratory symptoms, similar to those of byssinosis, amongst flax and hemp workers. In the early nineteenth century respiratory disease amongst British cotton workers was detailed by Jackson[76] and Kay[77] but the first report of the disease that is now accepted as byssinosis was by Greenhow[78] in 1860. The term 'byssinosis' itself was introduced by Proust in 1877. Although cotton dust is the commonest cause of byssinosis, the disease is also caused by flax and hemp dusts.

Byssinosis became a compensatable disease in England in 1941. The compensation scheme applied only to male operatives employed in the cotton chambers, blowing rooms or card rooms for at least 20 years, who were totally and permanently disabled. Over the past 40 years these compensation regulations have been extended to include women and to recognize partial disability (1948), to include the processes of spinning, winding and beaming (1974) and finally (1979) to allow qualification for compensation irrespective of length of employment.

The first large scale published epidemiological studies of byssinosis took place in the 1950s and were initiated by Schilling, McKerrow and others. As a consequence of these studies, recommendations for dust standards were laid down. Despite considerable improvement in dust levels in the cotton mills, a more recent prospective study[79] continues to show a significant problem with a total prevalence of byssinosis of 26·9 per cent of the workforce and the continuing development of new cases of byssinosis.


## Clinical features

The diagnosis of byssinosis is based on the presence of a work-related pattern of respiratory symptoms. These symptoms occur on the first day back after a break from work and either disappear or improve over subsequent days of the working week and over weekends. The cardinal symptoms of byssinosis, which are usually preceded for a period of weeks or months by exertional dyspnoea alone, are chest tightness, shortness of breath, cough and wheezing. These symptoms are always worst on the first working day, developing between 1–4 hours into the working shift, resolving slowly after ceasing work and occasionally recurring in the early hours of the morning, interrupting sleep. There is, with continued exposure to cotton dust, progression of the frequency of symptoms into the working week. In advanced disease symptoms are present on each working day and over weekends. This pattern of symptoms led Schilling[80] to recommend a clinical grading for byssinosis which has become the standard classification for epidemiological surveys:

*Grade ½:* occasional chest tightness on the first day of the working week.

*Grade 1:* chest tightness and/or shortness of breath on the first work day of the week only.

*Grade 2:* chest tightness and/or shortness of breath on the first and other days of the working week.

*Grade 3:* grade 2 symptoms accompanied by evidence of permanent respiratory impairment.

Later Bouhuys[81] proposed a functional grading system using baseline $FEV_1$ and changes in flow rate as the major criteria. This system has considerable practical limitations when applied to large epidemiological studies and consequently has not been widely used.

There is a latent period of a number of years between the commencement of employment in a cotton mill and the development of respiratory symptoms. The known factors determining the length of this period include intensity of exposure to cotton dust (strippers and grinders are most commonly affected) and the type of cotton — coarse or medium — being processed. Symptoms of byssinosis may occur within 4 years in 'coarse' mills and between 5 and 10 years in 'medium' mills.[79]

The physical and radiological examinations of individual workers with byssinosis initially reveal no specific abnormalities. When respiratory impairment develops, signs of airway obstruction are found on examination and the chest radiograph may show evidence of over-inflation of the lungs. There are no radiological changes specific to byssinosis. The acute symptoms of byssinosis are accompanied by changes in lung function indicating reversible airway obstruction. The measurement of lung function most commonly used is the $FEV_1$, but changes in $FEV_1$ across the working shift are frequently only small, 0·2–0·4 l, and measurement of maximum expiratory flow rates on flow volume curves has proved more sensitive.[82] Lopez-Merino et al.,[83] studying hemp workers, found decreases in arterial oxygen tension associated with often only small changes in $FEV_1$, suggesting the development of ventilation/perfusion abnormalities during acute responses to hemp dust. The physiological changes of grade ½ and grade 1 byssinosis are said to be reversible on ceasing exposure to cotton dust except for some workers with grade 2 disease who may remain with permanent respiratory impairment. However, there is some evidence of a chronic effect of dust on the lung. The mean annual decline in $FEV_1$ amongst cotton workers in comparison to workers handling man-made fibres is significantly greater: 54 ml per year in a cotton mill compared to 32 ml per year in man-made fibre mills.[84] Of particular importance are the findings of Bouhuys and Zuskin[85] in their follow-up study of hemp workers. They showed a more rapid decline in lung function amongst the hemp workers than amongst control subjects, even though the majority of hemp workers had ceased exposure to dust. This evidence of progression of disease, in a heavily exposed group of workers, after leaving the industry has important implications for the individual developing byssinosis. In order to prevent serious respiratory disability later in life he will need to leave the industry early in the course of his disease.

Pathological changes occurring in the lung in byssinosis are again not specific to this disease and are indistinguishable from those of chronic obstructive bronchitis. Edwards,[86] in a pathological study of 43 cases of byssinosis, described mucous gland hyperplasia and smooth muscle hypertrophy in large airways, and variable centrilobular emphysema; panacinar emphysema was uncommon (14 per cent). In their series there was no significant difference in ventricular weights between cases and controls. Unfortunately they did not relate smoking habits to lung pathology so the presence of emphysema is difficult to interpret. The fact that this type of pathological change is not a dominant feature in byssinosis is supported by the physiological study of Zuskin et al.[87] who found normal diffusing capacities amongst the group of workers they were studying. The conclusion that an individual initially develops byssinosis grade ½ and then with continued exposure to cotton dust progresses through the grades until he reaches 3, is an assumption based only on the histories of the workers and has not yet been investigated in long term prospective studies.

## Prevalence

Prevalence studies on byssinosis have shown the disease to be present in all countries with textile industries. Prevalence rates for byssinosis are determined by the quantity of respirable dust, the quality of this dust and the length of exposure of the individual, byssinosis being most prevalent amongst workers in the blowing room and card room and least amongst spinners. Molyneux and Tombleson,[79] in a survey of 1359 cotton workers in the Lancashire mills, found prevalence rates for byssinosis of 26·9 per cent in the total population and a prevalence of 58·2 per cent amongst card attendants. The rates were higher in workers exposed to coarse as opposed to medium cotton dust. Studies in other countries, including the United States and the Netherlands, have shown prevalence rates for byssinosis amongst card room workers of between 20 per cent and 40 per cent.

Most prevalence studies of byssinosis have been cross-sectional in design, representing studies of survivor populations and they thus probably underestimate the true prevalence of the disease. Community-based studies of both active and retired textile workers[88,89] have shown high prevalences of chronic respiratory symptoms and loss of lung function amongst these workers when compared to control subjects.


## Pathogenesis

The mechanisms by which textile dusts cause byssinosis are not fully understood, but certain facts are now clearly established. An aqueous extract of cotton dust, or cotton dust itself, inhaled by volunteers or by byssinotic subjects, produces symptoms and physiological changes characteristic of byssinosis.[90,91] These responses demonstrate marked tachyphylaxis, that is, those who react develop increasing tolerance to repeated inhalations. An aqueous extract of cotton dust also releases histamine from sliced human lung[92] and the histamine metabolite l-methyl-4-imidazole acetic acid is increased in the urine of volunteers following the inhalation of cotton dust extract.[93] The bronchoconstrictor substance or substances originate from the cotton bracts, which are the leaves around the stem of the boll. One, or what may be a number, of airway constrictor substances, have been shown to be of small molecular weight, less than 1000 daltons, and highly water-soluble.[94] It may be concluded from these experiments that cotton dust contains an agent capable of causing release of histamine in both workers and volunteers. The acute, as opposed to chronic, responses to textile dusts in man may be explained on this basis. This mechanism probably also explains the pattern of pulmonary response to textile dusts seen in individuals with bronchial hyper-reactivity or constitutional asthma developed either before, or while working in, the mills.

Such workers rapidly develop symptoms of airways obstruction on entering an environment containing textile dust and experience symptoms of increasing severity through the working week frequently leading to their early retirement from the industry.

Cotton dust is frequently heavily contaminated by bacteria and fungi, and the inhalation of endotoxin, arising from these bacteria, has been considered as a cause of byssinosis.[95] A number of epidemiological studies have demonstrated a correlation between exposure to Gram-negative bacteria and symptoms of byssinosis.[96,97] The inhalation of purified endotoxin causes a decrease in $FEV_1$ in normal subjects[98] and the highest airborne concentrations of endotoxin are found in the carding areas. However, if endotoxin were the cause of byssinosis, it would be difficult to explain how the same substance should cause two different syndromes — mill fever and byssinosis — in the same workforce.

The remaining mechanism to be considered in the pathogenesis of byssinosis is that it is mediated by an antigen–antibody reaction. There is now little evidence to support this hypothesis. The pattern of symptoms, with improvement on continued exposure to the allergen, is one not usually associated with occupational asthma. Taylor et al.[99] claimed to have identified the antigen as a condensed polyphenol based on leucocyanidin. This antigen produced the subjective sensation of byssinosis but no physiological changes. The serum of textile workers contains a number of different antibodies but none are specifically associated with the disease byssinosis.[100] There is, at present, no evidence that an antigen–antibody reaction is involved in the pathogenesis of byssinosis. The cause of the progressive airways obstruction associated with byssinosis remains unexplained. It is probable that the non-antigenic release of histamine is responsible for the acute symptoms occurring on exposure to textile dusts.

**Prevention and treatment**

The prevention of byssinosis should be primarily based on the reduction of the inhalable fraction of the cotton dust. This may be achieved by enclosure of the process, exhaust ventilation or by protection of the individual. However, even when low levels of inhalable dust are achieved those workers who have already developed byssinosis may continue to experience symptoms and further cases of byssinosis may occur.[101,102]

A different approach to the problem has been the treatment of the cotton itself before it enters the mill. Initial trials of various methods of cleaning cotton[102] suggested that steaming of the cotton effectively removed most of the biological activity without interfering with the cotton's spinning qualities. The subsequent intervention study of cotton steaming in a coarse cotton mill showed that while dust levels were reduced and respiratory flow rates improved in the early processes of opening and picking, in the later mill processes of spinning and winding the dust levels and the prevalence of byssinotic symptoms were increased.[103] A different type of intervention involving the early picking of cotton while green, before the bracts become friable and brittle, is at present undergoing study.

Finally, in those individuals who, usually for economic reasons, continue in the textile industry despite symptoms, medical treatment may be used to alleviate symptoms. Various studies have shown the prevention of the acute responses to textile dust by the administration of antihistamine,[104] $\beta$-adrenergic drugs, beclamethasone and disodium chromoglycate[105] before exposure to the dust.

## Conclusions

Despite the recognition of byssinosis as an occupational respiratory disease for approximately 200 years, and considerable improvement in the working conditions in the textile industry, an unacceptably high proportion of the workforce continues to experience symptoms. In many mills the current TLV for fine cotton dust, $0.5$ mg/m$^3$, is still exceeded and in the current economic climate it is unlikely that the necessary finance will be available to make the further improvement required. The previous heavy exposures to cotton dust are probably largely responsible for those individuals still developing byssinosis, most of whom will have worked many years in the industry. It requires careful prospective studies to evaluate whether the current recommended dust levels are low enough to prevent individuals entering the industry now from developing byssinosis in 10–20 years time.

## EXTRINSIC ALLERGIC BRONCHIOLO-ALVEOLITIS

### Introduction

The inhalation of organic dusts may produce, usually after prolonged or intense exposure, lung disease involving the lung interstitium, alveoli and bronchioles. The generic term applied to this group of diseases is 'extrinsic allergic bronchiolo-alveolitis' (EAB) or, as favoured in the United States, 'hypersensitivity pneumonitis'.

### Causative agents

An extensive literature is now established of occupational and avocational causes of EAB (*Table 10.2*). The antigens are variously derived from micro-organisms (fungi, actinomycetes, viruses and bacteria), avian proteins, small molecular weight chemicals and drugs. The antigenic exposure to many of these agents is characteristically heavy. The airborne spore concentration when moving mouldy hay has been measured at $1583 \times 10^6$ spores per cubic metre of air. It is estimated that a farmer working in such conditions may retain in his lungs 750 000 actinomycetes spores per minute. The spore diameter of actinomycetes is small ($< 1 \cdot 0$ $\mu$m per diameter) and the spores readily penetrate to alveolar level in the lungs. Studies attempting to characterize the antigens of actinomycetes have shown a complex picture of multiple antigens with varying biological activities. In a recent study of avian antigens,[106] four major antigens were identified by immunoelectrophoresis, including two glycoproteins with molecular weights suggesting that one was pigeon IgA and the other an Fab or Fc breakdown product of pigeon IgA. The full characterization of these antigens is of considerable importance since this may then lead to the development of specific diagnostic tests for each disease.

**Table 10.2.** Occupational allergic bronchiolo-alveolitis

| Disease/occupation | Exposure | Specific agent |
| --- | --- | --- |
| Bagassosis | Mouldy sugar cane | *Thermoactinomyces sacharii* |
| Birdfancier's disease | Avian dust | Avian proteins |
| Cheese washer's disease | Cheese mould | *Penicillium casei* |
| Farmer's lung | Mouldy hay | (*Micropolyspora faeni Thermoactinomyces vulgaris*) |
| Humidifier lung | Fungal spores | Thermophilic actinomycetes |
| Laboratory technicians | Animal experiments | Rat serum |
| | Biochemistry laboratory | Pauli's reagent |
| Malt worker's lung | Mouldy barley | *Aspergillus clavatus* |
| Maple bark stripper's disease | Mouldy maple logs | *Cryptostroma corticale* |
| Mushroom worker's lung | Mouldy compost | Thermophilic actinomycetes |
| Sequoiosis | Mouldy redwood dust | Craphium *Aureobasidium pullulans* |
| Suberosis | Mouldy cork | *Penicillium frequentans* |
| Woodtrimmer's disease | Mouldy logs | Fungal spores |
| Woodworker | Wood dust | Ramin |

## Prevalence

Despite the large number of occupational causes of EAB, relatively few of these affect industry in the United Kingdom and prevalence studies are restricted to the occupational groups of farmers, malt workers and bird breeders. The prevalence of farmer's lung varies with the local geographical and atmospheric conditions, ranging from 8 per cent of farmers in Orkney and Ayrshire[107] to 2 per cent of farmers in Somerset.[108] In a study of Scottish malt workers,[109] 5·2 per cent of the workforce were found to have symptoms of EAB. Avocational exposure to budgerigars, in terms of numbers of affected individuals, is associated with more disease, with 3·4 per cent of an exposed population having avian-related symptoms.[110] In the more heavily exposed pigeon breeders, disease is more frequent; Elgefors et al.[111] found 8 per cent of 180 pigeon breeders and Caldwell[112] found 6 per cent of 150 breeders with the disease. Recent interest in North America has been focused on workers exposed to actinomycetes colonizing humidifiers (humidifier lung). In this situation up to 15 per cent of the workforce may develop alveolitis.

## Pathology

The histopathological changes associated with the disease have now been well studied. In the acute form of the disease there is a cellular infiltration of the alveoli and bronchioles. The infiltrate characteristically consists of lymphocytes and plasma cells with large numbers of activated alveolar macrophages. Non-caseating granulomas and multinucleate giant cells, with clefts containing refractile material probably vegetable in origin, are frequently a prominent feature. In chronic disease the cellular infiltration and granulomas are replaced by fibrotic changes with destruction of lung architecture. Evidence of a pulmonary vasculitis is not usually found except in instances where lung biopsies have been performed shortly after exposure to the agent.[113]

## Immunopathogenesis

Some of the most exciting recent advances in the understanding of extrinsic allergic bronchiolo-alveolitis have been in the field of immunopathology. It has been traditional dogma that the disease results from a type III immune complex reaction occurring in the lung. The evidence for this is based on the presence of specific precipitating antibodies; a late skin reaction which, with some antigens, has shown histological changes of a necrotizing vasculitis; a pulmonary response on clinical exposure or provocation testing which has a similar time sequence to the skin reaction; and finally the lung biopsy finding of a pulmonary vasculitis with specific antibody and complement. There is, however, increasing evidence that other immunological mechanisms are involved. Both organic and inert dusts have been shown to activate the alternate pathway of complement[114] in animals, producing a picture indistinguishable from that of an allergic bronchiolo-alveolitis.

The histopathological findings of non-caseating granulomas of the lung initiated a number of studies to examine the possible role of type IV reactions in the pathogenesis of the disease. Studies in pigeon breeders have demonstrated both lymphocyte transformation and the production of macrophage inhibition factor on exposure to avian antigens in a high percentage of symptomatic breeders and in a smaller percentage of asymptomatic breeders.[115]

Bice et al.,[116] in experimental work on rabbits, showed that the transfer of lymph node cells, but not of serum, from sensitized to previously unexposed animals followed by the aerosol challenge of the recipient rabbit produced typical pulmonary lesions in that animal. Further evidence supporting a role for both type IV and type III hypersensitivity reactions in the pathogenesis of this disease comes from bronchial lavage experiments in men with extrinsic bronchiolo-alveolitis.[117] Reynolds found increased numbers of lymphocytes in the lavage fluid, with the majority being T lymphocytes, contrasting with a normal distribution of T and B lymphocytes in the peripheral blood. In addition, elevated levels of respiratory IgG and IgM were present. In two of the subjects bronchiolo-alveolar IgG precipitins to thermophilic actinomycete antigens were demonstrated.

In the past there has been little evidence for the participation of IgE-mediated reactions in this disease; the total IgE and blood eosinophil counts are usually normal. However, the late type skin reactions, seen to certain antigens, are always preceded by an immediate skin response, and both the lavage study[117] and the lung biopsy study[113] have shown a local accumulation of eosinophils in the lung. It is possible that a localized allergic reaction may be taking place. There is additional evidence from Schuyler[118] that the immunological response may be localized to the lung. They demonstrated lymphokine production from bronchiolo-alveolar cells but not from the peripheral blood lymphocytes, taken concurrently from individuals with pigeon breeder's disease. Current evidence suggests that the development of extrinsic allergic bronchiolo-alveolitis results from a number of different immunological and possibly non-immunological responses.

## Clinical features

The clinical presentation is determined to some extent by the type of exposure to the antigen. The acute variety usually results from intensive intermittent exposures, symptoms occurring about 6 hours following such an exposure and lasting for 24–48 hours. The reaction is characterized by 'flu-like symptoms of fever, malaise and myalgia with cough and breathlessness. Physical signs are localized to the respiratory tract, with fine inspiratory crackles heard on auscultation of the lung. The chest radiograph shows widespread, fine nodular shadowing predominantly affecting the middle and lower zones. The physiological changes occurring in the lungs are those of a restrictive ventilatory defect with impairment of carbon monoxide gas transfer. In chronic forms of the disease, as is seen in budgerigar fanciers, exposure to the antigen is usually regular and lower in intensity. This form of disease tends to present with exertional dyspnoea and evidence of advanced pulmonary fibrosis. The physiological changes are those of both a restrictive and obstructive ventilatory defect, and the chest radiograph shows evidence of fibrosis affecting mainly the upper lobes.

### Diagnosis and management

The diagnosis of extrinsic allergic bronchiolo-alveolitis is primarily based on a careful occupational history combined with appropriate clinical symptoms and the detection of precipitating antibodies to the implicated agent. In view of the tissue-damaging nature of the pulmonary response the early diagnosis and protection of the individual are of paramount importance. This protection may take the form of change of job, or of work practice, or the wearing of appropriate respiratory protection. The administration of oral steroids may be useful in accelerating the resolution of acute disease.

## HUMIDIFIER FEVER

### Introduction

In 1959 Pestalozzi[119] described an outbreak of respiratory symptoms in a group of woodworkers resulting from air humidifiers contaminated by moulds. Subsequently a variety of respiratory diseases have been described resulting from the inhalation of water contaminated by micro-organisms. The types of respiratory diseases described include bronchial asthma,[120] extrinsic allergic bronchiolo-alveolitis,[121] humidifier fever,[122] Legionnaire's disease[123] and Pontiac fever.[124] The aerosolization of water occurs in both domestic and industrial situations and may arise from a variety of different sources including central humidification systems,[121,125] portable and coolmist vaporizers,[120,126] vacuum pumps,[127] sauna baths,[128] evaporative condensers,[124] steam turbine condensers[129] and household hot water supplies.[130]

Banaszak et al.[121] described the development of a respiratory disease in a group of office workers due to the contamination, by thermophilic actinomycetes, of the office air conditioning unit. The clinical, physiological, immunological and radiological features were characteristic of extrinsic allergic alveolitis. Subsequent publications have described further outbreaks of disease occurring in both work and home environments.

In 1976 Pickering[125] identified a respiratory disease, similar to that described by Pestalozzi[119] in a group of printers, again caused by an air conditioning unit. This disease has some features of extrinsic allergic bronchiolo-alveolitis but differs in other important aspects and has been called 'humidifier fever'.

## Clinical features

The symptoms of humidifier fever vary from a mild, afebrile illness with headache, malaise and joint pains to an acute illness with a high fever, cough and breathlessness. These symptoms occur on the first day back at work after a weekend or on returning from holiday, developing 4–8 hours after the start of the working shift and resolving over 12–24 hours. The level of symptoms and the physiological changes improve through the working week and by Friday are absent. In the milder forms of the disease physical signs and respiratory physiological changes are not found. In acute disease a high fever and basal inspiratory crepitations are usually present. Finger clubbing has not been described in association with humidifier fever. There are no specific radiological features even at the height of the illness. In the study described by Friend et al.[127] when some individuals were hypoxic with diffusing capacities reduced by more than 50 per cent, chest X-rays remained normal. The respiratory physiological changes in acute disease are those of a restrictive ventilatory defect with a reduced diffusing capacity. A polymorphonuclear leucocytosis is commonly present at the height of the reaction.

## Prevalence

Although a number of outbreaks of humidifier fever have been described in industry (*Table 10.3*), no systematic examination of a

**Table 10.3.** Reported outbreaks of humidifier fever

| Reports | Exposures | Proportion of workforce affected |
| --- | --- | --- |
| Pestalozzi[119] | Joinery | 12/17 |
| Pickering[125] | Printing | 9/350 |
| Edwards[131] | Rayon manufacture | 20/50 |
| Friend[127] | Envelope manufacture | 24/560 |
| Newman Taylor[132] | Printing | 3/32 |
| Campbell[133] | Operating theatre | 11/60 |
| Ganier[134] | Office building | 26/50 |
| Parrote[135] | Printing | 7/26 |
| Horsefield[136] | Printing | 12/500 |

workforce exposed to humidified air has yet been carried out. The number of men affected has varied between 4 per cent and 52 per cent of workforces. In a number of outbreaks symptoms have been very mild and it is possible that this problem is more widespread than is at present suggested by the small number of reports.


## Pathogenesis

A common feature of all outbreaks of humidifier fever is the wide range of micro-organisms isolated from both the factory environment and the humidifiers themselves. These have included numerous bacteria and fungi, including thermophilic actinomycetes and protozoa. A number of publications have implicated specific organisms as causes of the disease, amongst which are acanthamoeba,[122] pullularia,[128] *B. subtilis*[135] and flavobacterium,[137] their conclusions being based on serological testing and not on bronchial provocation studies. The fact that the water contains the agent causing the disease has been confirmed by provocation testing,[127,134] but similar studies using individual organisms suggested by serological tests[125] have failed to identify the responsible organism. The universal finding of heavy precipitating antibodies to extracts from the humidifiers in symptomatic workers and the disease's similarity to an alveolitis suggest an immune-complex-mediated disease. The lack of radiological changes or persisting physiological abnormalities following prolonged exposures, usually features of allergic alveolitis, may simply result from the form in which the antigen is presented to the respiratory tract, in this instance being soluble, rather than particulate, as is the case in extrinsic allergic alveolitis.

An alternative theory has been proposed by Rylander and his colleagues.[137] They identified raised levels of bacterial endotoxin in the humidifier water and suggested that the symptoms resulted from the activation of complement, by endotoxin, via the alternative pathway. Since tolerance to endotoxin occurs with continuous exposure and is rapidly lost when exposure ceases, this theory is compatible with the clinical pattern of disease. However, efforts to demonstrate complement activation have so far been unsuccessful.[134]

The mechanism responsible for humidifier fever remains a matter for debate.


## Management

In most outbreaks attempts at keeping the water free of humidifier antigens have proved difficult and either changes in the humidification system itself or the total removal of the humidification unit have been necessary. The frequent cleaning of the units combined with the introduction of a biocide has met with variable success. Pickering[125] found this method ineffective, but more recently after a further outbreak of the disease in a different printing works the regular monthly cleaning of the humidifiers appears to have led to the virtual disappearance of symptoms amongst the workforce. The introduction of steam in place of water is an effective but expensive solution, unless a source of steam is available on the site. In small humidification units the addition of hydrogen peroxide is described as effective in controlling the growth of micro-organisms.[138]

The design of water spray humidification systems is such that contamination by micro-organisms is inevitable and indeed may be enhanced by the provision of nutrition in the re-circulated air, for example cellulose particles in printing works. In many of the outbreaks described in the literature the humidification systems are old, heavily contaminated with micro-organisms and extremely difficult to clean effectively. It may be that regular cleaning, using steam, from the time the humidifier units are fitted will reduce contamination by micro-organisms and prevent humidifier fever. The design of humidifiers should allow for easy access and for the removal, for cleaning or replacement, of the baffle plates.


## Air conditioning and *Legionella pneumophila*

The outbreak of Legionnaire's disease in Philadelphia in 1976 and the identification of the causative organism *Legionella pneumophila* led to a renewed interest in previously unexplained febrile respiratory infections. Subsequently two distinct syndromes caused by *L. pneumophila* have been identified. First, *Legionnaire's disease,* which is a multisystem disease involving the respiratory and gastrointestinal tracts and the renal and central nervous systems. The attack rate is between 1 and 5 per cent of people exposed, the incubation period averages 5–6 days and the mortality rate is approximately 15 per cent. The second syndrome has been named *Pontiac fever*[124] or more recently 'non-pneumonic, short incubation period Legionellosis'.[123] This is a benign, self-limiting disease characterized by fever, headache and myalgia; there is no evidence of pneumonia. The attack rate is high, over 90 per cent of exposed people, and the incubation period short, 24–48 hours. One of the modes of transmission appears to be by exposure to aerosols of water arising from air conditioning cooling towers, evaporative condensers or showers. The variation in disease produced by the same organism remains unexplained. While there are similarities between Pontiac and humidifier fever, they differ in important aspects. In humidifier fever the attack rate is lower, there may be marked lower respiratory tract symptoms and physiological changes and the antibody titres to *L. pneumophila* are not elevated.

At the present time there have been no cases of Legionnaire's disease arising from cold water spray types of humidification. *L. pneumophila* grows best at 37°C and it is probable that the water in water spray humidification systems is too cold to promote growth of this organism.

REFERENCES

1. Nickol KH. Some occupational respiratory disorders. In: Ward Gardner A (ed.) *Current Approaches to Occupational Medicine*. Bristol: Wright, 1979: 101–23.
2. Department of Employment and Productivity/HM Factory Inspectorate. *Problems Arising from the Use of Asbestos*. London: HMSO, 1968.
3. McVittie JC. Asbestosis in Great Britain. *Ann. NY Acad. Sci.* 1965; **132**:128.
4. Montague Murray H. Report of the Departmental Committee on Compensation for Industrial Diseases: minutes of evidence, appendices and index. London: HMSO, 1907: 127–8.
5. Cooke WE. Fibrosis of the lungs due to the inhalation of asbestos dust. *Br. Med. J.* 1924; **2**:147.
6. Seiler HE. A case of pneumoconiosis: result of the inhalation of asbestos dust. *Br. Med. J.* 1928; **2**:982.
7. Burton Wood W. Pulmonary asbestosis: radiographic appearances in skiagrams of the chests of workers in asbestos. *Tubercle* 1929; **10**:353.
8. Merewether ERA and Price CW. *Report on Effects of Asbestos Dust on the Lungs and Dust Suppression in the Asbestos Industry*. London: HMSO, 1930.
9. British Occupational Hygiene Society. Hygiene standards for chrysotile asbestos dust. *Ann. Occup. Hyg.* 1968; **11**:47.
10. Rossiter CE, Bristol LJ, Cartier PH et al. Radiographic changes in chrysotile asbestos mine and mill workers in Quebec. *Arch. Environ. Hlth* **24**, 388.
11. Turner-Warwick M and Parkes WR. Circulating rheumatoid and anti-nuclear factors in asbestos workers. *Br. Med. J.* 1970; **3**:492.
12. Turner-Warwick M. HLA phenotypes in asbestos workers. *Br. J. Dis. Chest* 1979; **73**:243.
13. Kagan E, Webster I, Cochrane JC et al. In: Walton WH (ed.) *The Immunology of Asbestosis in Inhaled Particles IV*. Oxford: Pergamon Press, 1977: 429–33.
14. Haslam PL, Lukosmek A, Merchant JA et al. Lymphocyte responses to phytohaemagglutinin in patients with asbestosis and pleural mesothelioma. *Clin. Exp. Immunol.* 1978; **31**:178.
15. Pierce R and Turner-Warwick M. Skin tests with tuberculin (PPD), *Candida albicans* and Trichophyton spp. in cryptogenic fibrosing alveolitis and asbestos related lung diseases. *Clin. Allergy* 1980; **10**:229.
16. Berry G, Gilson JC, Holmes S et al. Asbestosis: a study of dose–response relationships in an asbestos textile factory. *Br. J. Ind. Med.* 1979; **36**:98.
17. McMillan GHG, Pethybridge RJ and Sheers G. Effects of smoking on attack rates of pulmonary and pleural lesions related to exposure to asbestos dust. *Br. J. Ind. Med.* 1980; **37**:268.
18. Liddell FDK and McDonald JC. Radiological findings as predictors of mortality in Quebec asbestos workers. *Br. J. Ind. Med.* 1980; **37**:257.
19. Weiss W. Cigarette smoking: asbestos and pulmonary fibrosis. *Am. Rev. Respir. Dis.* 1971; **104**:223.

20. Samet JR, Epler GR, Gaensler EA et al. Absence of synergism between exposure to asbestos and cigarette smoking in asbestosis. *Am. Rev. Respir. Dis.* 1979; **120**:75.
21. Lynch KM and Smith WA. Pulmonary asbestosis III: carcinoma of lung in asbesto-silicosis. *Am. J. Cancer* 1935; **24**:56.
22. Doll R. Mortality from lung cancer in asbestos workers. *Br. J. Ind. Med.* 1955; **12**:81.
23. *Asbestos:* Final Report of the Advisory Committee, vol. 2. London: HMSO, 1979.
24. Elmes PC and Simpson MJC. Insulation workers in Belfast 3: mortality 1940–66. *Br. J. Ind. Med.* 1971; **28**:226.
25. Knox JF, Holmes S, Doll R et al. Mortality from lung cancer and other causes among workers in an asbestos textile factory. *Br. J. Ind. Med.* 1968; **25**:293.
26. Lumley KPS. A proportional study of cancer registrations of dockyard workers. *Br. J. Ind. Med.* 1976; **33**:108.
27. Edge JR. Incidence of bronchial carcinoma in shipyard workers with pleural plaques. *Ann. NY Acad. Sci.* 1979; **330**:289.
28. Turner-Warwick M, Lebowitz M, Burrows B et al. Cryptogenic fibrosing alveolitis and lung cancer. *Thorax* 1980; **34**:496.
29. Wagner JC, Berry G, Skidmore JW et al. The effects of the inhalation of asbestos in rats. *Br. J. Cancer* 1974; **29**:252.
30. Hueper WC. *Occupational and Environmental Cancer of the Respiratory Tract: Recent Results in Cancer Research, 3.* Berlin: Springer Verlag, 1966.
31. Kannerstein M and Churg J. Pathology of carcinoma of the lung associated with asbestos exposure. *Cancer* 1972; **30**:14.
32. Whitwell F, Newhouse ML and Bennett DR. A study of the histological cell types of lung cancer in workers suffering from asbestosis in the United Kingdom. *Br. J. Ind. Med.* 1974; **31**:298.
33. Coutts II, Gilson JC, Kerr IH et al. Mortality in asbestosis in relation to initial radiographic appearance. *Thorax* 1980; **35**:235.
34. Wright PH, Heard BE, Steel SJ et al. Cryptogenic fibrosing alveolitis: assessment by graded trephine biopsy compared with clinical radiographic and physiological features. *Br. J. Dis. Chest* 1981; **75**:61.
35. Selikoff IJ, Hammond EC and Churg J. Asbestos exposure, smoking and neoplasia. *JAMA* 1968; **204**:104.
36. McDonald JC, Liddell FDK, Gibbs GW et al. Dust exposure and mortality in chrysotile mining 1910–75. *Br. J. Ind. Med.* 1980; **37**:11.
37. Hammond EC, Selikoff IJ and Seidman H. Asbestos exposure, cigarette smoking and death rates. *Ann. NY Acad. Sci.* 1979; **330, 473**.
38. Saracci R. Asbestos and lung cancer: an analysis of the epidemiological evidence on the asbestos smoking interaction. *Int. J. Cancer* 1977; **20**:323.
39. Enterline PE. Attributability in the face of uncertainty. *Chest* 1980; **78**: Suppl., p. 377.
40. Wagner JC, Sleggs CA and Marchand P. Diffuse pleural mesothelioma and asbestos exposure in the North Western Cape Province. *Br. J. Ind. Med.* 1960; **17**:260.
41. Meurman LO, Kivilnoto R and Hakama M. Mortality and morbidity among the working population of anthophyllite asbestos miners in Finland. *Br. J. Ind. Med.* 1974; **31**:105.
42. Yamicioglu S, Ilcayto R, Balci K et al. Pleural calcification, pleural mesotheliomas, and bronchial cancers caused by tremolite dust. *Thorax* 1980; **35**:564.
43. Seidman MBA, Selikoff IJ and Hammond EC. Short-term asbestos work exposure and long-term observation. *Ann. NY Acad. Sci.* 1979; **330**:61.
44. Harington JS, Gilson JC and Wagner JC. Asbestos and mesothelioma in man. *Nature* 1971; **232**:54.
45. Jones JSP, Pooley FD and Smith PG. Factory populations exposed to crocidolite asbestos — a continuing survey. *IARC Sci. Pub.* 1976; **52**:117.

46. McDonald AD and McDonald JC. Mesothelioma after crocidolite exposure during gas mask manufacture. *Environ. Res.* 1978; **17**:340.
47. Whitwell F, Scott J and Grimshaw M. Relationship between occupations and asbestos fibre content of the lungs in patients with pleural mesothelioma, lung cancer and other diseases. *Thorax* 1977; **32**:377.
48. Sheers G and Coles RM. Mesothelioma risks in a naval dockyard. *Arch. Environ. Health* 1980; **35**:276.
49. Newhouse ML and Berry G. Predictions of mortality from mesothelial tumours in asbestos factory workers. *Br. J. Ind. Med.* 1976; **33**:147.
50. Elmes PC and Simpson MJC. The clinical aspects of mesothelioma. *Q. J. Med.* 1976; **45**:427.
51. Stanton MF and Wrench C. Mechanisms of mesothelioma induction with asbestos and fibrous glass. *J. Natl Cancer Inst.* 1972; **48**:797.
52. Wagner JC, Berry G and Timbrell V. Mesotheliomata in rats after inoculation with asbestos and other materials. *Br. J. Cancer* 1973; **28**:173.
53. Baris YI, Sabin AA, Onzesimi M et al. An outbreak of pleural mesothelioma and chronic fibrosing pleurisy in the village of Karain/Urgup in Anatolia. *Thorax* 1978; **33**:181.
54. Artvinli M and Baris YI. Malignant mesotheliomas in a small village in the Anatolian region of Turkey: an epidemiologic study. *J. Natl Cancer Inst.* 1979; **63**:17.
55. Keskinen H, Alanko K and Saarinen L. Occupational asthma in Finland. *Clin. Allergy* 1978; **8**:569–80.
56. Burge PS, Perks W, O'Brien IM et al. Occupational asthma in an electronics factory. *Thorax* 1979; **34**:13–18.
57. Perks WH, Burge PS, Rehahn M et al. Work related respiratory disease in employees leaving an electronics factory. *Thorax* 1979; **34**:19–22.
58. Mitchell CA and Gandevia B. Respiratory symptoms and skin reactivity in workers exposed to proteolytic enzymes in the detergent industry. *Am. Rev. Respir. Dis.* 1971; **104**:1–12.
59. Newman Taylor AJ, Myers JR, Longbottom JL et al. Immunological differences between asthma and other allergic reactions in laboratory animal workers. *Thorax* 1981; **36**, 229.
60. Gross NJ. Allergy to laboratory animals: epidemiologic, clinical and physiologic aspects, and a trial of cromolyn in its management. *J. Allergy Clin. Immunol.* 1980; **66**:158–65.
61. Juniper CP, How MS, Goodwin BFS et al. *Bacillus subtilis* enzymes: a 7-year clinical, epidemiological and immunological study of an industrial allergen. *J. Soc. Occup. Med.* 1977; **27**:3–12.
62. Dally MB, Hunter JV, Hughes EG et al. Hypersensitivity to platinum salts. *Am. Rev. Respir. Dis.* 1980; **121**:230.
63. Pepys J, Wells ID, D'Souza MF et al. Clinical and immunological responses to enzymes of *Bacillus subtilis* in factory workers and consumers. *Clin. Allergy* 1973; **3**:143–60.
64. Newman Taylor AJ, Longbottom JL and Pepys J. Respiratory allergy tourine proteins of rats and mice. *Lancet* 1977; **2**:847–9.
65. Bjorksten F, Backman A, Jarvinen KAS et al. Immunoglobulin E specific to wheat and rye flour proteins. *Clin. Allergy* 1977; **7**:473–83.
66. Cromwell O, Pepys J, Parish WE et al. Specific IgE antibodies to platinum salts in sensitised workers. *Clin. Allergy* 1979; **9**:109–18.
67. Zeiss CR, Patterson R, Pruzansky JJ et al. Trimellitic anhydride-induced airway syndromes: clinical and immunologic studies. *J. Allergy Clin. Immunol.* 1977; **60**:96–103.
68. Ahmad D, Morgan WKC, Patterson R et al. Pulmonary haemorrhage and haemolytic anaemia due to trimellitic anhydride. *Lancet* 1979; **2**:328–30.

69. Patterson R, Addington W, Banner AS et al. Antihapten antibodies in workers exposed to trimellitic anhydride fumes: a potential immunopathogenetic mechanism for the trimellitic anhydride pulmonary disease — anaemia syndrome. *Am. Rev. Respir. Dis.* 1979; **120**:1259-67.
70. Butcher BT, O'Neill CE, Reed MA et al. Radioallergosorbent testing of toluene di-isocyanate-reactive individuals using p-tolyl isocyanate antigen. *J. Allergy Clin. Immunol.* 1980; **66**:213-16.
71. Davies RJ, Butcher BT, O'Neill CE et al. The *in vitro* effect of toluene di-isocyanate on lymphocyte cyclic adenosine monophosphate production by isoproteronol prostaglandin and histamine. A possible mode of action. *J. Allergy Clin. Immunol.* 1977; **60**:233.
72. Burge PS, O'Brien IM and Harries MG. Peak flow rates in the diagnosis of occupational asthma due to colophony. *Thorax* 1979; **34**:308-16.
73. Burge PS, O'Brien IM and Harries MG. Peak flow rates in the diagnosis of occupational asthma due to isocyanates. *Thorax* 1979; **34**:317-23.
74. Chan-Yeung M. Fate of occupational asthma. A follow-up study of patients with occupational asthma due to Western Red cedar *(Thuja plicata). Am. Rev. Respir. Dis.* 1977; **116**:1023-9.
75. Adams WG. Long term effects on the health of men engaged in the manufacture of toluene di-isocyanate. *Br. J. Ind. Med.* 1975; **32**:72-8.
76. Jackson J. On the influence of the cotton manufactories on the health. *Lond. Med. Phys. J.* 1818; **39**:464-6.
77. Kay JP. Trades producing phthisis. *North Engl. Med. Surg. J.* 1831; **1**:358-63.
78. Greenhow EH. Report of Medical Officer of the Privy Council, with appendix, 1860. Third report appendix V. London: HMSO, 1861: 172-6.
79. Molyneux MKB and Tombleson JBL. An epidemiological study of respiratory symptoms in Lancashire mills 1963-66. *Br. J. Ind. Med.* 1970; **27**: 225-34.
80. Schilling RSF, Vigliani EC, Lammers B et al. A report on a conference on byssinosis (14th International Conference on Occupational Health, Madrid 1963). *Excerpta Med. Int. Congr. Ser.* 1963; **62**:137-44.
81. Bouhuys A, Barbero A, Lindell SE et al. Byssinosis in hemp workers. *Arch. Environ. Health* 1967; **14**:533-44.
82. Bouhuys A, Mitchell CA, Schilling RSF et al. A physiological study of byssinosis in colonial America. *Trans. NY Acad. Sci.* 1973; **35**:537-46.
83. Lopez Merino V, Llopis Lombart R, Flores Marco R et al. Arterial blood gas tensions and lung function during acute responses to hemp dust. *Am. Rev. Resp. Dis.* 1973; **107**:809-15.
84. Berry G, McKerrow CB, Molyneux MKB et al. A study of the acute and chronic changes in ventilatory capacity of workers in Lancashire cotton mills. *Br. J. Ind. Med.* 1973; **30**:25-36.
85. Bouhuys A and Zuskin E. Chronic respiratory disease in hemp workers. A seven year follow-up study 1967-74. *Ann. Intern. Med.* 1976; **84**:398-405.
86. Edwards C, Macartney J, Rooke G et al. The pathology of the lung in byssinosis. *Thorax* 1975: 612-23.
87. Zuskin E, Valic F, Butkovic D et al. Lung function in textile workers. *Br. J. Ind. Med.* 1975; **32**:283-8.
88. Bouhuys A, Barbero A, Schilling RSF et al. Chronic respiratory disease in hemp workers. *Am. J. Med.* 1969; **46**:526-37.
89. Bouhuys A, Schoenberg JB, Beck GJ et al. Epidemiology of chronic lung disease in a cotton mill community. *Lung* 1977; **154**:167-86.
90. Bouhuys A, Lindell SE and Lundin G. Experimental studies on byssinosis. *Br. Med. J.* 1960; **1**:324-6.
91. Battigelli MC, Fischer JJ and Gamble JF. Controlled human exposure to cotton dust. *Beltwide Cotton Production Research Conference* 1977: 73-5.

92. Bouhuys A and Lindell SE. Release of histamine by cotton dust extracts from human lung tissue *in vitro*. *Experimentia* 1961; **17**:211–15.
93. Edwards J, McCarthy P, McDermott et al. The acute physiological pharmacological and immunological effects of inhaled cotton dust in normal subjects. *Proc. Physiol. Soc.* March 1970: 63–4.
94. Buck MG and Bouhuys A. Constriction of airways by cotton bract extracts. *Beltwide Cotton Production Research Conference* 1980: 31–4.
95. Pernis B, Vigliani EC, Cavagna G et al. The role of bacterial endotoxins in occupational disease caused by inhaling vegetable dusts. *Br. J. Ind. Med.* 1961; **18**:120–9.
96. Cinkotai FF, Lockwood MG and Rylander R. Airborne micro-organisms and prevalence of byssinotic symptoms in cotton mills. *Am. Ind. Hyg. Assoc. J.* 1977; **38**:554–9.
97. Fischer JJ. Relation of airborne bacteria to endotoxin levels in cotton mills. *Proceedings of the Fourth Special Session on Cotton Dust Research, Beltwide Cotton Production Research Conference* 1980: 29–30.
98. Cavagna G, Foa V and Vigliani EG. Effects in man and rabbits of inhalation of cotton dusts or extracts and purified endotoxins. *Br. J. Ind. Med.* 1969; **26**:314–21.
99. Taylor G, Massond AAE and Lucas F. Studies on the aetiology of byssinosis. *Br. J. Ind. Med.* 1971; **28**:143–51.
100. Popa V, Gavrilescu N, Preda N et al. An investigation of allergy in byssinosis; sensitization to cotton, hemp, flax and jute antigens. *Br. J. Ind. Med.* 1969; **26**:101–8.
101. Berry G, Molyneux MKB and Tombleson JBL. Relationships between dust level and byssinosis and bronchitis in Lancashire cotton mills. *Br. J. Ind. Med.* 1974; **31**:18–27.
102. Merchant JA, Lumsden JC, Kilburn KH et al. Pre-processing cotton to prevent byssinosis. *Br. J. Ind. Med.* 1973; **30**:237–47.
103. Merchant JA, Lumsden JC, Kilburn KH et al. Intervention studies of cotton steaming to reduce biological effects of cotton dust. *Br. J. Ind. Med.* 1974; **31**:261–74.
104. Bouhuys A. Prevention of Monday dyspnoea in byssinosis. A controlled trial with an antihistamine drug. *Clin. Pharmacol. Ther.* 1963; **4**:311–14.
105. Fawcett IW, Merchant JA, Simmonds SP et al. The effect of sodium chromoglycate, beclomethasone diprorionate and salbutamol on the ventilatory response to cotton dust in mill workers. *Br. J. Dis. Chest* 1978; **72**:29–38.
106. Fredericks W. Antigens in pigeon dropping extracts. *J. Allergy Clin. Immunol.* 1978; **61**:199.
107. Grant IWB, Blyth W, Dardrop VE et al. Prevalence of farmer's lung in Scotland: a pilot survey. *Br. Med. J.* 1972; **1**:530–4.
108. Pether JVS and Greatorex FB. Farmer's lung in Somerset. *Br. J. Ind. Med.* 1976; **33**:265–8.
109. Grant IWB, Blackadder ES, Greenberg M et al. Extrinsic allergic alveolitis in Scottish maltworkers. *Br. Med. J.* 1976; **1**:490–3.
110. Hendrick DJ, Faux JA and Marshall R. Budgerigar fancier's lung: the commonest variety of allergic alveolitis in Britain. *Br. Med. J.* 1978; **2**:81–4.
111. Elgefors B, Belin L and Hanson LA. Pigeon breeder's lung. Clinical and immunological observations. *Scand. J. Respir. Dis.* 1971; **52**:167–76.
112. Caldwell JR, Pearce DE, Spencer C et al. Immunological mechanisms in hypersensitivity pneumonitis. *J. Allergy Clin. Immunol.* 1973; **52**:225.
113. Ghose T, Landrigan P, Kileen R et al. Immunopathological studies in patients with farmer's lung. *Clin. Allergy* 1974; **4**:119–29.
114. Edwards JH, Baker JT and Davies BH. Precipitin-test-negative farmer's lung activation of the alternative pathway of complement by mouldy hay dusts. *Clin. Allergy* 1974; **4**:379–88.

115. Fink JN, Moore VL and Barboriak JJ. Cell mediated hypersensitivity in pigeon breeders. *Int. Arch. Allergy Appl. Immunol.* 1975; **49**:831–6.
116. Bice D, Salvaggio J and Hoffman E. Passive transfer of experimental hypersensitivity pneumonitis with lymphoid cells. *J. Allergy Clin. Immunol.* 1975; **55**:71.
117. Reynolds HY, Fulmer JD, Kazmierowski JA et al. Analysis of cellular and protein content of broncho-alveolar lavage fluid from patients with idiopathic pulmonary fibrosis and chronic hypersensitivity pneumonitis. *J. Clin. Invest.* 1977; **59**:165–75.
118. Schuyler MB, Thigpen TP and Salvaggio JF. Local pulmonary immunity in pigeon breeder's disease. *Ann. Intern. Med.* 1978; **88**:355.
119. Pestalozzi V. Febrile Gruppenerkrankungen in einer Modellschzeinerei durch inhalation von mist Schimmelpilzen Kontaminierstem Befenchterwasser. *Schweiz. Med. Wochenschr.* 1959; **27**:710–13.
120. Solomon WR. Fungus aerosols arising from coolmist vaporizers. *J. Allergy Clin. Immunol.* 1974; **54**:222–8.
121. Banaszak EF, Thiede WH and Fink JH. Hypersensitivity pneumonitis due to contamination of an airconditioner. *N. Engl. J. Med.* 1970; **283**:271–6.
122. MRC Symposium. Humidifier fever. *Thorax* 1977; **32**:653–63.
123. Fraser DW, Tsai TR, Orenstein W et al. Legionnaire's disease: description of an epidemic of pneumonia. *N. Engl. J. Med.* 1977; **299**:1189–97.
124. Glick TH, Gregg MB, Berman B et al. Pontiac fever: an epidemic of unknown aetiology in a health department. *Am. J. Epidemiol.* 1978; **107**:149–60.
125. Pickering CAC, Moore WKS, Lacey J et al. Investigation of a respiratory disease associated with an air-conditioning system. *Clin. Allergy* 1976; **6**:109–18.
126. Hodges GR, Fink JN and Schlmeter DP. Hypersensitivity pneumonitis caused by a contaminated coolmist vaporizer. *Ann. Intern. Med.* 1974; **80**:501–4.
127. Friend JAR, Gaddie J, Palmer KNV et al. Extrinsic allergic alveolitis and contaminated cooling water in a factory machine. *Lancet* 1977; **1**:297–300.
128. Metzger WJ, Patterson R, Fink J et al. Hypersensitivity pneumonitis due to contaminated water in a home sauna. *JAMA* 1976; **236**:2209–11.
129. Fraser DW, Deubner DC, Hill DL et al. Non-pneumonic, short incubation period Legionellosis (Pontiac fever) in men who cleaned a steam turbine condenser. *Science* 1979; **205**:690–1.
130. Muittari A, Kuusisto P, Virtanen P et al. An epidemic of extrinsic allergic alveolitis caused by tap water. *Clin. Allergy* 1980; **10**:77–90.
131. Edwards JH, Griffiths AJ and Mullins J. Protozoa as sources of antigen in humidifier fever. *Nature* 1976; **264**:438.
132. Newman Taylor AJ, Pickering CAC, Turner-Warwick M et al. Respiratory allergy to a factory humidifier contaminant presenting as a pyrexia of unknown origin. *Br. Med. J.* 1978; **2**:94.
133. Campbell IA, Cockroft AE, Edwards JH et al. Humidifier fever in an operating theatre. *Br. Med. J.* 1979; **2**:1036.
134. Ganier M, Lieberman P, Fink J et al. Humidifier lung. *Chest* 1980; **77**:183–7.
135. Parrott WF and Blyth W. Another causal factor in the production of humidifier fever. *J. Soc. Occup. Med.* 1980; **30**:63–5.
136. Horsefield N, Pickering CAC, Auswick P et al. Respiratory diseases at work in a printing works. *Clin. Allergy* (in the press).
137. Rylander R, Haglind P, Lundhold M et al. Humidifier fever and endotoxin exposure. *Clin. Allergy* 1978; **8**:511–16.
138. Rosenzweig AL. Hydrogen peroxide in prevention of water contamination. *Lancet* 1978; **1**:944.

# 11. INJURIES AT WORK
## *J. G. P. Williams*

Injury at work is a major epidemic, accounting for substantial immediate losses and considerable long term disability. Much has been done by way of prevention but even in the best regulated circles injuries will still occur. Their prompt and effective management will reduce morbidity and promote rapid recovery of function and return to work.

## CAUSES OF INJURY

In general, injury is the sequel of the application to the body of local stresses to which the body cannot effectively adapt. These stresses may be more or less instantaneous, in which case the damage is acute and often severe, or the stresses may be more chronically applied giving rise to over-use injury. The nature of the stress varies; it can be applied externally by extrinsic factors such as other individuals, the environment or the equipment with which the work is done, or the excess stress may be developed from within. It is not enough to recognize that certain dangerous situations exist: if injury is effectively to be prevented an understanding of the mechanics of the application of stress to the body is essential. This is particularly so in chronic and over-use injury.

Any single tissue or combination of tissues may be involved in injury. Major injury often involves severe damage to bones with fracture or to joints with dislocation or subluxation causing instability due to rupture of supporting ligaments. These injuries being by their nature serious and dramatic attract particular attention and in general are the subject of vigorous preventive measures. The reduction of morbidity and, indeed, mortality from severe injuries in, for example, the mining and steel industries has been a direct consequence of the nature of those injuries and their tendency to lead to long term permanent disability. By contrast the much more common and far less dramatic but often very disabling minor soft tissue injuries tend to pass relatively unrecognized. These injuries include muscle tears and bruises, over-use injuries around tendons (such as tenosynovitis), low grade traumatic synovitis and capsulitis as a result of minor stress to joints and damage to skin, including abrasions, lacerations and minor puncture wounds.

## PRINCIPLES OF MANAGEMENT

In all cases the aim of injury management should be the same: an early return to work and productivity. (Return to a normal working life is usually regarded as a fair indication of recovery.) It may be based on the principle that functional recovery may predate anatomical healing. In many instances the pathological condition causing the disability which prevents work is not so much the presence of damaged tissue as the secondary effects it may provoke. For example, in muscle bruising a few torn muscle fibres are largely irrelevant in functional terms; what causes the clinical problem is raised intramuscular tension due to the presence of extravasated blood and oedema fluid. Treatment designed to reduce intramuscular tension in such cases by the absorption of fluids will drastically reduce symptoms, and therefore disability, even before healing of the torn tissue elements has been completed.

## RETURN TO WORK

In many instances the most effective rehabilitation following injury is return to normal activity, or as near normal activity as possible. In the context of work an early return to normal work is the best possible outcome of injury. Where this is not possible the job may be modified or adapted so that the patient, while carrying out his normal occupation as nearly as possible, is relieved of certain particular elements of it of which he is not capable. When the injury is more severe, selected work may have to be provided within the context of the employment situation as a whole but involving rather different activities. For example, the patient who normally works with machine tools requiring the use of both hands would have to be given a different type of employment during the rehabilitation phase following a significant injury to one hand. If the worst comes to the worst and the patient remains significantly disabled, some form of sheltered employment may be necessary. There is no doubt that in both social and economic terms sheltered employment, even if the work is of a relatively low grade and menial character, is preferable to unemployed disability.

Management of injury at work begins with first aid, the objectives of which are: first, to save life; secondly, to stop deterioration in the patient's condition; and thirdly, to initiate recovery. Effective first aid drastically reduces subsequent disability. For example, the rapid application of cold, compression and elevation to a patient with severe bruising can reduce the post-injury disability period by 50 per cent or more and thus promote an early return to work. It is, of course, essential that first aid should be administered only by properly qualified and experienced individuals: untutored meddling can too readily make an injury infinitely worse.

## TREATMENT PLANNING

It is essential in planning the management of injury to see that the treatment is appropriate to the specific needs of the individual as a patient, as well as to the requirements of his local pathology. Too often at all levels programmes of treatment are prescribed which are totally irrelevant to the problems of the individual patient as a person. As an example, an individual with back pain may be seen in a hospital orthopaedic department and prescribed physiotherapy in a hospital department to which it is impossible for him to travel because neither personal nor public transport is readily available!

The specific objectives of treatment can be considered at both tissue level and in relation to the patient as a whole. At tissue level they are to re-align and encourage healing of divided tissues, reduce haematoma, promote absorption of oedema and tissue fluids, stretch tight tissues and scars and restore flexibility to joints and extensibility to muscles and soft tissues. In addition, supportive measures such as those required in the first instance to combat shock and later perhaps to combat infection and to achieve appropriate levels of nutrition may be required. At patient level the objectives are to minimize disability — and this includes psychological as well as physical disability — to prevent further damage and to restore function rapidly.

Too often soft tissue injuries particularly are thought to be of a minor nature and almost unworthy of serious medical attention. (The best advice given to the new consultant on taking up his appointment was that of the old general practitioner who said, 'My boy, the patient who comes to you in your hospital clinic is coming because he has a problem'.) The problem may not be immediately apparent; thus in the context of an injury the actual tissue damage may be minimal and the problem itself may be some effect of this injury on the individual in his life or work. There is little doubt that vigorous and interested treatment of even relatively minor injuries drastically reduces morbidity and disability and, far from generating additional work, will in fact reduce the load on the medical services. There seems to be a kind of 'mean consultation time' required before patients with certain conditions, for example, low back pain due to posterior facetal joint strain, can recover. In such an instance it could be that the patient would require 15 minutes of informed counselling and advice with perhaps a 5-minute follow-up to get him back to normal fitness, or ten 2-minute interviews spread over a correspondingly longer period before the same result is achieved! There is little doubt that acute conditions become chronic because inadequate attention is paid to them in the first instance, particularly inadequate counselling in respect of simple home treatment and personal management. In part this may be due to the medical attendant's lack of understanding, perhaps even a form of fear of the unknown in tackling apparently complicated yet common conditions. Low back pain is a classic example: the very fact that low back pain is often given as a diagnosis rather than referred to as a symptom (as it really is) is an indication of the extent to which problems presenting this symptom cause confusion.

## DIAGNOSE ACCURATELY, PRESCRIBE LOGICALLY

While treatment should never in any sense become mere régime (it is sad how often patients presenting with low back pain are rigidly diagnosed 'P.I.D.' (prolapsed intervertebral disc) and sent for the same type of physiotherapy regardless of the specific clinical findings), there is, nevertheless, much to be said for the development of a scheme, skeleton or structure into which the symptom complex can be fitted as a means towards determining an appropriate treatment policy (*Table 11.1*). To some extent this will depend upon a proper understanding of the modalities of treatment available and the effect they are likely to have on an individual patient.

**Table 11.1.** Examples of an outline scheme for relating treatment to clinical presentation (in this case for low back pain)

| Presentation | Diagnosis | Management |
| --- | --- | --- |
| (Back pain +ve)<br>Sciatica +ve<br>Neurological signs +ve | Direct pressure due to disc prolapse (or space-occupying lesion) | Bed rest — if no rapid improvement then explore |
| (Back pain +ve)<br>Sciatica +ve<br>No neurological sign | Root irritation — posterior spinal joints inflamed | Rest, anti-inflammatory drugs, traction, rehabilitation |
| Plain X-rays positive<br>  Spinal instability | Spondylolisthesis<br>Wedge fracture<br>Spondylolysis | } Corset or fusion |
| No instability | Congenital abnormality, e.g. hemivertebra<br>Degenerative joint disease — early | } Spinal muscular re-education |
| | Degenerative joint disease — late, osteoporosis | } Corset |
| Back pain only<br>No radiological change<br>  Midline<br>  Lateral | Ligament sprain<br>Muscle strain | } Initial rest, then mobilization and rehabilitation |
| If muscle spasm +ve | | Traction |

Heat is one of the most commonly prescribed forms of physiotherapy yet it is remarkable how few of its prescribers really understand its effect and, therefore, the logical therapeutic uses to which it may be put. In fact heat has a number of possible effects ranging from direct tissue destruction when intense (as in cautery) to gentle sensory stimulation when used sparingly. In the management of injury, and particularly in rehabilitation, heat is used for two main purposes: first, by its pleasant sensory effects, to induce relaxation and secondly, by its direct heating effects on the surface, to promote local circulation. Commonly the desired effect is sought through promoting a reflex response in deep tissues rather than by the direct action of heat on the surface itself. Some specific methods of heat production directly facilitate the development of heat deep in the tissues. Examples include short-wave diathermy (where heat is generated by the passage of ultra-high frequency alternating electric current which does not reach stimulating levels) and ultrasound (by the absorption of kinetic energy and its conversion into heat). Since ultrasound involves kinetic energy it clearly offers an additional therapeutic modality in also producing what is effectively deep micro-massage. It is the difference between the effects of short-wave diathermy and ultrasound which would determine the choice in dealing with, for example, a deep haematoma.

Given adequate understanding of the effects of different treatment modalities it becomes possible to use them logically in the management of injury. In the example of treatment of patients suffering from low back pain, heat (usually radiant heat) will be helpful in alleviating the secondary muscle spasm that is so commonly a feature in these patients. Insofar as it is thought that tension may be a significant factor in symptom production, the relaxation engendered by heat may secondarily assist in the relief of symptoms. Clearly, however, heat, unless it is deep heat at the site of the tissue lesion, will not necessarily affect the underlying disease process. However, since it can be seen that many disabling conditions are as much due to the secondary effects provoked by the primary pathology as by the latter itself, treatment may be effective in terms of symptom relief when directed at these secondary effects.

It is essential in dealing with any injury to have a clear three-dimensional mental picture of the primary pathology and to have a clear concept of the secondary effects which it may produce. It is not necessary that these concepts should be based on scientifically proved histopathological findings, but quite clearly they should fall in with general principles of physiology and physiopathology and, further, should in their likely consequences tie in accurately with observed effects. Thus, it is reasonable to reserve the diagnosis 'prolapsed intervertebral disc' for those patients in whom there is evidence of nerve root pressure unassociated with other possible causes (such as fracture, tumour or infection). If such a concept is applied then it is clear that in patients so diagnosed the objective of treatment must be the relief of pressure on the nerve root, and appropriate steps taken to achieve that objective. It can of course be argued that some prolapsed intervertebral discs may actually be symptom-free (post-mortem studies have shown that disc prolapses can be demonstrated in patients with no previous history, although this is not necessarily conclusive). In such asymptomatic disc prolapses decompression is not required. This introduces the controversial concept that pathological processes by themselves are irrelevant; they become relevant only when they interfere with some aspect of patient function. Even malignant tumours are asymptomatic early on, and while so cause no concern to the patient. It is when they interfere with function, especially threatening the ultimate interference of function (that is, death) that they demand treatment. In practical terms, however, it is acceptable that in planning treatment the method chosen should relate not only to the primary pathology but also to its secondary effects with the specific objective of obliterating anti-functional elements in the clinical picture. This approach to injury management, while simplistic, has the merit of relating treatment to the way in which the patient's particular difficulties are caused by the pathological process. As a practical basis it is highly effective and is readily applied.

A regularly structured process for the planning of treatment from first aid to final rehabilitation (and including prevention of injury) clearly leads to the minimizing of disability and loss of working capacity (*Fig. 11.1*). It should stimulate thought and not merely provide a rigid routine into which all patients are squeezed willy-nilly. The basis of all effective medical management is summed up in the adage 'Diagnose accurately, prescribe logically'. So far so good, but how is this put into practice?

In the first aid situation the niceties of treatment programming take second place to dealing with the immediate life- or health-threatening situation. Once, however, the stage of primary treatment is reached (that is, the patient's condition has been stabilized and attention can now be given towards promoting improvement rather than preventing deterioration), it becomes necessary to reach a clear understanding of the nature of the injuring stress in order to assess the damage which it has produced. In minor trauma, the first aid stage is often an essentially do-it-yourself affair with the patient himself taking whatever steps seem appropriate, while in the case of the over-use injury there is no first aid stage as such.

It is axiomatic that in a work situation the damaging stress and its mechanism should as far as possible be identified and excluded. It is not always possible to exclude this stress, particularly in some

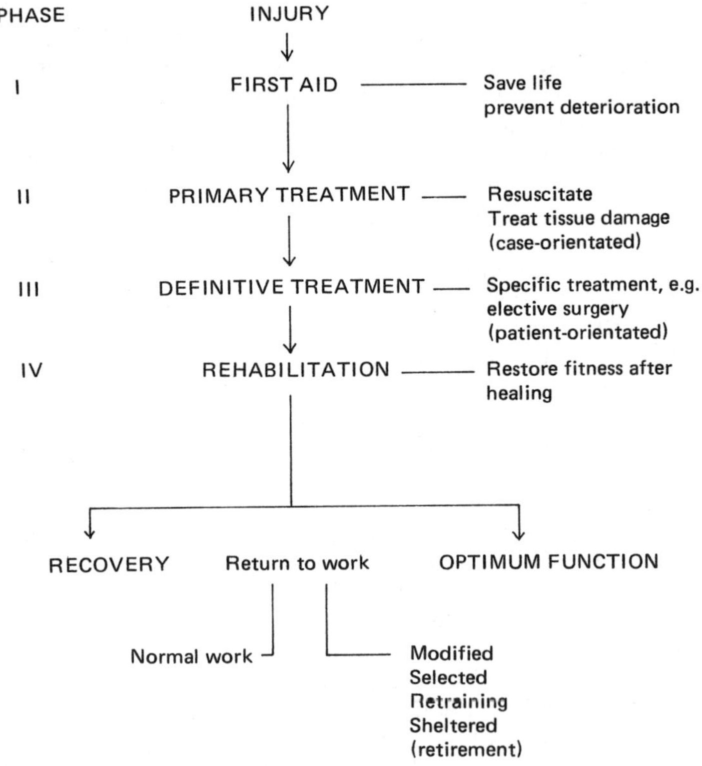


*Fig. 11.1.* Structured process for treatment planning. Phases may overlap (e.g. primary and definitive) or may be left out (e.g. first aid in over-use injury). It is important to identify the mechanism of injury to prevent recurrence.

instances of over-use injury (it is less possible in sport where there is virtually no scope for alternative techniques and methods). However, in the industrial context elimination of unnecessary stress to prevent recurrence is an essential component of injury management to be carried out *pari passu* with definitive treatment and rehabilitation, once the cause of the injury has been identified.

Understanding of the mechanism is often directly relevant in the management of the specific injury. Thus, for example, where rotational stress has been applied it becomes particularly essential to prevent rotatory movements during healing. This is applicable both in bone injury, as in spiral fractures of the tibia, and in soft tissue injury, as in anteromedial rotatory instability of the knee. In the latter example, dealing only with the meniscus tear will be inadequate and the knee will remain unstable unless action is taken to stop subsequent rotation, for example by pes anserinus transfer. At a much less clinically technical level the same applies in minor injuries. Inadequate or unsuitable footwear is often a cause of ankle and lower limb injury. Inadequate support of mildly sprained joints and inadequate muscular re-education will predispose to recurrence or relapse.

While various factors, including the specific nature of the injury and the consequences to the patient (in clinical, social and economic terms) will influence the outcome of definitive treatment, perhaps the key factor is the extent to which the responsible clinician works within his particular limitations. These limitations may be either in technical expertise or availability of facilities. It is interesting that there are two main reasons for referral for a second or consultant opinion: first, for diagnosis when the nature of the clinical problem appears obscure and secondly, for expert or particular treatment where diagnosis is clear but the facilities immediately available to the referring practitioner are inadequate.

No doctor should ever feel embarrassed at referring a patient for a further opinion whether for diagnosis or management. By definition no one can be expert at everything. Indeed a practitioner must be morally bound to seek help when confronted by a clinical problem with which he is not familiar or where he has any doubt as to his diagnostic ability. It must here be emphasized that diagnostic ability is not merely a matter of labelling a patient with a suitable tag relating to his complaint (as has already been shown in the example of cases of backache diagnosed as prolapsed intervertebral disc); it involves a positive understanding of the relevant pathological processes. Given that an accurate diagnosis can be made on this basis, the second half of the adage — 'prescribe logically' — readily follows. It may not always be possible for the clinician first seeing the patient to provide the necessary treatment from resources available to him. For example, a patient with a significant hand injury with fractures, tendon or nerve damage clearly cannot be managed at factory surgery level — only in exceptional cases will the necessary clinical facilities be available. Usually the patient must be transferred to hospital for appropriate management. This will be true to a greater or lesser extent with the whole spectrum of injuries.

## TREATMENT TECHNIQUES

Various techniques of treatment are available for dealing with injuries and it is essential that the clinician deploying them should be fully conversant with their rationale (*Table 11.2*).

### Rest

Perhaps the simplest and most readily applied treatment is rest. This can apply to the body as a whole or simply to the affected part. It may

**Table 11.2.** Outline of methods available for injury management

| | |
|---|---|
| Rest | General (bed rest)<br>Local (plaster-of-Paris, splints, foot orthosis, supports) |
| Medication | Analgesics<br>Anti-inflammatories<br>   Systemic<br>   Topical (including injection)<br>Others (antibiotics, enzymes, rubifacients etc.) |
| Physiotherapy | Thermal<br>   Heat (radiant heat, SWD)<br>   Cold<br>   Contrast baths<br>Thermal/mechanical (ultrasonics)<br>Mechanical<br>   Massage, vibration<br>   Manipulation<br>   Traction<br>Electrical<br>   Faradism<br>   Galvanism<br>   T.C.N.S., interferential<br>   Others |
| Invasive | (Acupuncture)<br>Aspiration<br>Surgery |
| Re-education | Exercise<br>   General<br>   Specific<br>      Strengthening<br>      Mobilizing<br>      Co-ordinating<br>      Increasing endurance |
| Psychological | Realistic objectives |

consist of the complete cessation of activity (as, for example, in the prescription of bed rest for the patient with an acute spinal strain) or it may simply be a reduction in activity (as in advising the patient with a mild over-use injury to reduce the level of activity to that at which symptoms are no longer reproduced). Local rest may be secured by a number of means including rigid or semi-rigid splinting or simple support from elasticated and/or adhesive bandages. In prescribing local rest it is essential to remember that if the patient's general level of working capacity is to be maintained, or at least if the minimum deterioration is to be permitted, general activity for the rest of the body is essential. The fact that the patient has a bandage or splint on, say, one wrist does not necessarily mean that the rest of the body should not be exercised. Rest, whether local or general, allows for the reparative processes to proceed unimpaired and this will be important when damaged tissues must be kept in apposition. However, rest can be counter-productive at local as well as general level if, for example, adhesions are allowed to form or scar tissue to develop in a shortened contracted mode. Thus it is necessary to restore mobility at the earliest possible time consistent with proper tissue healing, and in planning this the need for clear understanding of the nature of the repair processes is obviously essential.

## Medication

One of the most simply deployed treatment methods available is chemical therapy, that is, the use of drugs. A number of types are available for injury management. Analgesics to reduce pain are of obvious value, particularly in the early stages. Pain with associated anxiety and, in the chronic case, depression are potent factors in the maintenance of disability. In general, however, it must be remembered that pain is essentially a symptom and that wherever possible pain relief should be sought by dealing with the primary clinical problem rather than by a direct use of analgesics.

The second type of useful medication is the non-steroidal anti-inflammatory. Since the primary cause of disability from injury is often the ungoverned inflammatory response, anti-inflammatory medication, particularly in minor injury, may be of great value in reducing disability and restoring function. A wide variety of different preparations is available and patient response is variable. In general where the patient fails to respond to an anti-inflammatory preparation of a particular type, for example, a propionic acid derivative, it is better to use an alternative of a different type (such as piroxicam) rather than to persist with different variations of the same type. Anti-inflammatory preparations are normally administered systemically and there is much to be said for prescribing the highest tolerable dose for the shortest possible time. (This contrasts with the way in which these drugs are used in management of chronic rheumatic conditions in which the lowest dose producing the required therapeutic effect is used over a longer period.) Clinical evidence suggests very strongly that where non-steroid anti-inflammatory medications are used they are most effective if given as soon as possible after injury. Non-steroid anti-inflammatories are not entirely without side-effects although these vary enormously from one drug to another and indeed from one individual to another. They should, therefore, be used with care but not with excessive caution except in patients with previous history of intolerance or of gastrointestinal ulceration.

In some instances, particularly in chronic injury, the lesion is sufficiently well localized for local as opposed to systemic chemotherapy to be appropriate. In this context the drug most often used is a steroid, although local anaesthetic with or without a spreading agent such as hyaluronidase may be appropriate in some instances, as in acute teno-synovitis. There is, however, little justification for the practice common in some quarters of using local anaesthetic and spreading agents in acute soft tissue injury such as muscle tears. Where steroids are employed they should be used with caution and should never be injected into tendon due to the risk of causing a rupture.

Some topical agents may be of value but there is considerable difficulty in the objective clinical assessment of their efficacy. In this context may be included rubifacients, which have the effect of producing a 'chemical heating' by promoting increased local tissue surface circulation, and anti-inflammatory agents, which are absorbed directly through the skin overlying the lesion. A number of preparations are available and it is often difficult to exclude the placebo effect. It may sound cynical, but it is nevertheless not unreasonable to use them on the basis that they do no harm and may do some good even if it is only by the placebo effect!

Other forms of drug treatment are perhaps less relevant but the role of antibiotics and of appropriate tetanus inoculation should be remembered in appropriate cases.


## Physiotherapy

By contrast with chemical therapy, which is usually readily available, physical therapy often necessitates trained staff, although there is much that patients can do in the way of home physical treatment, particularly in minor injuries. The spectrum of physiotherapeutic techniques is wide and is increasing with the introduction of new types of electrical treatment, many of which show promise but are not yet fully evaluated. The main types of physical treatment available include heat and cold, the application of kinetic energy and various types of electrical stimulation.

Heat will include the use of radiant heat from appropriate lamps, induced heat specifically by short-wave diathermy and microwave and as a spin-off from ultrasonics, and direct conduction to the surface using hot packs, muds and hot water. The general effect of heat is to increase local tissue circulation and as such it is of value in promoting the absorption of local tissue fluids and oedema. In addition, the sensory effect when heat is applied to the skin has a relaxing value. By contrast cold shuts down the local circulation both in the skin and in the deep tissues (it is thought by an axon reflex). Cold is used, therefore, in the first aid situation to diminish bleeding. It is also used for its stimulatory sensory effect on the skin and appears to relax muscle spasm, both protective spasm as a result of injury and neurological spasm in upper motor neurone lesions. Cold must be applied with care since in excess, as with heat, tissue damage can be caused. Some patients are particularly intolerant to local cold and may produce quite marked weals.

Cold and heat used alternately for their opposing effects form the basis of the contrast bath technique (in which the patient is encouraged to immerse the injured part in water that is alternately as hot and as cold as he can tolerate). This has a particularly stimulating effect on local tissue circulation, and is of especial value in the management of painful foot and ankle conditions. It has the additional advantage that as a treatment technique it is amenable to the 'do-it-yourself' approach and patients can therefore be encouraged to give themselves contrast baths as part of home treatment.

Kinetic energy is commonly applied either as the micro-massage of ultrasonics or in more gross forms by direct massage or manipulation. The exact way in which ultrasound works is not known; there is considerable conjecture and experimental evidence is inconclusive. The situation is made more difficult in that it is virtually impossible to establish dose levels as transmitted to the patient. Certainly output is often quoted in terms of watts per square centimetre but it is virtually impossible to determine what proportion of this power is actually transferred to the patient and indeed what exact effect it may have.

Kinetic energy applied through massage is equally unquantifiable and its efficacy is very much dependent on the skill of the individual therapist. It has a variety of possible effects both mechanical and sensory. Being a pleasant form of passive treatment it is said to be 'addictive', and for that and other reasons has tended to fall into disrepute. However, in appropriate instances massage can be extremely effective as, for example, in mobilizing oedema fluid, stretching tight muscle and reducing the symptoms in that most pernicious and elusive condition, fibrositis.

Other forms of kinetic energy involve vibrations which can be applied at various frequencies, all relatively low, by vibrating equipment or, in the most extreme form, in 'impact therapy'. As is so often the case the effectiveness of these forms of treatment is dependent less on their inherent properties than on the way in which they are used, specifically in the selection of cases and in the skill and assiduity with which the therapist plies his or her trade. As such they should not be ridiculed but used with the appropriate degree of selection.

Another and fascinating area of physical treatment involves the use of electricity. The peculiar values of faradic stimulation and direct current (galvanism) have long been known and regularly used; both have had their detractors. Faradism now appears to be effective less to give 'electrical exercise' than as a means of producing feed-back, allowing the patient to *feel* his muscles contracting and, hence, enabling him to re-educate them. Galvanism, that is direct current, has had an even more chequered career and its therapeutic use is still not clearly defined. Transcutaneous nerve stimulators for the relief of pain use direct current and there appears to be some value in the use of half-wave rectified currents in treating chronic over-use injury.

Interferential therapy has long been regarded by therapists experienced in its use as a valuable method of pain relief. Other methods of electrical treatment have been introduced recently but formal evaluation has yet to be made and their roles in the management of injury therefore are as yet undefined.

Manipulation is included in physical treatment and involves the manual passive movement of joints and tissues as well as certain specific treatments such as traction. In general the objective is to mobilize joints with restricted movement. In the case of other (soft) tissues the distinction from massage becomes extremely fine. Traction is most commonly used in spinal complaints, particularly those associated with nerve root irritation causing brachialgia or pain in the leg (sciatica). It is thought to work partly by the direct mechanical relief of tissue tension at the posterior spinal joint and disc and partly by overcoming secondary muscle spasm. There are many schools of thought as to the value of manipulation and the techniques that are used and they are frequently at loggerheads! The situation is further confused by the extent to which this type of treatment (involving as it does the laying-on of hands and often accompanied by much mumbo-jumbo) has been adopted by practitioners on the fringe of medicine. Another fringe medical practice is acupuncture, which in appropriate cases may be helpful in pain relief. It is, however, no more than a means of achieving a particular type of analgesia; in the management of injury its role is strictly limited.

**Surgery**

Invasive treatment involves surgical methods varying from formal elective surgery to simple procedures such as joint aspiration for effusions interfering with joint function.

There is considerable argument as to whether an effused joint should be aspirated. Effusion of rapid onset following trauma is due to haemorrhage into the joint (haemarthrosis). A serous effusion develops more slowly, for example the patient will complain that the joint is swollen on the day following the injury. In general the former is large and tense, the latter less large and relatively soft. In the acute phase the aspiration of a haemarthrosis may be required for diagnostic purposes since the tense swollen joint is difficult to examine. Generalizations are often made regarding the possibility of complete ligament rupture in patients with a haemarthrosis but they are frequently inappropriate in individual cases. As a diagnostic aid aspiration of a haemarthrosis is of real value but must be carried out under the most rigid possible aseptic conditions. Nowadays the availability of pre-packed, pre-sterilized needles and syringes makes this easier. Aspiration is generally unnecessary other than when required for diagnostic purposes, unless the effusion is actually embarrassing joint excursion and/or inhibiting the contraction of muscles supporting the joint. Aspiration of other cystic lesions is of doubtful value. The encysted haematoma is best drained surgically rather than aspirated and this is certainly true of an abscess, although a small subcutaneous abscess can sometimes be drained effectively through a wide-bore needle. An inflamed bursa such as an olecranon or prepatellar bursa should not be treated by aspiration since it tends to refill and repeated aspiration tends to cause loculation which is difficult to treat.

The need for formal surgery in major trauma is self-evident and the indications are usually clearly defined. A fruitful field for surgery in recent years has been over-use injury, more common perhaps in the sporting rather than industrial context but a source of very real disability nonetheless. The surgical procedures most commonly used in these cases involve decompression of compartments or individual structures. Over-use causes an acute oedematous inflammatory reaction which if persistent leads on to a chronic fibrous reaction. Increased tissue tension in anatomical compartments (of which the anterior tibial is perhaps the best known) may cause symptoms which can be crippling. Decompression procedures, while invasive and aggressive (and indeed carrying all the risks of surgery), nevertheless produce dramatic reduction in disability and are of real value in the management of patients for whom recovery of function at the earliest possible time is essential. Surgical decompression in extensor tenosynovitis of the wrist has reduced the mean disability period from 3 weeks to 3 days.


## REHABILITATION

The mere treatment of injury is not enough. It must be accompanied by the active restoration of the patient's fitness (that is, performance capacity). This means the retraining of wasted muscles and redevelopment of motor skills and coordination, and this is achieved by exercise — repeated exercise. The value of therapeutic exercise cannot be overestimated. In an industrial context it is particularly important to recognize that the patient debilitated by injury or illness often finds it a struggle to return to his previous employment. The whole objective of a rehabilitation programme is so to restore the level of fitness that the patient is in fact enabled to 'go back to work for a rest'. The patient who is able to return to work without difficulty or strain is unlikely to develop long term problems.

Too often the specific management of damaged tissue is regarded as the prime concern of the clinician, without reference to the context of the patient as a whole. This leads to case-orientated as opposed to patient-orientated medicine. For the caring clinician failure to return the patient to a normal life in all its aspects should be a positive affront. Clearly in some instances restoration of normality is impossible, as in the irremediable loss of digits in a factory accident, in which case the aim of treatment becomes to maximize the effective function of the rest of the hand and to enable the patient to adjust to his new body image, with all that it implies. It is essential that the clinician sets rehabilitation targets that are realistic as well as optimistic and perseveres in their attainment. False hopes are as unproductive as clinical neglect.

Hippocrates declared 'that no head injury is so severe that it need be despaired of nor so trivial that it should be neglected' — this dictum is applicable to all injury. An energetic and aggressive philosophy in the management of injury, particularly minor injury, will do much to restore the working capacity of the patient and minimize long term complications. At the same time it will contribute materially to the health and well-being of the community as a whole.

# 12. PRINCIPLES, PRACTICE, PROBLEMS AND PRIORITIES IN TOXICOLOGY
*James W. Bridges and Susan A. Hubbard*

## PRINCIPLES OF TOXICOLOGY

Toxicology is concerned with identifying and characterizing the adverse effects of chemical and physical agents on biological systems. Information on the toxic effects of chemicals must be combined with data on the dosage, duration and form of exposure in order to assess potential toxicity. Thus toxicity can be considered to represent the potential of a chemical to cause harm. The hazard to man provides the likelihood that it will do so and is usually assessed from toxicity data and data on human exposure.

### Interactions of chemicals with biological systems

The events leading to toxicity can be considered to take place in three phases: a chemical phase, a distributive phase and an interactive phase (*Fig. 12.1*).

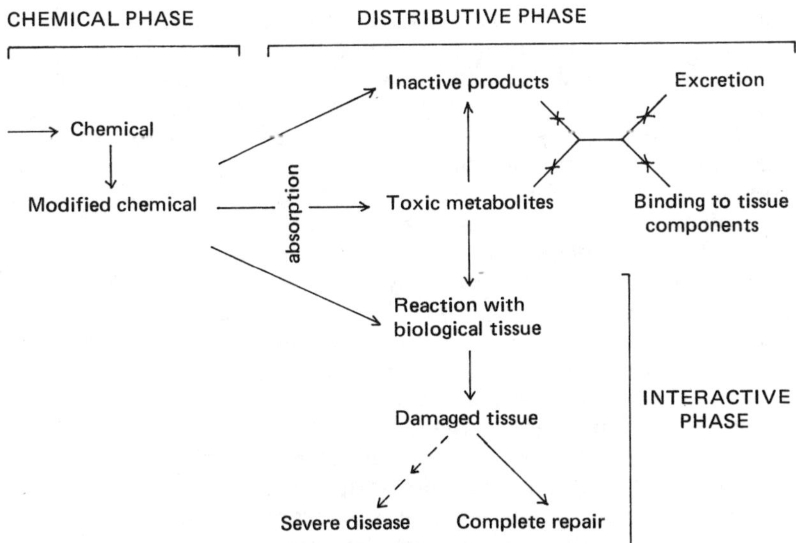


*Fig. 12.1.* Phases involved in chemically mediated toxicity.

## The chemical phase

In the chemical phase the chemical prior to absorption may break down spontaneously or interact with other materials in the environment or diet. Examples include the hydrolysis of many organophosphorus pesticides, photochemical oxidation of alkyl hydrocarbons and the reaction of nitrite with dietary amines in the stomach to form nitrosamines. In order to exert their toxic effects most chemicals, other than caustic agents and irritant particles, must be absorbed. The majority of low molecular weight materials pass into the body via the lung, intestine and/or skin to at least a limited extent. As all animal cell membranes are predominantly lipoid they favour the uptake of fat-soluble (lipophilic) non-ionized compounds such as the non-polar organic solvents.

## The distributive phase

Absorption is the first stage of the distributive phase. Chemicals absorbed via the intestine pass initially almost entirely to the liver whereas those penetrating the lungs and skin pass directly into the general systemic blood circulation. In most cases absorption occurs by passive diffusion. Under normal circumstances this is a non-saturable process. Only compounds resembling naturally occurring substances are likely to be absorbed to a significant extent by active, and therefore saturable, processes. Following absorption a chemical may undergo one or more of the following fates: binding to tissue constituents, metabolism by the so-called 'drug metabolizing enzymes' and/or excretion via the urine, bile, other secretory fluids (tears, saliva, sweat, seminal fluid, milk), faeces, or, for volatile compounds, via the expired air. On entering the systemic circulation the initial distribution of lipophilic chemicals throughout the body is rapid, the rate of distribution in a particular body organ being determined by the blood flow in that organ and the ability of the chemical to cross the capillary bed and enter the various cell types which comprise the organ. Water-soluble molecules or compounds which are very extensively bound to plasma proteins may have a much more restricted distribution. With time the pattern of distribution of a chemical between organs may change to reflect the affinity of the cell components of the different organs. Commonly a chemical is concentrated in one or more specific organs. The site(s) of accumulation may or may not also be the primary sites (target organ(s)) for the chemical's toxic effect.

Organ(s) concentrating the chemical but not experiencing the toxic effects can be regarded as reservoirs of the chemical and the concentration of the chemical by the organ as a buffering system lessening toxicity. Examples of this situation are the accumulation of lead in the bone-marrow, the concentration of chlorinated biphenyl compounds in the fat deposits and the binding of dieldrin to blood albumin. Examples of chemicals which concentrate at their target site include carbon monoxide which binds selectively to haemoglobin, paraquat which is concentrated by active transport in the type 2 cells of the lung and cadmium which is bound in the kidney and liver to metallothionein.

Lipophilic compounds usually require to be metabolized to more water-soluble forms before they can be excreted, because the excretory systems of the body are designed to clear water-soluble substances. Thus not only are the excretory fluids themselves aqueous, but the cells lining the excretory ducts have lipoid membranes thus encouraging retention in the ducts of water-soluble substances, except those which are actively re-absorbed, but permitting re-uptake of lipophilic substances. Passage of compounds into the excretory ducts may occur either by filtration (a separation based on molecular weight) or by active transport. These active transport systems are geared to dealing with polar materials particularly anions.

The conversion of lipophilic chemicals to more water-soluble, less lipophilic, products is carried out by a group of enzymes known collectively as the 'drug metabolizing enzymes'. These enzymes are most active in the liver but are also located at the portals of entry into the body (intestine, lung and skin) and in a number of other organs. Metabolism of lipophilic compounds usually occurs in two stages: a preconjugation (phase 1) reaction in which an electrophilic substituent is inserted or revealed in the molecule and a conjugation (phase 2) reaction in which the new substituent group is conjugated with an endogenous substance such as glucuronic acid, sulphate or an amino acid (*Table 12.1*). The majority of these enzymes are characterized by relatively high Kms (the dose which causes the enzyme K to be half saturated with substrate). Under most circumstances they will, therefore, not be saturated by substrate. In addition to their metabolism by tissue enzymes, orally administered chemicals, or biliary excreted materials, may also be metabolized by the intestinal microflora. In contrast to tissue drug metabolism, with the microflora hydrolytic and reduction reactions are very common while oxidation is rare. The metabolic fates of benzene and parathion may be used to illustrate some of the capabilities of the drug metabolizing enzymes (*Fig. 12.2*).

Oxidation is the most common phase 1 reaction. In the case of benzene an unstable epoxide is formed (cytochrome P-450) which is rapidly broken down, spontaneously to form phenol and enzymically to form a dihydrodiol (epoxide hydrolase) or a glutathione conjugate (glutathione transferases). The primary urinary excretion products are phenylglucuronide and phenylsulphate.

Parathion undergoes both oxidative and hydrolytic attack. Some of the ethyl groups enter intermediary metabolism pathways eventually

**Table 12.1.** The principal drug metabolism reactions

| Preconjugation (phase 1) | Reaction | Enzyme(s) involved |
|---|---|---|
| Oxidation | Aromatic hydroxylation | Cytochrome P-450* |
| | Aliphatic hydroxylation | |
| | N, O or S Dealkylation | |
| | Desulphuration | |
| | Sulphoxidation | } Cytochrome P-450 |
| | N hydroxylation | } Amine oxidase |
| | Oxidation of alcohol | |
| | Oxidation of aldehydes | |
| Reduction | Nitro reduction | } Cytochrome P-450 |
| | Azo reduction | } Cytochrome C reductase |
| | Dehalogenation | |
| | Aldehyde and keto reduction | |
| Hydrolysis | Esters | Esterases* |
| | Amides | Amidases* |
| | Epoxides | Epoxide hydrolase |

| Conjugation (phase 2) | Substrates | Enzymes and reaction |
|---|---|---|
| Glucuronidation | Hydroxyl, sulphydryl, carboxyl and amino groups | Glucuronyl transferase* (transfer of glucuronic acid group) |
| Sulphation | Hydroxyl and amino groups | Sulphatransferase* (transfer of sulphate group) |
| Glutathionation | Various electrophiles, e.g. epoxides | Glutathione transferase* (transfer of glutathione) |
| Mercapturic acid formation | Glutathione conjugates | Glutathionase + peptidase + N-acetyl transferase (loss of glycine and glutamic acid then acetylation) |
| Amino acid conjugation | Carboxylic acids | Amino acid conjugase (transfer of an amino acid such as glycine) |
| Acetylation | Aromatic amines | *Acetylase (transfer of acetyl group) |
| Methylation | Catechols, some N and S containing groups | Methyl transferase (transfer of methyl group) |

*Represents a family of enzymes.

being excreted from the body as $CO_2$. The $p$-nitrophenyl glucuronide and $p$-nitrophenyl sulphate are the major urinary excretion products. Small amounts of amine are formed.

Most compounds are metabolized by more than one route. The drug metabolizing enzymes operate largely independently of one another and are therefore in competition for substrate. Since the body

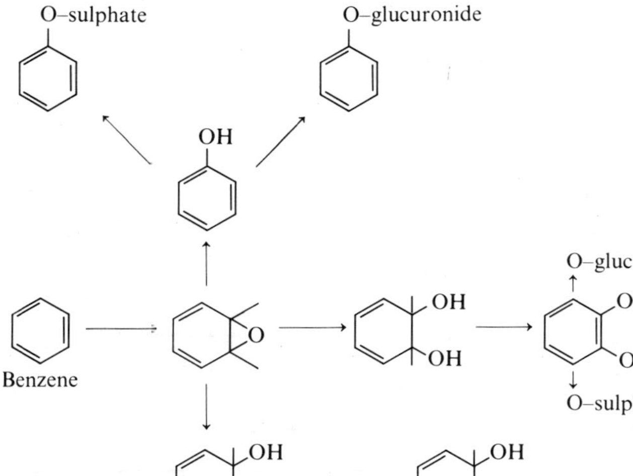


*Fig. 12.2.a,* Metabolic fate of benzene.

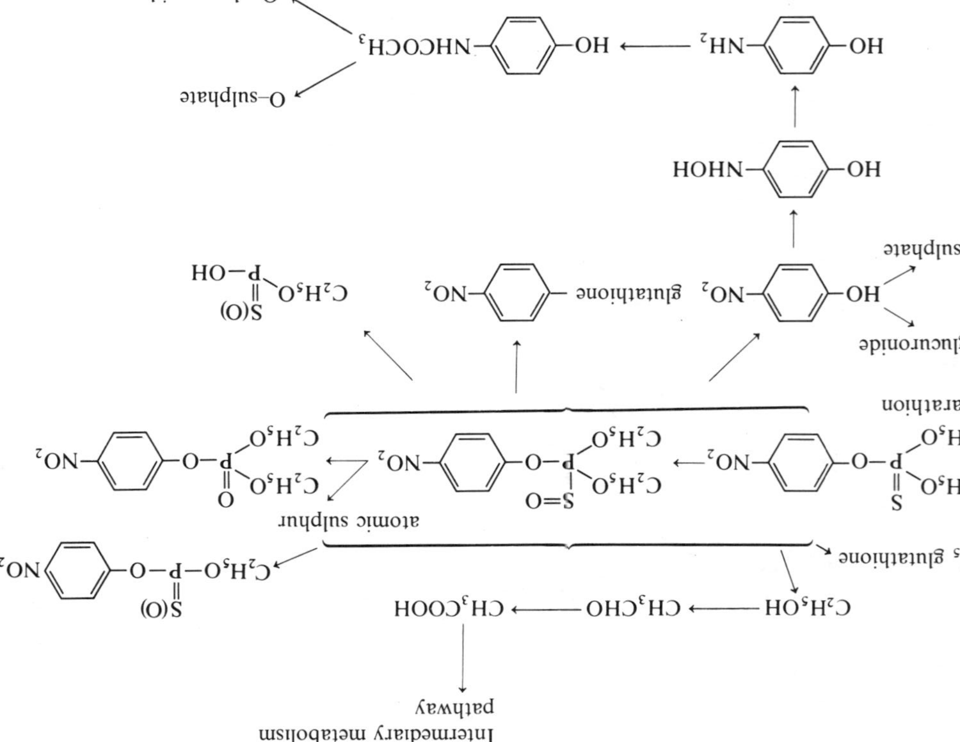


*Fig. 12.2.b*, Metabolic fate of the insecticide parathion.

is unable to limit uptake of lipophilic chemicals, metabolism is often an essential and rate-limiting step in terminating exposure. The bulk of an absorbed chemical is usually detoxicated by metabolism. Although following metabolism the general trend is towards increased water solubility, this is by no means invariable. For example, the nephrotoxicity of high doses of ethylene glycol is due largely to its metabolism to oxalate which precipitates (as calcium oxalate) in the kidney tubules. Acetylation and methylation quite frequently lead to less water-soluble products. Reactive intermediates are also commonly produced during phase 1 (preconjugation) metabolism. These metabolites must be regarded as potentially highly toxic since they may covalently bind to essential cell organelles such as nucleic acids, enzymes, cofactors, membrane components and so on. Whether they produce a deleterious effect or not is dependent on their relative rates of formation and breakdown, whether they are able to leave the enzyme forming them, their half-life in biological membranes and fluids and the types of endogenous molecule available for them to react with. Active metabolites with very short half-lives may only exert an effect in the organs in which they are formed. For example, the free radical intermediate of carbon tetrachloride causes severe liver damage in rats. Active metabolites with longer half-lives may provoke a toxic reaction in several organs. Since the oxidation enzymes are particularly concentrated in the endoplasmic reticulum of the liver, the liver is a particularly common target organ for metabolite-mediated toxicity. The lung, which has high concentration of the oxidation enzyme cytochrome P-450 in one particular cell type (clara cells), and the proximal tubule portion of the kidney are other frequent target sites. There is a steadily growing number of examples of chemicals which exert their toxic effects partly or entirely through the aegis of active metabolites (*Table 12.2*). For benzene the formation of an epoxide appears to play a key role in its toxicity whereas for parathion the sulphoxide, elemental sulphur and paraxon are all contributors to the observed toxic effects.

## *The interactive phase*

It is generally considered that toxicity arises through the reversible or irreversible interaction of a chemical and/or its metabolites with one or more molecular or receptor sites. A list of some common types of toxic effect is given in *Table 12.3*. The nature and degree of toxic effect are thus related to the type of molecular or receptor sites involved, the extent of the interaction(s) and its persistence. Chemicals are unlikely to mediate their toxic effects on cells by creating new functions; rather they modify existing ones, often rendering apparent functions that were formerly latent. As a consequence the observed

**Table 12.2.** Examples of chemicals which solely or partly exert their toxic effects via the formation of reactive metabolites

| Type of intermediate | Example | Toxic effect |
| --- | --- | --- |
| Epoxide | Bromobenzene | Hepatocellular necrosis |
| | Benzene | Leukaemia |
| | Ipomeanol | Lung cell necrosis |
| | Aflatoxin | Liver cancer |
| | Allylisopropylamide | Porphyria |
| | Benzpyrene | Lung and skin carcinomas |
| N-Hydroxy (ultimately active compound may be an N-hydroxy conjugate) | 4-Amino biphenyl | Intestinal cancer |
| | $\beta$-Naphthylamine | Bladder cancer |
| Free radical | Carbon tetrachloride | Hepatocellular necrosis |
| | Paraquat | Lung fibrosis |
| | Alloxan | Pancreatic necrosis |
| Elemental sulphur and/or S-oxides or S-dioxides | Carbon disulphide | Hepatocellular necrosis |
| | Parathion | Hepatocellular necrosis |
| | $\alpha$-Naphthyl thiourea | Pulmonary oedema |
| | Thioacetamide | Hepatocellular necrosis |
| Quinolimine or semiquinone | Paracetamol | Hepatocellular necrosis |
| | Anthracycline | Anti-tumour agent |
| Carbene | Safrole | Hepatocellular necrosis |

tissue changes frequently resemble those occurring in 'natural' disease. Indeed, in chronic studies it may be difficult to distinguish between chemically mediated toxicity and 'natural disease' processes. The toxic effect which is measured may be quantal (that is, all or none), such as the death of the animal, or graded (that is, continuous), such as the progressive inhibition of an enzyme. For either situation it is important to establish the relationship between the exposure dose and response. A typical dose–response relationship is shown in *Fig. 12.3*. Determination of blood levels with time of chemical and/or metabolites of drug often provide a much more reliable indicator of the concentration of drug available to the sites of action than does a knowledge of the extent of exposure to the chemical. (NB, The study of the quantitative aspects of the distributive phase is usually referred to as pharmacokinetics or toxicokinetics.) Data on blood levels may provide information on the exposure levels at which one or more aspects of the distributive phase are saturated. Such information is valuable in ascertaining the relevance of a toxic effect to the likely normal human or environmental situation[1] (*Fig. 12.4*).

Many of the reactive species indicated in *Table 12.2* are electrophilic (that is, they have a partial positive charge) and are therefore most likely to react with nucleophilic cellular components. In the liver, especially with low doses, the most abundant and therefore the major

**Table 12.3.** Some examples of common types of toxic effect

| Type of toxic effect | Possible causes |
| --- | --- |
| Cell death (necrosis) | Anoxia may arise, for example by occlusion of blood flow<br>Serious damage to plasma membranes (often arising from lipid peroxidation?)<br>Mechanical trauma or ionizing radiation |
| Modification of plasma membrane fraction | Membrane damage, e.g. due to lipid peroxidation |
| Endoplasmic reticulum changes | Loss of ribosomes, loss of integrity, excessive proliferation of smooth endoplasmic reticulum |
| Lysosomal changes | Proliferation, engorgement with damaged organelles |
| Other organelle changes | Nuclear enlargement, peroxisome proliferation |
| Lipid accumulation | Disturbance in lipid metabolism |
| Glycogen depletion | 'Uncoupling' of link between carbohydrate and lipid metabolism |
| Other major disturbances in intermediary metabolism | Uncoupling of oxidative phosphorylation, loss of cofactors, e.g. ATP, reduced nicotinamide, protein synthesis inhibition, reduction in active transport |
| Replacement of 'normal' cells by another cell type or proliferation of a second cell type | |
| Metaplasia | Replacement of glandular or transitional epithelium by squamous epithelium<br>Repeated trauma |
| Granuloma formation | Repeated or prolonged damage — macrophage infiltration and granulated tissue formation |
| Fibrosis | Repeated damage |
| Cell mutation and/or tumour development | Interaction of reactive chemical or its metabolite or ionizing radiation with nuclear DNA. Effect of promotors |
| Immune complex disorder | Production of antibodies against host protein |
| Formation of thrombi, stones and so on | Blockage of fine vessels and/or ducts |

target for electrophilic reactive metabolites is the tripeptide glutathione. Since some loss of glutathione can be tolerated without
any untoward effect, toxicity may only be observed at relatively high
doses when the glutathione protective mechanism is overwhelmed.
Glutathione occurs at much lower levels in most extrahepatic organs
and is therefore less effective in other organs as a protection
mechanism. The interaction of reactive chemicals and/or their
metabolites with nuclear DNA is thought to be the initiating event in
producing cell mutations which may lead to tumour development.

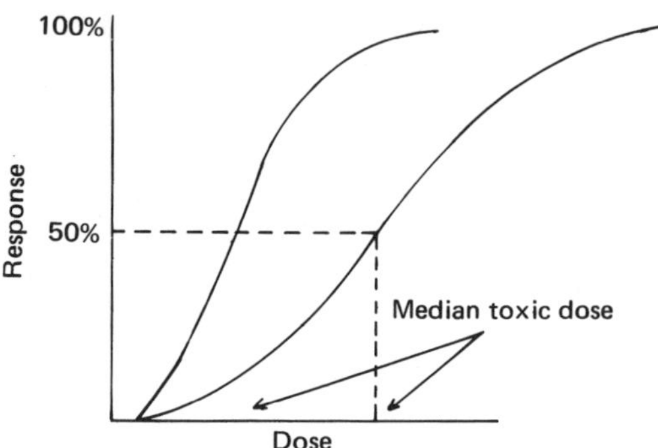


*Fig. 12.3.* Dose–response relationships.

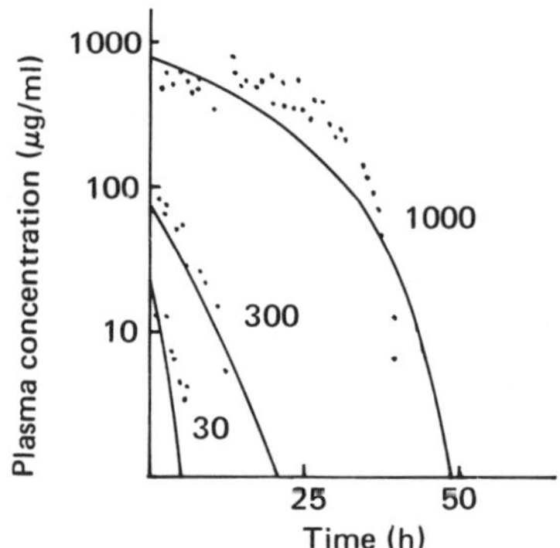


*Fig. 12.4.* Blood plasma concentration of 1,4-dioxane in rats administered intravenous doses of from 30 to 1000 mg/kg.

This concept provides the basis for the use of many of the short term tests for carcinogens.[2]

The magnitude of the observed toxic effects depends not only on the initial interaction of a chemical and/or its metabolites but also on the persistence of the damage. This is related both to the persistence of the interaction and to the body's ability to repair effectively any damage caused (*Fig. 12.1*). Efficient restoration of tissue to normal after any form of damage is likely to be rate-limited at high chemical exposure levels. The same is true for drug metabolism. This raises the question of the relevance of extrapolating information obtained using high doses to low exposure situations.

## Factors affecting toxicity

The nature and extent of toxicity depend on a variety of other factors (*Table 12.4*). These factors contribute to inter-individual differences in response to apparently identical exposure to a chemical. It is obviously not possible in a chapter of this nature to discuss each of these factors individually, so we have selected two as illustrations of their possible influence on the toxicity of individual chemicals. (NB, A third, species differences, is considered later.)

## Other chemicals

In real, as opposed to experimental laboratory, situations it is rare to be exposed to only one chemical at a time. Chemicals may modify the toxicity of one another in the chemical, distributive and/or interactive phases. Most attention has been focused on the ability of one chemical to modify the metabolism of another by either inhibiting particular drug metabolizing enzymes or causing selective induction, that is bringing about *de novo* synthesis. Induction of drug metabolizing enzymes is a very common phenomenon. Over one thousand drugs, industrial and naturally occurring chemicals have been reported to act as inducers. However, these chemicals fall into a rather small number of categories. The major class of inducers, which is typified by phenobarbitone, manifests its inductive effects on many of the major drug metabolizing enzymes, particularly some types of cytochrome P-450, glucuronyl and glutathione transferases and epoxide hydrolase. However, the effects are largely confined to the liver and intestine. A second important class of inducers, typified by 2,3,7,8-tetrachlorodibenzodioxan (dioxin, TCDD) and many polycyclic aromatic hydrocarbons, is more restricted in the types of enzymes induced but affects many more tissues, including skin, kidney, steroidogenic organs and brain. Alcohol, safrole and other benzodioxoles, clofibrate and pregnenolone carbonitrile represent other classes of inducers. Induction of 'drug metabolizing enzymes' leads to enhanced metabolism of many chemicals. The resulting effect of this enhanced metabolism may be to increase detoxication pathways or to favour formation of toxic metabolites. Thus induction increases the hepatotoxicity of carbon tetrachloride but decreases the neurotoxicity of *n*-heptane. The magnitude of change in toxicity may be considerable; for example, the liver necrosis observed in rats after treatment with carbon tetrachloride is greatly increased when rats are pretreated with alcohols.[3]

There are many possible ways in which one chemical may affect the toxicity of another in the interactive phase. A particular source of concern is the ability of some chemicals to 'promote' the carcinogenic (cancer-causing) action of others. The effects of such promotion may be dramatic; for example, an initial application of a polycyclic aromatic hydrocarbon to mouse skin does not in itself produce tumours but may do so if the animals are treated weeks, months or even a year later with phorbol esters (in castor oil).[4]

Fortunately synergistic toxic effects such as those indicated above

**Table 12.4.** Possible factors affecting toxicity of a chemical

| Factor | Comments |
| --- | --- |
| **Animal-related factors** | |
| Species | *See* text |
| Strain | Very large differences may occur in rat and mice strains, some known, which have a total absence of particular drug metabolizing enzymes or are deficient in certain receptors |
| Sex | On maturity sex differences in toxicity common in rats and mice. Large differences may arise if a chemical affects hormonal status |
| Age | Very young and old animals more vulnerable to many toxins partly because of a deficiency of many drug metabolizing enzymes |
| Hormonal status | May cause differences in response to drugs between the sexes |
| Disease | Stress caused by infection can alter response to chemicals. Infection can also affect metabolic status of body |
| **Environmental-related factors** | |
| Temperature | Differences in toxicity of chemical may occur, e.g. some pesticides (parathion, DDT) more toxic to mice at 27°C than at 1°C and 30°C. Drug response can be affected by temperature but the body temperature can be affected in different ways by drugs at various ambient temperatures |
| Humidity | Humidity closely aligned with temperature. Physiological hydration of animal can affect response |
| Barometric pressure | In man effects of digitalis and ethanol affected by altitude. All such changes attributed to changes in environmental oxygen tension rather than direct pressure changes. However at equal tension the effect of morphine and amphetamine are less at less pressure. Several pressure changes cause stress changes |
| Time of day/month | Little evidence of differences in response due to time rhythms alone. Light and dark exposure affects drug metabolism in rat liver. Mortality due to methadone in rats affected by circadian cycle |
| Caging conditions | Overcrowding can lead to increase in toxicity of amphetamines and other CNS stimulants in rats |
| Stress | Stress can be caused by many factors (e.g. environment, drug treatment). Chemical distribution in body can be affected considerably. Toxicity can also be affected |
| Exposure to other chemicals | *See* text |
| Diet | *See* text |
| **Compound-related factors** | |
| Physical form of chemical | May influence aqueous solubility and therefore intestinal absorption |

**Table 12.4** (*cont.*)

| Factor | Comments |
| --- | --- |
| Presence of impurities | A number of examples known in which toxicity partly due to impurities, e.g. saccharin, 245T |
| Stability of chemical | A number of chemicals are largely degraded prior to absorption, e.g. thalidomide |
| Route of exposure | Route can affect distribution, metabolism and excretion of chemicals |

tend to be the exception rather than the rule. Unless evidence is available to the contrary, the toxic effects of a mixture of chemicals must be presumed to be equivalent to the effect which would be produced by simply adding the toxicities of each of the components of the mixture at the levels at which each is present. This concept is largely ignored in practice. For example, in chemical plants the emphasis is on ensuring that the level of each chemical is not above its TLV; insufficient consideration has been given to how the presence of one chemical at about its TLV should influence the calculation of the TLV for a second chemical.

## Diet

In experimental toxicology animals normally have continuous, immediate access to an unchanging but nutritionally rich diet, and being confined in a small space, segregated from animals of the opposite sex and often any other animals, maintained under conditions of tightly controlled humidity and temperature, it is not surprising that the animals spend most of their waking hours eating. Not unexpectedly, young animals show a very rapid, probably maximal, weight gain. Failure to gain weight as rapidly when animals are exposed to chemicals can be an important parameter of toxicity, although it may be misinterpreted if it is not considered in relationship to diet palatability and intake. Recent findings have cast increasing doubt on the relevance of using these overfed animals in many toxicology studies. Even rather slight dietary restriction has been shown in rats and mice to cause a very dramatic reduction in liver and mammary tumours while mineral imbalance in the diet has been demonstrated to increase significantly urinary calculus formation and risk from bladder tumours[5].

## THE PRACTICE AND PROBLEMS OF TOXICOLOGY

### Selection of animal models for toxicity testing

The magnitude and frequency of species differences in toxicity are obviously important because species of animals have to be selected for practical and ethical reasons as models in which to identify and characterize likely toxic effects in man (*see below*). In experimental animals one can of course only obtain information on signs as opposed to symptoms of toxicity. The animal species most widely used in toxicity testing are mice, rats and dogs. There is also a growing tendency to use monkeys. In addition, rabbits are used for teratology studies and guinea-pigs for skin irritancy testing. It is important to realize that these species are not selected because they have been demonstrated to be especially relevant models for man. Rather their choice is based on the ease of breeding and maintaining these animals, on economic considerations and on the fact that toxicologists have long accumulated experience of distinguishing between chemically induced and 'natural' disease in these species. However, these species show a number of dissimilarities both from man and each other (*see Table 12.5*). It is apparent that no single species is a better universal model for man than another. Considerable differences in toxicity between species are common; they may vary both in the target organ affected and/or in the observed dose-response relationship. For example, in a recent survey of toxicological studies on some fifty chemicals carried out by Dr Ralph Heywood (unpublished data) in which the principal target organs affected in rodents were compared

**Table 12.5.** Some species differences which may lead to differences in toxicity

| | Mouse | Rat | Dog | Man |
|---|---|---|---|---|
| Gastric pH | 2–4 | 2–4 | 1–4·5 | 1·5–2·5 |
| Microflora in stomach | Present | Present | Absent | Absent |
| Molecular weight threshold for biliary excretion | $\simeq 325$ | $\simeq 325$ | $\simeq 325$ | $\simeq 525$ |
| Vomiting reflex | Absent | Absent | Present | Present |
| Gallbladder | Present | Absent | Present | Present |
| Hepatic mixed function oxidase activity | Fast | Fast | Moderate Fast | Slow |
| Diet | Omnivore | Omnivore | Carnivore | Omnivore |
| Sex differences in toxicity | Common | Common | Probably uncommon | Probably uncommon |
| Basal metabolic rate (kcal $g^{-1}$ $h^{-1}$) | $7·9 \times 10^{-3}$ | $4·2 \times 10^{-3}$ | $1·6 \times 10^{-3}$ | $1·0 \times 10^{-3}$ |
| Average lifespan | 2–3 yr | 3–4 yr | 9–11 yr | $\sim 70$ yr |
| Body weight : surface area ratio | 0·43 | 0·62 | 2·08 | 3·89 |
| Hepatic arylamine acetylase | Present | Present | Absent | Present |
| Hepatic glycine conjugase | Present | Present | Absent | Present |

with those in non-rodents, in 34 per cent of cases the same organ was affected, in 20 per cent a similar organ was involved and for 46 per cent of chemicals the target organ was distinctly different (*Table 12.6*). Even using animals of the same sub-strain, age and so on, and housed under apparently identical conditions, there will be differences in the individual toxic response to chemicals. It is relatively uncommon for repeat studies on a chemical to be performed. Where such investigations have been carried out the reproducibility has often been less than impressive. Several interlaboratory studies have been carried out on toxicity tests of short duration. For example, an interlaboratory trial of the Draize eye irritancy test involving some 24 laboratories showed a very considerable variation in scoring the severity of irritancy (*Table 12.7*), even when protocols for the investigations were issued and apparently adhered to. Species differences in the acute lethal dose ($LD_{50}$) have been most widely studied. Variations of tenfold or more are fairly common and for some chemicals much larger differences are observed. For example, in the case of TCDD (dioxin) the guinea-pig is a particularly sensitive species while the dog is about one thousand-fold more resistant.

**Table 12.6.** Examples of failure of present methods of toxicity testing

| Chemical | Effect | Species |
|---|---|---|
| *Effects in man not detected in one or more of the species commonly used in toxicity testing* | | |
| Thalidomide | Fetal abnormalities | Most rat strains |
| Practalol | Skin, eye and intestinal lesions | All species examined |
| Halothane | | All species examined |
| Oral contraceptive pill | Thrombosis | All species examined |
| Dinitrophenol | Cataracts | All mammals examined |
| Azauracil | CNS effects | All mammals examined |
| Clioquinol | Eye | All mammals examined |
| TCDD | Chloroacne | All except hairless mouse |
| Benzene | Leukaemia | All mammals examined |
| β-Naphthylamine | Bladder cancer | Rats and rabbits |
| *Effects detected in one or more common laboratory animals not observed in man* | | |
| Oral contraceptive pill | Bleaching of retina | Monkey |
| | Cancer | Mice |
| DDT, dieldrin and phenobarbitone | Liver tumours | Mice |
| Penicillin | Rapid death after single dose | Guinea-pig |
| Aspirin | Fetal abnormalities | Rats |

**Table 12.7.**Results of an interlaboratory comparison of the *in vivo* Draize test for assessing ophthalmic toxicity in 24 laboratories

|  | Score range from 24 laboratories | | |
| Test compound* | Lowest score | Medium score | Maximum score |
| --- | --- | --- | --- |
| 95% ethanol | 2 | 21·2 | 50 |
| 20 volume cream peroxide | 7 | 48·5 | 83·5 |
| Methyl ethyl ketone | 2 | 19·2 | 55 |
| Ethylene glycol monoethyl ether | 1 | 20·8 | 60 |
| Decaethoxyoleyl ether | 4 | 14·2 | 79 |

*Chemicals applied directly to the rabbit eye. Maximum score 110 based on 80 marks for the degree of corneal opacity and area of damage, 10 marks for damage to the iris, 20 marks for conjunctival redness, chemosis and discharge.

## TESTING FOR TOXIC EFFECTS

### Basis of test selection and design

The major purpose of toxicity investigations is to enable the prediction of hazard. It is obviously impossible to detect every adverse effect of chemicals on biological systems. Lesions which are considered to be common and for which methodology is available for detection and characterization are particularly sought as endpoints. Others, although probably very important — for example behavioural effects — tend to be given scant attention. The acute and chronic effects of chemicals may be distinctly different. For example, benzene produces narcosis after a large single dose but smaller continual doses may cause leukaemia. Acute toxicity data are required in order to assess the hazard of a high but very short term exposure such as may occur following a chemical spillage. Acute findings, however, can only rarely be used to predict the likely toxic effects of continual exposure to moderate doses of a substance which is by far the most common form of exposure both in the work place and in the environment. In order to obtain this type of information, subacute and chronic investigations must be embarked on (*see below* and *Table 12.8*).

It is impossible for purely practical reasons to examine the contribution which each of the factors listed in *Table 12.4* (and briefly discussed above) makes to the nature and magnitude of the toxicity engendered by an individual chemical. The compromise is usually therefore adopted of investigating the effects of just three of these factors — species, sex and development — and standardizing as rigorously as practicable the other possible variables.

The reason for selecting these three variables is that they are felt to represent a wide spectrum of possible biological variability in response to toxic chemicals among mammalian systems. The underlying philosophy is that man is likely to fall somewhere within the spectrum of response. If the variability range is relatively small between species and sexes, then it is frequently reasonable to use the data with some confidence to predict the likely hazard to man. However, if large differences are observed, then the problem arises of which, if any, of the animal models is most relevant to man. This is an especially difficult problem if the large variations between models are related to a chronic toxic effect because cost and time scale may preclude many relevant experiments which could be used to clarify the situation (*Table 12.8*). If, in addition to ascertaining the likely toxic effect of a chemical in man, it is necessary to derive information on its possible environmental impact (ecotoxicology), toxicological investigations on additional exposed species, e.g. fish and birds, may be required. (NB, Ecotoxicological studies may also involve studies on the effect of the chemical on model ecosystems and may require a rather detailed analysis of the 'chemical phase' of the chemical's behaviour in the environment.) In order to control other variables it is normal practice to use, wherever possible, genetically well-defined, healthy animals, of a particular species, of a common age, fed on a diet of known composition and housed in carefully controlled humidity, temperature and lighting, and as far as possible in stress-free conditions. The chemical to be administered should be well defined in terms of physical form, solubility, impurities and stability. The likely route(s) of exposure when the chemical is in use must also be taken into account in considering the tests to be carried out. The most common routes are oral administration either in the food or by gavage, by inhalation (an expensive and technically demanding means of chemical administration) or topical application.

It is unfortunate that the term 'test' has been introduced to describe various components of toxicological investigations because it implies that a specific endpoint is involved. Because in acute and chronic toxicity investigations it is usually the case that it is not known whether a toxic effect will occur and if it does what the form of toxicity will be, the term 'test' is misleading. The concept that there is a 'test' for toxicity has undoubtedly contributed to much of the current misunderstanding about the nature of toxicology work. Unfortunately the word is now established terminology and will therefore be referred to in subsequent sections, although the word 'study' would be preferable.


## ACUTE TOXICITY TESTING

The term 'acute' is used by most toxicologists to mean exposure on one occasion of short duration only. Occasionally it is also applied to multiple exposure over a short period of time, that is up to 24 hours.

**Table 12.8.** Types of routine toxicity tests*

| Types of test | Species normally examined | No. of animals/ dose/test group | Duration | Details | General cost range based on three doses and a minimum number of animals |
|---|---|---|---|---|---|
| *In vivo* mammalian test | | | | | |
| 1. Acute ($LD_{50}$) | Rats, mice | 5–10 | 7–10 days | Single exposure<br>$LD_{50}$<br>Acute effects<br>Clinical signs<br>Gross and histopathology | £600–£1200 |
| 2. Short term | Rats, mice | 5–10 | 7–28 days | Multiple exposure<br>Range–finding study<br>Clinical signs<br>Clinical chemistry<br>Gross and histopathology | £15 000–£30 000 |
| 3. 90-day | Rats, mice, non-rodents (e.g. dogs) | 20–30<br>4–8 | 90 days | Regular dosing (daily)<br>Clinical chemistry<br>Clinical signs<br>Gross and histopathology | £25 000–£50 000 |
| 4. Long term | | | | | |
| A | Rats, mice, non-rodents (e.g. dogs) | 50–100 | 6 mth–1 yr | As 90-day study | £50 000–£150 000 |
| B (carcinogenicity) | Rats, mice, non-rodents (e.g. dogs) | 50–100 | Life span (e.g. 2 yr rats 7 yr beagles) | Tumour development and as 90-day test | £100 000–£350 000 |

| 5. Reproduction | Rats, mice, rabbits | Starting with 20 females | Up to 1 yr | Offspring born by caesarean section; general abnormalities on fetus and progeny; toxicological studies on dosed animals; followed through 3 generations; effects on males may also be studied |
| 6. Irritancy | Skin — guinea-pigs<br>Eyes — rabbit | | Up to 3 weeks | Visual examination and histochemistry |
| *In vitro* tests<br>Bacterial mutation tests (e.g. Ames tests)<br>Cell transformation<br>DNA repair<br>Mammalian mutagenicity | | | 3 days–4 weeks | £500–£2000 |

*An *in vivo* study for toxicity testing generally consists of tests 1, 2 or 3 and test 4A. For carcinogenicity, tests 1, 2 or 3 and test 4B are generally carried out. The cost of the various tests depends on the route of administration, species and the number of doses tested. The cost for a study of 3 doses and a minimum number of animals is given.

Acute toxic effects may be lethal or non-lethal. The signs and symptoms of a non-lethal effect may be overt or subtle. For example, the insecticide Mipafox has been shown to cause paralysis in man while some anticholinesterase pesticides have been reported to affect attention and short term memory through their effects on the hippocampus. Certain organotin pesticides cause cerebral oedema while provoking only very minimal histologically observable changes.

To aid interpretation, statistical inferences based on population effects are often employed rather than individual risks. The symbol ED is used by convention to indicate 'effective dose', and if the endpoint to be measured is death, then the symbol employed is LD (lethal dose). The percentage of animals affected is expressed by a subscript, hence $LD_{50}$ is the lethal dose (usually expressed in mg/kg) to 50 per cent of the animals. Sometimes the term MLD is used instead of $LD_{50}$ (median lethal dose). In acute toxicity tests involving whole atmosphere exposure, for example aquatic and inhalation toxicology, the effect levels are expressed in terms of concentrations rather than doses and so the terms effective concentration (EC) and lethal concentration (LD) are used in preference to ED and LD for such situations.

Results are normally plotted in graphical form to quantitate the toxicity. If linear : linear and linear : log scales are used, sigmoidal curves are normally obtained (*Fig. 12.3*). The term probit is well known and is widely used in statistics. It is derived from the word 'probability unit'. In essence it is a mathematical trick to linearize a sigmoid curve. It is inappropriate to go into details in this chapter as to how it is derived.[6] Increasingly, probit mortality and log dose plots are employed in order to obtain linear presentations which by extrapolation theoretically enable the highest non-lethal dose to be identified (*Fig. 12.5*). It should be realized that two chemicals with an identical $LD_{50}$ may give quite different slopes. To extrapolate $LD_{50}$ information to possible acute hazard in low dose situations it is clearly very important to have a knowledge of the slopes of the graph as well as the $LD_{50}$ value. Measurement of $LD_{50}$ values is commonly carried out as part of the determination of the acute toxicity properties of chemicals, but it must be emphasized that it is only *part* of such a study (*see below*). The dose causing 50 per cent of deaths rather than some other percentage figure is selected largely because it can be measured with the most precision and is strongly favoured by legislative authorities. Because many variables may affect the $LD_{50}$ value the term should always be qualified, not only by the species, but also by information on the other conditions under which it was obtained. Two- to threefold differences in $LD_{50}$ often occur between sexes, between animals of different ages and between strains of animal of the same species. In addition, the route of administration and the time at which death is recorded must be defined. For example, in the

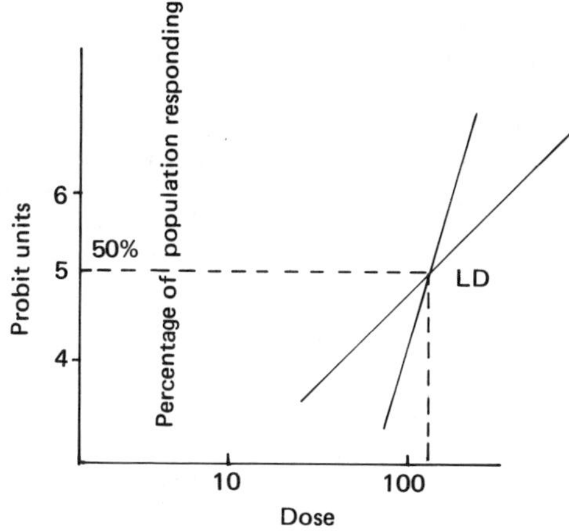


*Fig. 12.5.* Determination of LD$_{50}$ (lethal dose).

case of TCDD (dioxin) guinea-pigs appear to survive fairly high levels if deaths are recorded within a few days of exposure. However, between 20 and 40 days many animals die and the LD$_{50}$ reduces to $\sim$ 0·6 $\mu$g/kg body weight, that is, to extremely toxic levels. The main uses of acute toxicity testing are:

To predict acute hazard to accidentally exposed humans and animals.
To help in deciding the exposure levels which should be used in subacute and chronic toxicity investigations.
To satisfy legislative authorities.
To check for toxic contaminants in biological products such as vaccines.

The number of animals used by investigators to ascertain acute toxicity varies considerably. The number of rats and mice per dose level for an acute toxicity study is commonly five. Except for rats and mice it is relatively rare to find significant sex differences in toxic response; the value of insisting on studies in both sexes is therefore problematical. In addition, it is not sensible to insist on LD$_{50}$ data via the parenteral route for most industrial chemicals. (NB, Information on differences in toxicity between oral and parenteral routes may be helpful in defining rates of intestinal or skin absorption.) For expensive species an accurate LD$_{50}$ figure is both impractical and in most cases unnecessary. In addition to counting the number and carrying out a pathological examination of dead and moribund animals at regular intervals, careful observation of live animals should be continued and untoward behaviour recorded until the observers are confident that the surviving animals are no longer showing obvious effects from exposure to the chemical; typically this period is around 7 days. At the end of this time the animals should be sacrificed and at least some of them examined for gross and histopathological changes. (NB, Necropsies on animals that have died after a number of days of treatment are usually much more informative than those on animals that have died rapidly.) It is important to realize that even when the variables mentioned above are very closely defined the $LD_{50}$ value is still very far from being an absolute figure. Repeat experiments under apparently almost identical conditions often produce variations of $> 40$ per cent. If large species differences occur in the $LD_{50}$ value the problem arises of which is the most relevant to the human situation. In an attempt to identify whether one species is more predictable for effects in man than another, Krasovskij[7] has analysed the data on the acute toxicity of some 260 chemicals in man and other mammals. Krasovskij concluded that man was not much more sensitive than other species examined. In *Table 12.9* this is illustrated through a comparison of acute toxicity data in man and rats. Interestingly, Krasovskij found that no one non-human species was generally more sensitive than another. Thus for acute toxicity, sensitivity is not generally directly relatable to anatomical or physiological characteristics, to body weight or to metabolic rate. If large species differences in acute toxicity are found, further investigation of the mechanism(s) involved in the toxicity are often needed to resolve such problems. Frequently the differences may be explained largely by discrepancies in the distributive phase between species. There is an increasing tendency among legislative authorities for $LD_{50}$ values to be isolated from acute non-lethal toxicity data and used directly to compare the relative safety of quite different types of chemical. This approach is questionable for in a number of cases it is those chemicals with high $LD_{50}$ values that have highly undesirable non-lethal acute effects which may even be irreversible, while compounds with lower $LD_{50}$ values may show no such sub-lethal effects. Acute toxicity data must be examined

**Table 12.9.** Ratios of acute toxicity of various chemical classes in man compared with the rat

| Chemical class | No. of compounds | Sensitivity ratio man : rat* |
|---|---|---|
| Organochlorine compounds | 20 | 1·8 ± 0·3 |
| Organophosphorus compounds | 16 | 1·9 ± 0·46 |
| Inorganic compounds | 40 | 4·2 ± 0·66 |
| Drugs | 62 | 10·5 ± 5·8 |

*From* Brown.[6]
*Where rat = 1.

as a whole if significant assessment of hazard is to be achieved. $LD_{50}$ values alone give little idea of the 'safe' levels of acute exposure in either animals or man.

## SUB-ACUTE (SUB-CHRONIC) AND CHRONIC TOXICITY INCLUDING *IN VIVO* CARCINOGENICITY TESTING

Repeat dose studies may vary in duration from around 14 days to lifetime investigations. Before embarking on long term studies, short term range-finding studies are needed to determine the appropriate dose levels to be used. The dose which causes some tissue damage or produces plateau blood levels or reaches a maximum of about 5 per cent of the diet or provokes a significant reduction in weight gain is particularly sought. This dose is termed the 'maximum tolerated dose'. Three dose levels and a control are normally selected. One group of animals is given the maximum tolerated dose, a second group, half the maximum tolerated dose or geometric spacing and the third group, a small multiple of the likely human daily exposure if this is known. If the likely human exposure levels are not known, the lowest dose should be one which is not expected to produce a toxic effect. The number of animals required in a toxicity study is dependent on the required sensitivity (*Fig. 12.6*). Broadly speaking, the longer the study the greater the number of animals used. Fifty animals per group are commonly used in a carcinogenicity study. The longer the duration of a study the greater is the prospect that the control group will show a

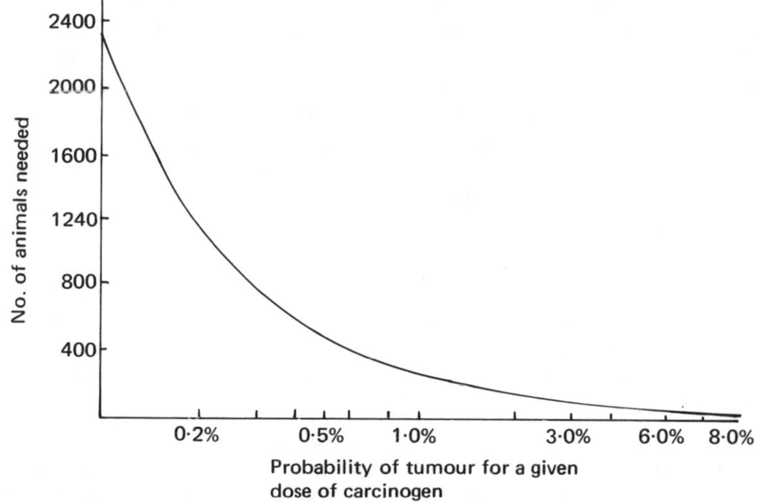


*Fig. 12.6.* Numbers of animals needed to be 90 per cent certain of finding one animal with a tumour, assuming control animals have no tumours.

significant amount of 'natural' disease, for example in lifetime studies in rodents some members of the control group are likely to show tumours in one or more organs. This incidence of 'background disease' may influence the sensitivity of a test by interfering with the identification and characterization of the effects due to the chemical. Even when a significant incidence of 'natural' disease is present, effects due to the chemical may be distinguished from it first because the particular lesion is not seen in the controls and secondly because the same lesion occurs but develops earlier and/or with increased frequency in the test animals.[2]

Most carcinogenicity studies are confined to rats and mice. Chronic studies may be performed on dogs and monkeys. Good quality animals and very high standards of animal husbandry are vital for long term toxicity studies. Many of the factors listed in *Table 12.4* potentially have a much greater effect on long term toxicity tests than on acute tests and this may worsen still further the problem of distinguishing and characterizing differences between changes in the test animal and those in control animals. Statistical tests of the significance of a difference between control and test groups are often misused. Evidence for a dose–response relationship should be sought and particular attention given to whether a threshold dose for production of the lesion exists. Such information is vital for extrapolation purposes. Whereas few toxicologists would disagree on whether an animal was dead or not, the diagnosis of many other lesions is much more controversial. This is particularly the case for diagnosis of tumours. As a consequence, the interpretation of the same lesions by acknowledged expert pathologists may differ considerably and lead to differences in the type of advice a toxicologist would give. This may result in an impasse because the cost of a carcinogenicity study in a single species (*Table 12.8*) may be £100 000 or more which, except in special cases, may rule out a repetition of the experiment. It is therefore extremely important in long term toxicity tests to pay very careful attention to design in order to derive the maximum, relevant information from the study. Many methods in addition to pathology may be used to seek and characterize the development of toxic effects, including haematology, clinical chemistry of blood and urine, careful physical examination of the live animals including ECG and respiration rate, measurement of body weight gain and food intake. The animals should be observed for the development of behavioural changes. Behavioural investigations, although extremely important, are often poorly carried out. The selection of variables to monitor is an essential aspect of test design. The choice of variables to measure must be based on the consideration of the toxic properties of similar chemicals and of the organs most affected in the acute toxicity and dose range-finding studies.

A common protocol for all chronic toxicity tests is quite inappropriate and, despite its appeal to administrators, it should be resisted. The problem of extrapolation of results to the human or environmental situation is greatly exacerbated by the fact that, on economic grounds, the number and nature of the species investigated is very limited. Studies, even those of questionable quality, cannot usually be repeated. It is becoming increasingly apparent that certain types of lesion in particular species have little or no relevance to the likely toxic effects of chemicals in man. For example, liver tumours are produced in the mouse by chemicals such as DDT, phenobarbitone and chloroform, whereas epidemiological evidence based on humans exposed to these chemicals for many years demonstrates that is very unlikely in man. Information on situations in which animal models give invalid warnings of possible hazards to man from chronic exposure to chemicals is, of course, rather limited because good data on levels and duration of exposure and toxic effects of particular chemicals in man are rather rare. New chemicals which produce untoward effects in animals are not usually viable commercially.

## Reproductive toxicity tests

Before the early 1960s, it was generally believed that the placenta provided an effective protection for the fetus against the toxic effects of chemicals and that chemicals which did cause toxic effects in the fetus would also produce toxic effects in the mother at similar doses. The thalidomide disaster and later the demonstration of transplacental carcinogenicity by diethylstilboestrol provided tragic proof that neither assumption was correct. In the aftermath of these events, legislation has been introduced requiring the testing of chemicals for their effects on the reproductive organs and the development of the fetus. Sex discrimination legislation has added impetus to the needs to examine the effects of many industrial chemicals on reproduction because the approach of restricting women of child-bearing age from certain types of working environment is becoming less acceptable to our society. Furthermore, although the numbers of spontaneous abortions and malformed newborn in the human population each year is quite large, our knowledge of the aetiology of these effects is only fragmentary. Survival of a species depends on the integrity of its population of germ cells and therefore chemicals which might affect the human reproductive system are a particular cause for concern. A number of chemicals may affect sperm production to a greater or lesser degree including toluene, benzene, xylene and 1,2-dibromo-3-chloropropane. Cadmium, methylmercury and hexafluoracetone are toxic to both male and female reproductive systems. Apart from these effects on the development of the sperm and ova, chemicals might also interfere with fertilization, formation of the blastocyst, implantation, embryo development by either a direct or indirect effect — for example, by influencing maternal nutrition to the fetus via placental effects or to the newborn via lactation — birth and adjustment to the postnatal environment or postnatal growth and maturation. A composite testing procedure has been adopted in an attempt to detect all these effects. The basic approach is that the compound is administered at three different dose levels to rats, mice and/or rabbits during two to three generations by the route to which the human population is likely to be exposed. Undoubtedly this approach would detect another thalidomide but may miss more subtle effects, for example modification in brain development. Species differences occur, for example thalidomide is teratogenic in rabbits and man but not in rat strains, while aspirin is a teratogen in rats but apparently not in man. As in other areas of toxicology there is no ideal model for the human situation.

**Irritancy testing**

Skin and eyes frequently come in contact with chemicals and effects on them are usually visible. It is therefore not surprising that irritancy and inflammation of these organs is by far the most common form of toxicity in the industrial environment and an important cause of loss of production. A number of tests have been devised to identify the irritant and/or immunological effects of chemicals on these organs. Irritant effects range from subjective discomfort without any obvious inflammation through all grades of inflammation up to gross tissue destruction. The effects are generally dose-related and although they often occur after a single dose, several exposures may be needed before inflammation is detected. The usual technique for detecting irritants is to apply known doses of the chemical under test to a small patch of lint or gauze to the shaved backs of animals, the patch being secured by adhesive tape for the required period of exposure.

This test is by no means ideal. The occlusion of the application site may lead to accumulation of any solvent that is used and may also cause changes in the hydration state of the skin, thus altering its permeability characteristics to the chemical and any solvent used. The method and care with which the hair is removed may also influence the observed irritancy. Skin of hairy animals tends to be much thinner than that of man and therefore more sensitive to irritants. Fortunately, in contrast to the tests described above, for many chemicals patch tests can be applied to man himself thus providing direct toxicity information. If clear irritant effects occur they are likely to be experienced by many individuals. In contrast, sensitization effects are far less predictable. Sensitization, once developed, may persist for many years. This poses ethical problems in the use of human volunteers to detect sensitizers; moreover it is very difficult to verify whether a safe level exists for sensitizers. Guinea-pigs are normally used as the animal model, the chemical being applied at levels below those which cause an irritant effect. Pre-conditioning of the skin by irritants is often employed. Other effects on the skin such as changes in pigmentation intensity may also occur and should be looked for. Extrapolation of animal results to man varies according to the type of lesion: effects such as chloracne (caused by many aromatic chlorinated chemicals in man) are not detected by our present animal models.

Irritant effects on the eye may also require assessment, the rabbit eye being the favoured model. The test is far from ideal, it is difficult to reproduce and there is little mechanistic basis for many of the observed effects. Considering the importance of skin and eye effects in the industrial environment, surprisingly little is known about them. A particularly important area of ignorance is the factors which render some individuals much more susceptible to the topical effects of chemicals than others. Rapid advances in the field of immunology may soon provide answers to this problem. For both sensitization and irritancy *in vitro* models are both theoretically possible and highly desirable; however, as yet these 'tests' are in their infancy.

## Environmental impact testing (ecotoxicology)

Establishment of valid models for assessing the environmental impact of chemicals is a very difficult proposition. There is insufficient scientific basis for identifying which organisms to study, let alone what effects to look for and species : species interactions must be considered.

Pesticide environmental impact studies have become quite sophisticated, but for water and airborne industrial effluents, where studies are carried out at all, they typically consist of rather primitive field studies of easily identifiable fauna and/or flora in the immediate environment of the effluent. Good chemical analysis is an integral part of investigating the adverse effects of chemicals on the environment. Such investigations must include studies on the ease of chemical and biological degradation of the product under scrutiny. A particular problem in environmental impact studies is that in the wild the environment is continually changing naturally and therefore comparison of the test and 'control' situations may be complicated. It is important for toxicologists to realize that the adverse effects of chemicals on animals in the wild may be much more subtle than simply overt toxic signs. For example, even a relatively small reduction in the mobility of an animal may inhibit its ability to catch its food and/or render it a ready prey to predators. For further information on this topic the reader is referred to books by Duffus[8] and Butler.[9]

**Short term testing**

The very high cost and length of conventional carcinogenicity tests has led to very great interest in the development and use of rapid screening tests for the detection of chemical carcinogens. Many tests have been developed (*Table 12.10*).[10,11] Most are based on the ability of chemicals to induce gene or chromosome damage in somatic or germ cells. The underlying presumption in these tests is that there is a relationship between DNA damage and chemical carcinogenesis, hence certain types of DNA-damaging agents are likely to be carcinogens. Although there is much evidence in support of this, some chemicals could be carcinogenic by affecting other cellular constituents, for example, RNA and proteins.

The short term test that has made the biggest impact so far is the bacterial mutation assay, commonly referred to as the Ames test. In this test the ability of chemicals to induce reverse mutation in specific strains of *Salmonella typhimurium* in the presence of a rat liver microsome fraction is assessed. Gene mutation can also be assessed in cultures of yeast (for example, *Saccharomyces cerevisae),* in insects (for example, *Drosophila melanogaster*), or in mammalian cells in culture (for example, mouse lymphoma cells, Chinese hamster cells and human fibroblasts or lymphocytes). Other tests have been designed to measure DNA repair rather than actual gene mutation. DNA repair may be assessed in bacteria, yeast, Drosophila and also in mammalian cells in culture. Carcinogen-invoked cell transformation *in vitro* has appeal as an indicator of malignant changes because it appears to parallel the development of cancer *in vivo* in a number of respects. One objection to the use of cell transformation as a test is that the cell line employed is usually a partially transformed fibroblast line. As few human cancers arise from fibroblast transformation the relevance of results obtained using these cell lines for the human situation is questionable.

*In vivo* and *in vivo/in vitro* tests may also be used to detect carcinogens. Included in this group of tests are cytogenetic studies, cell transformation and bacterial mutations. In one form of test (host-mediated assay) mutated bacteria are inserted into the peritoneal cavity of a rat and the test chemical administered. After 24 hours the bacteria are removed and cultured to detect a change in mutation frequency. The claimed advantage of this type of test is that the drug metabolizing enzymes are in their normal relationship to one another and therefore the results are more relevant to man. Since no single test is likely to detect all potential carcinogens a battery of short term tests has been proposed as the most appropriate screening system. The problem is to decide which tests should comprise this screening battery. All short term screening tests have drawbacks. One disadvantage of many *in vitro* tests is that they do not provide adequate drug

| Organism | Gene mutation | DNA repair | Chromosome damage |
| --- | --- | --- | --- |
| Bacteria | *Escherichia coli*<br>　Fluctuation test<br>　Host-mediated assay<br>　Induction of prophage<br><br>*Salmonella typhimurium*<br>　Ames test<br>　Fluctuation test<br>　Host-mediated assay | *Escherichia coli*<br>Bacillus | |
| Fungi | *Saccharomyces cerevisae*<br>*Neurospora crassa*<br>Aspergillus<br>　Fluctuation test<br>　Mutation assays<br>　Host-mediated assay | *Saccharomyces cerevisae*<br>Aspergillus<br>　Mitotic gene conversion | *Saccharomyces cerevisae*<br>　Mitotic non-disjunction<br>Aspergillus<br>　Meiotic non-disjunction |
| Plants | Tradescantia | | Tradescantia<br>*Vicia faba*<br>*Allium capa*<br>　Chromosome aberrations |
| Insects | *Drosophila melanogaster*<br>　Induction of sex-linked recessive<br>　lethals | *Drosophila melanogaster*<br>　Non-disjunction<br>　Deletions<br>　Dominant lethal mutations<br>　Translocations | |
| Mammalian cells | Mouse lymphoma L5178Y cells<br>　Host-mediated assay<br>　Fluctuation test<br>Chinese hamster cells, V79, ovary,<br>　lung cells<br>Human fibroblast cells<br>Human lymphoblast cells | HeLa cells<br>Hepatocytes<br>Human fibroblasts<br>　Unscheduled DNA synthesis | Chinese hamster cells<br>Human and rat lymphocytes<br>　Chromosome aberrations<br>Chinese hamster cells<br>　Lymphocytes<br>　Sister chromatid exchange |
| Whole animals | Mice<br>　Coat colour spot test<br>　Host-mediated assay | | Mice<br>　Micronucleus<br>　Sperm abnormalities<br>　Sister chromatid exchange<br>　Chromosome aberration |

metabolizing facilities. Other facets of the distributive phase of a chemical that occur *in vivo* may also not be mimicked in an *in vitro* test, for example tissue binding. It is therefore very difficult to extrapolate from *in vitro* studies to the human situation; for example, methyl orange causes mutations *in vitro,* but *in vivo* it is unlikely to reach the relevant drug metabolizing enzymes and be converted to reactive metabolites due to its highly lipophobic nature. Most of the present short term tests for carcinogens are based on detecting gene mutation *in vivo*. They are therefore unlikely to detect chemicals that may be carcinogenic through a non-genetic mechanism. With the exception of asbestos and possibly the carcinogenic metals the number of human carcinogens in this category is uncertain. The present methods are also not suitable for identifying promoters, which a number of workers feel may be particularly important for the development of human cancer. A two-tier screening system for carcinogens has been proposed in which *in vitro* tests would constitute the first tier and *in vivo* tests the second. However, it is probable that some form of short term *in vivo* test should be included in the initial screening battery of tests.

The problem of extrapolation of results obtained in short term tests to the likelihood of a chemical causing cancer in man is obviously crucial. A positive result in a test indicates that the agent has the potential to react with DNA (it may need to be metabolically activated to do so). In the interpretation of most of these tests no account is taken of the contribution of differences in distribution or balance of activation or detoxification pathway *in vivo,* nor of the fact that only genetic damage is being detected. It has been suggested that a potency correlation between carcinogenicity and mutagenicity is also possible, at least for structurally related chemicals. This would appear unlikely to be the case for the *in vitro* tests as they are currently performed, unless appropriate pharmacokinetic data are available and taken in conjunction. A particular use of *in vitro* tests is in situations in which a chemical similar to a known human carcinogen is under investigation. In such cases a measure of prediction is often possible based on *in vitro* results alone.

It is now possible to conclude that there are several good screening tests, with much data to back them up, which are quite economical and reproducible, but these tests must be used in conjunction with other toxicological data to estimate a likely risk to man.

## PRIORITIES IN TOXICOLOGY

Estimates have been made that between 30 000 and 50 000 chemicals currently in use in significant amounts require some further toxicological examination. (NB, For the majority no toxicology data or acute data only are available.) In addition, between 500 and 2000 new products are introduced each year which require some toxicological investigation and the toxic properties of a number of naturally occurring compounds ought also to be evaluated. Even if we could automate our present test procedures we could not examine a significant proportion of these chemicals. Moreover, it is clear from the previous sections that prediction of likely hazard to man and to the environment from exposure to a particular chemical using our present approaches is expensive, requires very considerable technical expertise and may sometimes produce findings of questionable value. In view of these problems it is disturbing that the legislative authorities in many countries are concentrating their efforts on laying down rather rigid protocols as to how each test should be conducted. Furthermore, these protocols may vary from one country to another. Such emphasis on the details is ill-placed when the overall approach to evaluating the toxic properties of chemicals requires an urgent re-examination. In order to improve the predictability of hazard from toxicological investigations much more effort must be devoted to devising more appropriate methods of investigating the toxicity of both pure chemicals and mixtures of chemicals. The assessment of toxicology properties of mixtures requires the development of methods that can be used directly in the field, that is an updating of 'the canary in the mines' approach. To improve our methods requires:

1. Much more reliable information on the relationship of chemical structure and toxicological activity for particular types of toxic effect, an understanding that will only develop by improvements in our knowledge of mechanisms of toxicity. In this regard it is worth noting that the mutagenicity tests for carcinogens have arisen from research investigations of the initiating events for cancer development.

2. A critical re-evaluation of both the successes and failures in ascertaining hazard to man and other species of our present models for identifying and characterizing the toxicological properties of chemicals. Major progress in this quest could be achieved by a retrospective examination of the data on those chemicals which have been properly studied in animals and for which significant amounts of data on the effects in man exist. Only by such a study will the pros and cons of individual models be properly established. To achieve success in this field will require close collaboration between industry, research establishments and universities. It needs a new more flexible attitude from industry on the confidentiality of toxicology data on its products. The potential gains of more valid, more cost-effective toxicological studies with a concomitant and significant reduction in the loss of valued products, which at present can arise through the generation of toxicological data of dubious relevance in respect of hazard, will provide an incentive for such collaboration.

3. The thorny issue of obtaining some forms of toxicological data in man *in vivo* and/or in human tissues has to be tackled. Clearly there are ethical problems to be overcome; however, the ethics of our present approach which involves subjecting large numbers of animals to a lifetime exposure to high levels of chemicals must also be questioned if only because the scientific validity of such a procedure for assessing the likely hazard to man of exposure to low levels of the chemical may be doubtful.

It is apparent that selection of priorities in toxicology must involve a much wider population than toxicologists alone, for it is necessary to establish what types of product and which forms of toxicity should be focused on first. This latter aspect is particularly pertinent. At present the priority is to pay particular attention to the cancer-causing properties of chemicals. It is important to realize that giving detection of cancer-causing agents priority must entail, because of the limitation of resources, the relative neglect of other types of chronic toxicity such as effects on the newborn or on the central nervous system. Those concerned with occupational medicine have a very important role to play in explaining these problems in a lucid manner so that society can make informed decisions on the priorities.


## Acknowledgements

The authors thank Professor P. Grasso for his useful discussions and Mrs W. Horwood for preparing the typescript.


REFERENCES
1. Bridges JW and Chasseaud LF. *Progress in Drug Metabolism,* 5. Chichester: Wiley, 1980; 311–43.
2. Dayan AD and Brimblecombe RW. *Carcinogenicity Testing.* Lancaster: MTP Press, 1978.
3. Doull J, Klassen CO and Amdur MO. *Toxicology,* 2nd ed. New York: Macmillan, 1980; 206–31.
4. Boutwell RK. The function and mechanisms of promoters of carcinogens. *CRC Crit. Rev. Toxicol.* 1974; 2:419–69.
5. Roe FJC. Food and cancer. *J. Hum. Nutr.* 1979; 33:405–15.
6. Brown VK. *Acute Toxicology.* Chichester: Wiley, 1980.
7. Krasovskij GN. Extrapolation of experimental data from animals to man. *Environ. Hlth Perspect.* 1976; 13:51–8.
8. Duffus JH. *Environmental Toxicology.* London: Arnold, 1980.
9. Butler GC. *Principles of Ecotoxicology: Scope 12.* Chichester: Wiley, 1978.
10. Paget GE. *Mutagenesis in Sub-mammalian Systems.* Lancaster: MTP Press, 1979.
11. Stich HF and San RHC. *Short Term Tests for Carcinogens.* New York: Springer-Verlag, 1981.

# 13. MORTALITY IN A SMALL INDUSTRIAL TOWN:
## Problems of Analysis and Interpretation
*Owen Ll. Lloyd*

**INTRODUCTION**

The evaluation of an isolated and unusual finding is always difficult in epidemiology. The recourse to replication studies — that instinctive stand-by of scientists in other disciplines — can often be impracticable, unrewarding or unconvincing in epidemiological research. The reason for that handicap is, of course, that the environmental, demographic and genetic differences between different areas and populations may be both innumerable and cryptic. In these circumstances, 'control' and 'replicant' populations are so defined only with a large measure of hope.

In some instances, a pathogenic stimulus in the environment of a community can be readily identified. At Seveso, for example, a loud explosion was followed by the appearance of a cloud of vapour. Contact of this vapour with the eyes and skin respectively caused lacrimation and a stinging sensation; a disfiguring ailment, chloracne, followed shortly thereafter. Thus, both the stimulus and the early responses were obvious. Hence the causal link of chemical exposure was easily established.

In other instances, also, the pathological sequelae of a hidden toxin are so bizarre that questions about more insidious environmental poisons become inevitable, sooner or later. The cases of Minamata and Itai-Itai diseases were examples, as was hepatic angiosarcoma.

However, a major problem arises in epidemiology when it is a common disease whose incidence or prevalence is found to be abnormally high in a community, and where the environmental toxin is not patently obvious. When that community is a small one, furthermore, the statistical bogy of small numbers distils its 'postulates of impotence'[1] into the apprehensive cogitations of the epidemiologist.

Yet, 'humane physicians should be set to explore the nature, condition and constitution of the tiniest village. They should investigate its diseases and their causes in the most precise detail . . . In this way they would prepare for each district a kind of special geography.'

These encouraging words were written almost exactly two centuries ago by one of the pioneers of public health, Johann Peter Frank.[2] Further encouragement, too, for the contemporary 'humane physician' comes from the recent developments in statistical methodology for detecting 'clusters' of disease, despite the problems of small numbers.[3] And while using such methods it is to be hoped that physicians will also remain aware that yet more methods await discovery or application in problems of environmental epidemiology.

In this chapter an epidemiological episode will be described. This episode possesses the familiar epidemiological handicaps: a small community, an epidemic of common disease and no glaringly obvious toxic factor in the environment.

Also described will be various analytical techniques. Either they are relatively unfamiliar to doctors (although well proved in other disciplines) or they are in the early stages of evaluation. In the latter case, they are presented merely in order to stimulate critical thought and, thereby, to generate further research towards improved techniques for analysing problem epidemiology and problem environments.


## TOWN V, THE SOCIOECONOMIC BACKGROUND*

Town V is a long-established community in central Scotland. It was founded in 1790. It remained a wayside hamlet for only a few years until the Industrial Revolution stimulated its growth into a small burgh, through the exploitation of local coal deposits. This growth, however, has remained moderate. In 1971 the burgh V had a population of about 7100 — just 1000 more than in 1961. Since 1971 there has been little further increase. Only a thin 'shell' of housing has expanded the area of town V beyond the boundaries of the original burgh. Since regionalization in 1974, the burgh has ceased to exist as an administrative entity.

For many decades the major industries in town V were the traditional ones, brickworks, steel foundries and (until its closure in 1964) a colliery. Nowadays, many workers travel to neighbouring towns for employment, mainly at collieries or in a large vehicle factory.

In its socioeconomic structure the population of town V has remained homogeneous. Private house ownership, for example, was 6 per cent in 1968. The percentages of inhabitants in social classes I and II have been negligibly small, according to recent census data.

For all relevant parameters of social deprivation (for example, the parameter of domestic overcrowding), town V was unexceptional in recent decades.

*The terms 'burgh' and 'town' used herein are not precisely synonymous. In brief, town V comprises burgh V plus small areas of adjacent housing which lie outside the administrative area of the burgh proper.

So far as health care of town V was concerned, for many years the medical needs of that working class population have been met by a stable group practice of dedicated general practitioners. More elaborate medical services have been available at the district general hospital, about six miles away. Thus, there have been no sudden changes in diagnostic or therapeutic facilities in the recent history of town V.

Retrospectively, therefore, town V has shown itself to be unexcitingly humdrum in everything demographic and socioeconomic. It has reserved its epidemiological surprises for its record of mortality.


## RESPIRATORY CANCER IN TOWN V

The history of this disease in town V is shown in the number of cases described in various sources (*Table 13.1*).

(*a*) numbers of deaths from respiratory cancer as noted for the burgh (that is most of town V) by the Registrar-General for Scotland in his Annual Reports; for the years when that source was unproductive, the relevant death certificates for town V were counted in the parish registers;

(*b*) numbers of registrations of respiratory cancer for town V in the local cancer registry;

(*c*) numbers of deaths or discharges with the diagnosis of respiratory cancer for the parish of town V, from the records of the local hospitals.

From the data in *Table 13.1,* it can be seen that the annual numbers of deaths from respiratory cancer averaged around 2 until 1967. But thereafter, the numbers were never less than 6.

Using Scottish experience as the 'norm', the mean standardized mortality ratios (SMRs), with twice the standard error of the means, were calculated for the two 7-year periods 1961–7 and 1968–74. The mean SMRs for the 3-year periods on either side of 1961–74 were also calculated (*Fig. 13.1*). (The SMRs for 1975–7 were approximations only: the Registrar-General's Reports published after regionalization provided neither the observed deaths for burgh V nor the estimated populations of the small burghs, from which the expected deaths for burgh V could be calculated. Hence for these years, the observed deaths were those for the slightly larger area of the town while the expected deaths were derived on the assumption of a continuation of the previous demographic trends in the burgh.)

The distinctive pattern, established for the absolute numbers of cases, reasserted itself in this part of the mortality study. The mortality from respiratory cancer was generally low before 1968, but consistently high thereafter.

**Table 13.1.** Numbers of cases of respiratory cancer in town (or burgh) V, 1951–77, by source

|  | 1951 | '52 | '53 | '54 | '55 | '56 | '57 | '58 | '59 | '60 | '61 | '62 | '63 |
|---|---|---|---|---|---|---|---|---|---|---|---|---|---|
| Deaths | 0 | 0 | 0 | 6 | 2 | 1 | 0 | 4 | 1 | 1 | 2 | 1 | 5 |
| Registrations | - | - | - | - | - | - | - | - | - | - | 1 | 2 | 2 |

|  | '64 | '65 | '66 | '67 | '68 | '69 | '70 | '71 | '72 | '73 | '74 | '75 | '76 | '77 |
|---|---|---|---|---|---|---|---|---|---|---|---|---|---|---|
| Deaths | 3 | 2 | 1 | 2 | 7 | 7 | 8 | 7 | 7 | 10 | 9 | 6 | 9 | 7 |
| Registrations | 2 | 0 | 1 | 3 | 6 | 6 | 8 | 7 | 6 | 9 | 7 | - | - | - |
| Deaths/discharges | 3 | 5 | 3 | 6 | 21 | 18 | 14 | 12 | 13 | 8 | 14 | - | - | - |

**Table 13.2.** Annual SMRs and Cu-sum values for respiratory cancer in town V, 1958–77

|  | 1958 | '59 | '60 | '61 | '62 | '63 | '64 | '65 | '66 | '67 |
|---|---|---|---|---|---|---|---|---|---|---|
| SMR | 156 | 37 | 35 | 71 | 33 | 155 | 89 | 57 | 28 | 53 |
| Difference from target | +56 | −63 | −65 | −29 | −67 | +55 | −11 | −43 | −72 | −47 |
| Cu-sum | +56 | −7 | −72 | −101 | −168 | −113 | −124 | −167 | −239 | −286 |

|  | '68 | '69 | '70 | '71 | '72 | '73 | '74 | '75 | '76 | '77 |
|---|---|---|---|---|---|---|---|---|---|---|
| SMR | 179 | 172 | 183 | 158 | 154 | 214 | 183 | 126 | 179 | 142 |
| Difference from target | +79 | +72 | +83 | +58 | +54 | +114 | +83 | +26 | +79 | +42 |
| Cu-sum | −207 | −135 | −52 | +6 | +60 | +174 | +257 | +283 | +362 | +404 |

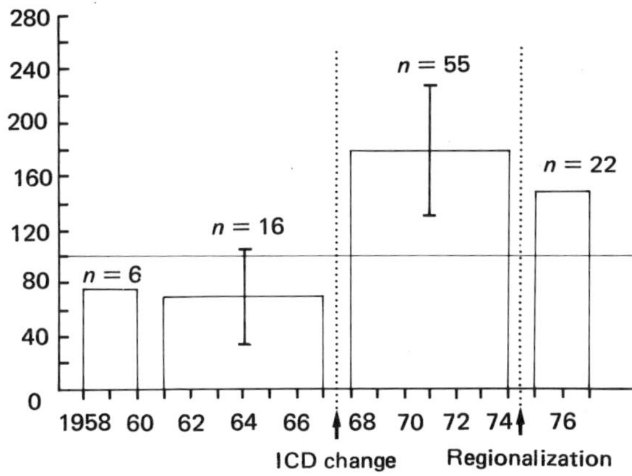


*Fig. 13.1.* Lung cancer in town V, mean SMRs 1958–77.


## THE ANALYSIS OF RESPIRATORY CANCER MORTALITY BY THE CU-SUM TECHNIQUE

'Cu-sum' stands for 'cumulative sum'. Its application can be followed in the example provided in *Table 13.2*. There, numerical differences appear between a series of values and a 'target' or 'reference' value. A cumulative sum is made from those differences.

As an analytical procedure, Cu-sum is well established in commercial and industrial spheres of statistical analysis.[4] However, it is rarely mentioned in medical papers.[5] Its use conveys two benefits. First, it can reveal subtle changes in the trend of a series of values, changes which might not be identified by conventional quality-control measures.[4] Secondly, an examination of the types of changes in the slope of the Cu-sum graph might indicate the nature of the underlying cause of that change. For example, a sharp 'point-change' in Cu-sum values would imply that a distinctively new stimulus had been applied to the process. In the context of biological experimentation, above all, a point-change is found where the stimulus has been applied 'shortly' before the observed response. A point-change is not normally compatible with the existence of a long latent period between stimulus and response. For long latency affords time for the effects of biological variability in the affected tissues to be manifested; in which case the Cu-sum values would then display a gradual change instead of a point-change.

The Cu-sum statistics were derived from the SMRs (1958–77) for respiratory cancer (*Table 13.2*). The SMR of 100 for Scotland, that is the reference population, was taken as the 'target', or normal value.

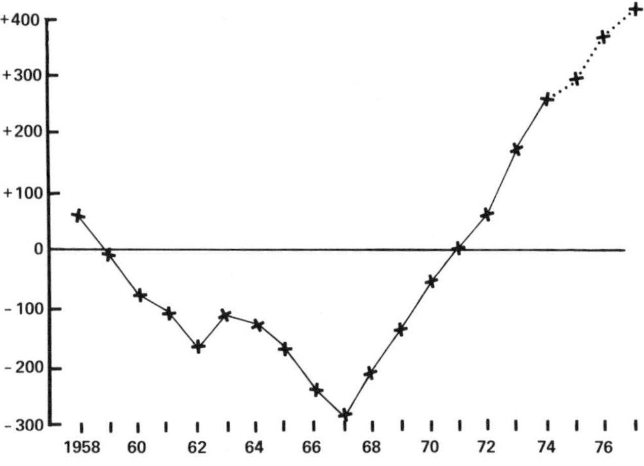



*Fig. 13.2.* Cu-sum of SMRs for lung cancer 1958–77.

The SMR for each year was evaluated against this target number, and the consecutive differences were summed cumulatively.

The resulting Cu-sum graph revealed a clear-cut point-change at 1967 (*Fig. 13.2*). Statistical methods exist[4] whereby the statistical significance of this change in trend could be confirmed.

Provided that the assumptions underlying the interpretation of a point-change are correct, and provided that no clerical error or other artefact was responsible (none has been detected as yet), the inference must be that the population of town V was exposed to an environmental carcinogen, or carcinogenic promoter, or 'trigger-factor' shortly before the point-change at 1967. A latent period shorter than 5 years might be indicated. The concept of a short latency for cancer does not form part of 'majority wisdom' for human carcinogenesis. Yet there are suggestions from pathologists, occupational epidemiologists and toxicologists that the cancer process may not always move with the ponderous momentum of a slow-moving juggernaut (see relevant references).[5] Indeed, latencies of 6 years and less have been predicted and recorded in the occupational field.[6,8] Certainly, there is now an increasing awareness and acceptance of ideas such as multifactorial causation, initiators and promoters and 'trigger-factors' in carcinogenesis.

## OTHER MORTALITY IN TOWN V

If an environmental agent had been present in sufficient strength to have promoted an epidemic of respiratory cancer, it would be surprising if other categories of disease had remained unaffected.[6]

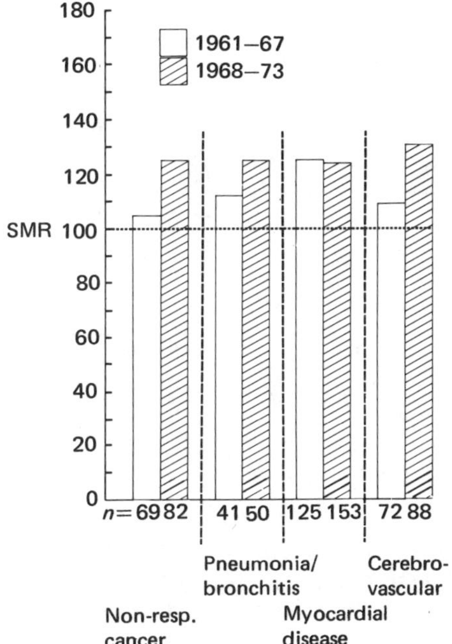


*Fig. 13.3.* SMRs for other major causes of death, 1961–7 and 1968–73.

From data provided in the Annual Reports of the Registrar-General for Scotland, the mean SMRs were calculated for other common causes of death in town V during 1961–7 and 1968–73 (*Fig. 13.3*).

All major causes of death were found in town V more frequently than expected. The increases were statistically significant in the cases of myocardial (1961–7, 1968–73) and cerebrovascular (1968–73) categories. Thus mortality from other categories of degenerative disease was indeed abnormal.

The general practitioners in town V had commented that, in their experience, the frank onset and diagnosis of respiratory cancer in many patients was preceded by respiratory infections. As a sequel to this observation, the trends of the annual SMRs for non-malignant chest disease (pneumonia, bronchitis and so on) and for respiratory cancer were compared during 1961–73 (*Fig. 13.4*).

It was clear that the graphs for these two categories of mortality were remarkably similar. What was less clear was how this phenomenon should be interpreted. Although toxicologists have shown that viruses can act as co-carcinogens in experimental animals, it would be a bold epidemiologist who would apply that explanation confidently in the case of town V.

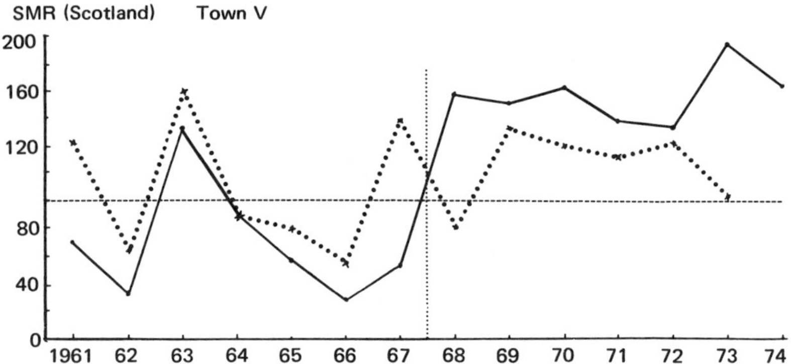


*Fig. 13.4.* Annual SMRs for respiratory cancer (solid line) and bronchitis/pneumonia (dotted line) in town V, 1961–74.

## MORTALITY IN TOWN V COMPARED WITH OTHER SCOTTISH COMMUNITIES

The annual mortality rates from all causes, standardized for age and sex, for town V were compared with those rates for the other communities in the same county. The years studied were from 1955 to 1973 (*Fig. 13.5*).

Three distinct phases appeared in the graph for town V's standardized death rate: (i) a normal level, relative to its neighbours; (ii) a high plateau; (iii) a very high and less uniform plateau. It was of interest that the high mortality from respiratory and non-respiratory cancers

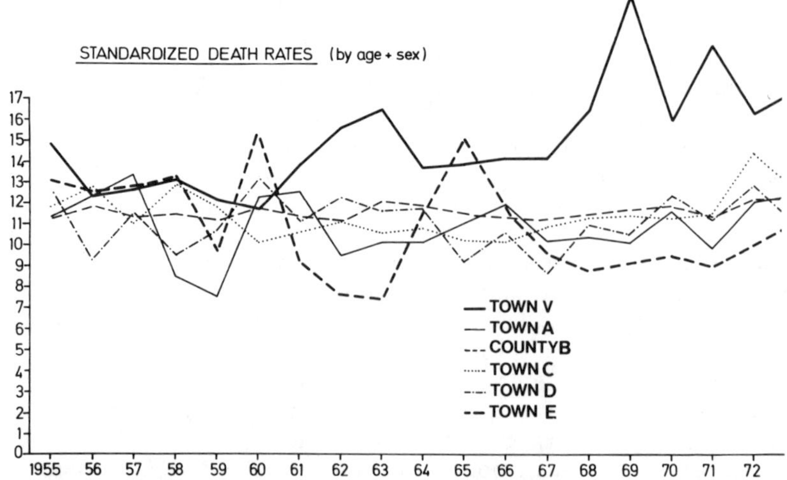


*Fig. 13.5.* Standardized death rates of town V and neighbouring communities, 1955–72 (rates per thousand population). (From the Annual Reports of the Registrar-General for Scotland.)

only appeared consistently in phase (iii). Hence the high plateau in phase (ii) was associated with non-malignant disease only, for example, with myocardial disease.

Of equal importance was the finding that phase (i) was quite normal. That finding made it less likely that the later high mortality was of genetic origin, was the result of long standing environmental factors, or was the pathological expression of traditional peculiarities in town V's social or cultural practices, such as cigarette smoking.

Having demonstrated that the standardized mortality rate in town V during 1968–73 was consistently higher than the rates of its neighbours, it was considered appropriate to evaluate town V in the widest, national context. Hence the mean standardized death rate from all causes in town V during 1969–73 (that is, 2 years on either side of the 1971 census) was compared with the similar mean rates for all communities in Scotland, as named in the Annual Reports of the Registrar-General — 175 other small burghs, 21 large burghs and 4 cities. The mean of the standardized death rates of all small burghs put together was calculated, as was the value for 3 standard deviations above that mean. The value for town V is indicated by the arrowed column in *Fig. 13.6.*

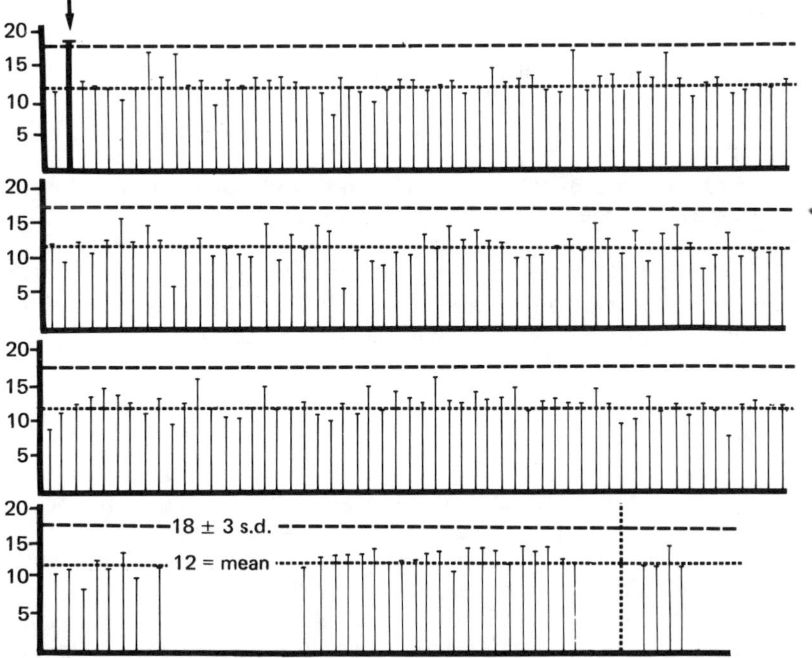


*Fig. 13.6.* Mean standardized death rates in named Scottish communities, 1969–73 (rates per thousand population).

From *Fig. 13.6* it was clear that the mortality record for town V was outstandingly bad, even in the context of the whole of Scotland. For its rate was the only one to transgress the line representing 3 standard deviations above the mean for all small burghs.

The other small burghs with high rates (*Fig. 13.6*) were of minor interest. On closer inspection, these burghs proved to be mainly very small communities, some having populations as low as 1000. Hence their eccentricity was probably attributable to the statistical instability associated with very small numbers of deaths and populations.

## INDIVIDUAL CIRCUMSTANCES IN TOWN V

The attributes of the individuals who had died of respiratory cancer during 1968–74 were examined. Personal details were obtained from death certificates, hospital case-notes and family doctors.

### Sex group

There were 9 death certificates for females with respiratory cancer in the town during that period. The resulting SMR was 172. Because of small numbers, however, this increase was not statistically significant. Nevertheless, this finding did suggest that both sexes might have been affected similarly. This deduction would indicate a carcinogenic factor in the general, rather than occupational, environment.

### Age group

The mean age-specific rates for males in town V during 1968–74 were contrasted with the male age-specific rates for Scotland in 1971. The ratio of town V to Scotland for each age group was calculated (*Fig. 13.7*). The evidence was that only the usual age groups of males were being affected in town V, and that they were suffering more than elsewhere. Thus there was a quantitative increase in mortality in 'at-risk' age groups, but no qualitative change in the age grouping.

### Occupation

Of the factories present in town V, only the steel foundry and the brickworks fell into the category of 'heavy industry'. When the occupations of the respiratory cancer victims were extracted from the death certificates, neither of those two industries was frequently mentioned. The failure of occupational mortality data to implicate heavy industry was corroborated by occupational morbidity data, that is the discharges/deaths of residents of town V from regional hospitals (*Table 13.3*).

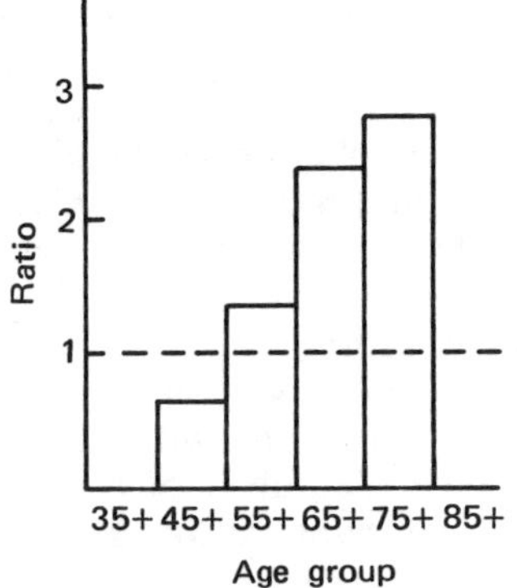


*Fig. 13.7.* Ratios of male age-specific rates for respiratory cancer.

**Table 13.3.** Occupations of persons dying from respiratory cancer in town V, 1964–74

| Hospital cases, town V, respiratory cancer | 1964–7 | 1968–74 |
| --- | --- | --- |
| In foundry workers | 0 | 4 |
| In coalminers | 6 | 24 |
| In all other males | 11 | 53 |

**Table 13.4.** Smoking habits of respiratory cancer victims in town V, 1968–74

| | |
| --- | --- |
| Heavy smokers, 20 or more cigarettes daily | 19 |
| Moderate smokers, 5–19 cigarettes daily | 11 |
| Light smokers, less than 5 cigarettes daily or 'occasional pipe' | 5 |
| Non-smokers | 5 |
| Not known | 4 |

Nevertheless, the possibility remained that an investigation of previous occupations might reveal some significant common factor. Such an investigation is currently under way.


## Smoking histories

Information was gathered concerning the smoking habits of 44 males with a 'confirmed' diagnosis of respiratory cancer (*see below*) during 1968–74 (*Table 13.4*).

Thus, 10 of the 40 males appeared to have had little or no exposure to tobacco smoke of their own making. This proportion of minimal smokers is slightly higher than in other series where smoking habits of respiratory cancer victims were investigated.

The opinion of the manager of the local chain store was that cigarette sales since 1960 had followed two trends: first, a massive switch from 'plain' to 'filter' and secondly, a reduction in total sales with each increase in the level of taxation of tobacco. In general, therefore, there were no indications that the epidemic of respiratory cancer in town V was either entirely or largely due to a sudden and excessive indulgence in tobacco by its population. Hence this finding in a general population is consistent with observations made in occupational epidemiology on the failure of cigarette smoking alone to explain high cancer rates in many occupational groups.[6]

**The reliability of diagnosis**

The medical evidence for the diagnosis of respiratory cancer in residents of town V was examined (*Table 13.5*). The numbers of deaths and autopsies for respiratory cancer in Scotland were obtained from the Registrar-General's Annual Report during the same period, 1968–74.

Thus, 26 per cent of the cases in town V were confirmed by autopsy, compared with only 15 per cent of the cases in Scotland as a whole. Hence there was evidence that the diagnoses of respiratory cancer in town V were made on valid criteria. Added to the knowledge of stable medical facilities in town V, it seemed unlikely that diagnostic artefact could explain the phenomenon of respiratory cancer there.

A review of the histology of the pathological specimens revealed the usual preponderance of squamous and oat cell types.

**THE ENVIRONMENT OF TOWN V**

An investigation of air pollution in town V and in three neighbouring towns revealed that in the monthly values for suspended particulates town V consistently outranked its neighbours. For the year May 1976 to April 1977, the mean value in town V was 47 $\mu$g/m$^3$, compared to values of 30, 30 and 21 in the adjacent towns. Indeed, the monthly values in this semi-rural town V were usually higher than the contemporary values in the city of Edinburgh and in the large industrial town of Motherwell.

Closer investigations showed that the air pollution in town V was associated only with light easterly winds and with calm days, indicating that the eastern half of the town was the probable source of this pollution. This supposition was confirmed when the level of pollution was shown to have fallen during the period of industrial holiday there, that is during the July of each year studied; most of town V's industry was on its eastern side.

**Table 13.5.** Validation of diagnosis of respiratory cancer in town V

| | No. deaths diagnosed respiratory cancer | Death certificate diagnosis later *confirmed by hospital staff* | | | | |
| --- | --- | --- | --- | --- | --- | --- |
| | | *either disproved or not confirmed* | only clinically and by X-ray | also by sputum cytology or direct observation | also by other histology | also by autopsy |
| Town V 1968–74 | 57 | 6 | 8 | 11 | 17 | 15 |
| Scotland 1968–74 | 20 729 | | | | | 3 097 |

Within town V the most noticeable source of dust and fume was the steel foundry. This factory, situated in the eastern half of the town, was very closely surrounded by housing. The fume-stacks of the foundry were low, lower than the factory roof. Consequently, the emerging fumes were rapidly returned to ground level by the eddy currents generated by the edge of the roof. For many years, local residents had complained about the airborne dusts and fumes from the foundry. Paradoxically, however, since the introduction of technological changes in the steel-making process in the foundry during the mid-1960s, there had been less *visible* pollution, and hence fewer complaints from the residents.

## MOSS BAG STUDY OF METALLIC AIR POLLUTANTS IN THE ENVIRONMENT OF TOWN V

Moss bags have been used in several large scale investigations in order to determine sources of metallic air pollution in industrial communities.[9,10] Their deployment in town V was considered advisable, to determine objectively the extent of major air pollution from the foundry.

Each moss bag was a square, flat, nylon-mesh packet containing Sphagnum moss; this moss has a valuable capacity to entrap and adsorb metallic particles (*Fig. 13.8*).

The moss bags were fixed horizontally to bamboo canes. They were then distributed within town V, at various distances from the foundry.[11] After exposure to the ambient air for 30 days, the moss bags were removed from their sites. The concentrations of some metals in the moss were measured by means of atomic absorption spectrophotometry.

The results (*Fig. 13.9*) demonstrated a 'hot-spot' of metallic air pollution close to the foundry. Beyond a distance of 500 m, however, little gross pollution was detected by this technique. Relatively high concentrations of nickel were also detected.[11]

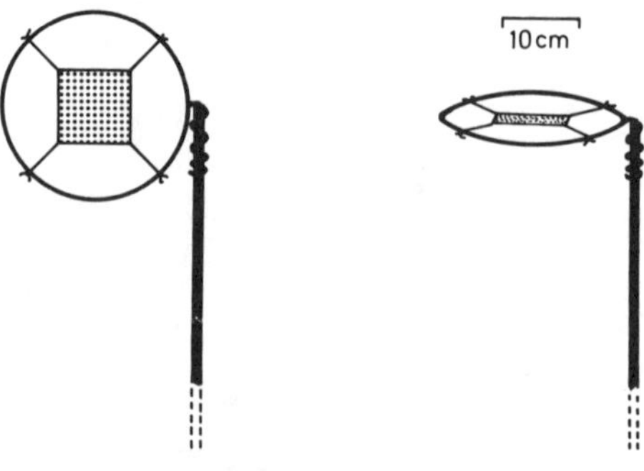


*Fig. 13.8.* Moss bags.

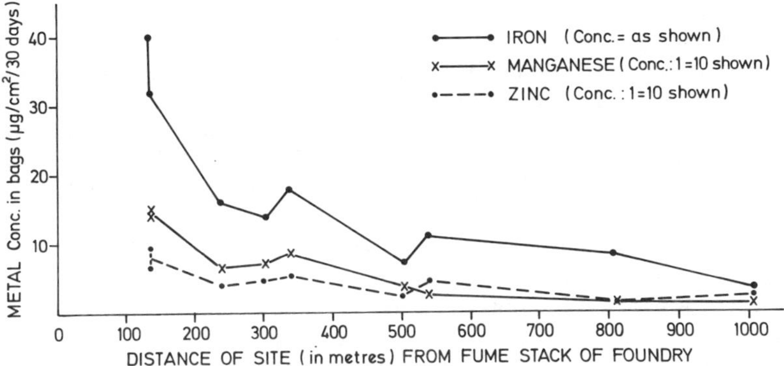


*Fig. 13.9.* Moss bag study: concentrations of metallic air pollutants near foundry.

The moss bags thus corroborated the subjective evidence provided by smell and sight, that the major source of metallic air pollutants was the steel foundry. They also demonstrated the circumscribed nature of the gross pollution from that source.

Easy and inexpensive to make, and unobtrusive when installed, moss bags can be recommended as aids to diagnosis of pollution sources and of areas of 'fall-out' when circumstances are similar to those in town V.

The anatomy of the zonal air pollution in town V, as portrayed by means of moss bags, has been corroborated recently by the use of a different technique. Lichens are sensitive to many air pollutants. Hence their scarcity reflects the long term presence of air pollution.[12] By analysing the pattern of lichen distribution in town V, a member of the Epidemiology Unit for Environmental Cancer (Fiona A. Yule) has confirmed the presence of the 'hot-spot' of air pollution, geographically associated with the foundry.

## CLUSTERING OF RESIDENCES OF VICTIMS OF RESPIRATORY CANCER WITHIN TOWN V

The moss bags and the lichens had identified a 'hot-spot' of air pollution. How many of the victims of respiratory cancer had resided in this 'hot-spot'? That was the next question.

The identification of a significant degree of clustering of disease can be an elusive goal.[3] In this study, several approaches to this problem were used.

### Enumeration districts

Enumeration districts (EDs) of a census may subdivide a community appropriately, thereby providing information about the numbers and age–sex structure of the populations in its districts. With this information, the rates of various diseases in an area of particular interest can be calculated and evaluated statistically. The presence or absence of a 'cluster' can thereby be determined.

There are drawbacks to this technique. First, the area of suspected clustering of disease may be spread across the borders of several EDs, so the cluster becomes diluted. Secondly, this technique may not allow the radial gradient of mortality around a point-source of toxin or carcinogen to be demonstrated. Thirdly, when the rates of disease are calculated over a number of years on either side of a census year, it may be rash to assume that there has been no dramatic change in the size and structure of the population before or after that census year.

In the case of town V, however, the population had remained relatively stable. There had been little re-development of the housing areas. Fortunately, too, aggregates of EDs provided suitable boundaries; within some of these boundaries lay those areas of housing at risk from metallic air pollution from the steel foundry. Those 'at-risk' areas were to the east and west of the foundry, and not further than 1 km from it.[11] A 'control' area of town V was also selected, that is, an area of similar size with comparable housing and almost as close to the foundry while not downwind from it.

The residences of the victims of clinically confirmed respiratory cancer 1968–74 were plotted within town V; where a change in residence during the 18 months prior to death was discovered, the previous address was substituted. Also plotted were the residences of individuals who had died of other common diseases. The rates, per 1000 population aged over 60 at the 1971 census, were calculated for those categories of mortality. The rates of the 'at-risk' area of town V were contrasted with those of the control area (*Figs. 13.10, 13.11*).

A substantial excess mortality for respiratory cancer alone was demonstrated in the 'at-risk' area, when compared with its control area.

The 'at-risk' area was readily divisible further into two equal components situated one on either side of the foundry. In the high-risk area to the west of the foundry, which was at risk when the pollution-associated easterly winds were blowing, there were 15 male deaths from clinically confirmed respiratory cancer between 1968 and 1974. The expected number, derived by indirect standardization from the rest of the burgh, was 6. This difference was statistically highly significant.

Hence, by using this approach, the deaths from respiratory cancer were shown to be clustered in the 'hot-spot' of metallic air pollution.

### Radial clustering

In archaeological fieldwork the site of a pre-historic centre of commerce can be deduced from the centripetal gradient of the spatial

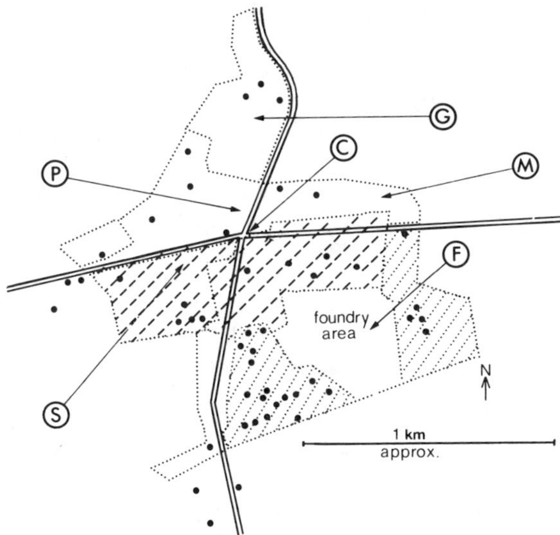



*Fig. 13.10.* Map of town V showing residences of victims of (confirmed) respiratory cancer: 'at risk' (dotted) and control (dashed) areas; hypothetical sources of pollution. G. Gasworks; P, swimming pool; M, old mineworks; C, cross-roads; F, foundry; S, school.

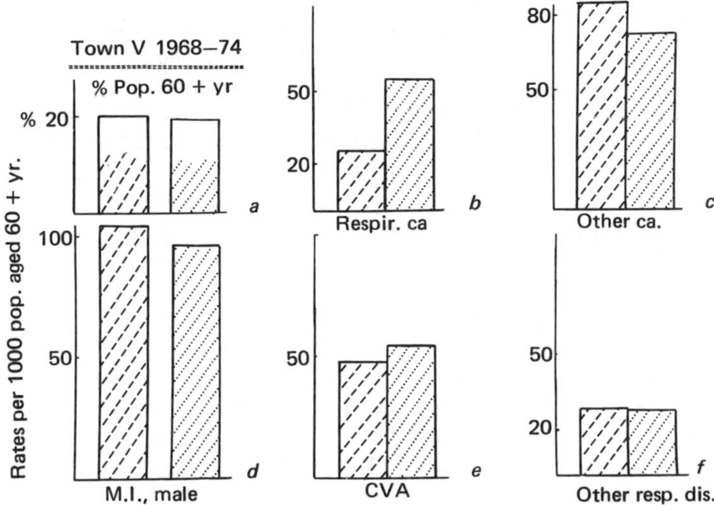



*Fig. 13.11. a,* Percentage of the population aged 60+ in 'at risk' and control areas. *b–f,* Rates per 1000 population aged 60+, for respiratory cancer (*b*), other cancer (*c*), male ischaemic heart disease (*d*), cerebrovascular heart disease (*e*) and other chest disease (*f*).

concentration (that is the radial clustering) of coin-finds.[13] This concept was applied to determine whether or not radial clustering of mortality from respiratory cancer was demonstrable around the centre of air pollution, the foundry.

The demographic data available from the census tabulations were not suitable for the purposes of analysis for radial clustering. Hence, mortality from a 'control' disease had to be chosen. Thus, deaths from this disease acted as a 'marker' of the demographic characteristics of the population in various areas of town V.

Acute myocardial disease in males was chosen as the 'control' cause of death. Only male deaths were noted, since most victims of respiratory cancer were also males. Like respiratory cancer, myocardial disease affects middle-aged as well as retired age groups. Likewise it is linked causally with cigarette smoking, and possibly with general air pollution. There are similar social class gradients for the two diseases. Lastly, being a very common cause of death under normal conditions, the total deaths from myocardial disease in town V were large enough to facilitate statistical comparisons with the numbers of deaths from respiratory cancer. Indeed the large numbers of 'normal' deaths from myocardial diseases were an advantage: they would render of no importance any minor contributions (to the total male deaths from heart disease) which might have resulted from cardio-tropic actions of respiratory carcinogens in a chemically polluted environment. No other cause of death appeared to have similar advantages as a 'control' for respiratory cancer.

The addresses of the 49 victims of respiratory cancer and of the 74 men who had died from myocardial disease were plotted on maps of town V (scale: 6 inches per mile, one map for each cause of death). On each map, a series of 17 concentric circles was drawn, the centre of the circles being the foundry area in town V (*Fig. 13.12*). The fractional radius or width of each annulus was the same, 0·5 cm. Thus the 17 concentric circles on each map had radii of between 2 cm and 10 cm. (The central circle was excluded because it was open ground round the factory. In *Fig. 13.12* only some circles have been shown, to enhance clarity.)

The numbers of deaths from cancer and from myocardial disease within each annulus were noted. These numbers were then expressed as percentages of the total numbers of the relevant disease; in turn, these percentages were summed in centripetal sequence for each disease. Thus, for each of the two diseases, the cumulative percentages were found of the deaths outside the circumference of each of the 17 concentric circles on the two maps (*Fig. 13.13*).

The differences between these two cumulative percentage distributions were evaluated statistically by means of the Kolmogorov-Smirnov test[14] (*Table 13.6*). When compared with other tests such as the $x^2$ test, the Kolmogorov-Smirnov test has inherent advantages for

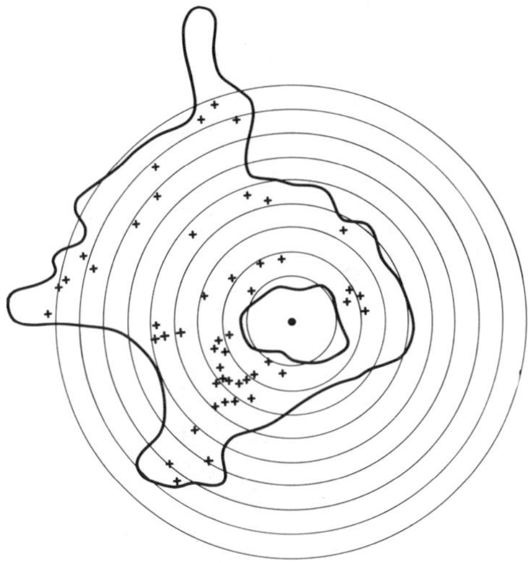

*Fig. 13.12.* Concentric circles around foundry and deaths from respiratory cancer.

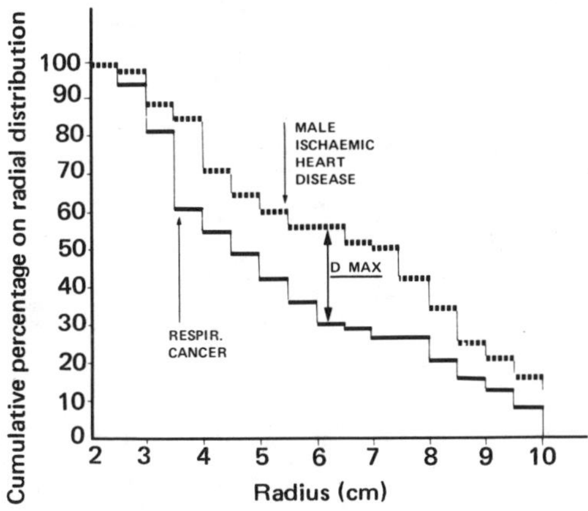


*Fig. 13.13.* Cumulative percentages around the foundry for respiratory cancer and male heart disease. Maximal difference ($D_{max}$) between graphs is given in *Table 13.6.*

the statistical evaluation of geographically dispersed data related to environmental pollution from a point source. Practically, its outstanding feature is its utilization of data in the form of percentage distributions, when making comparisons between sets of values.

From the graphs in *Fig. 13.13,* it was clear that the deaths from myocardial disease were distributed relatively evenly in town V. By contrast, there was an increasing concentration of deaths from respiratory cancer as the foundry area of town V was approached, that is as the cumulative percentage approached 60 and as the radii of

**Table 13.6.** Radial clustering: percentage distribution of deaths from respiratory cancer (1968–74) around factory

| | Percentages (and numbers) of deaths outside circles with radii (cm) | | | | | | | | | | | | | | | | | Total deaths |
|---|---|---|---|---|---|---|---|---|---|---|---|---|---|---|---|---|---|---|
| | 10 | 9·5 | 9·0 | 8·5 | 8·0 | 7·5 | 7·0 | 6·5 | 6·0 | 5·5 | 5·0 | 4·5 | 4·0 | 3·5 | 3·0 | 2·5 | 2·0 | |
| Respiratory | 2·04 | 8·16 | 12·24 | 16·33 | 20·41 | 26·53 | 26·53 | 28·57 | 30·61 | 36·73 | 42·86 | 48·98 | 55·1 | 61·22 | 81·63 | 93·88 | 100 | 49 |
| cancer | (1) | (4) | (6) | (8) | (10) | (13) | (13) | (14) | (15) | (18) | (21) | (24) | (27) | (30) | (40) | (46) | (49) | $= N1$ |
| Myocardial | 12·16 | 16·22 | 21·62 | 25·68 | 35·14 | 43·24 | 51·35 | 52·7 | 56·76 | 56·76 | 60·81 | 64·86 | 71·62 | 85·14 | 89·19 | 98·65 | 100 | 74 |
| | (9) | (12) | (16) | (19) | (26) | (32) | (38) | (39) | (42) | (42) | (45) | (48) | (53) | (63) | (66) | (73) | (74) | $= N2$ |

This distribution was compared to that for male myocardial disease. $D_{max}$ was 0·2615 (that is 26·15 per cent) at radius $= 6·0$ cm; this value for $D_{max}$ exceeded the lower of the 95 per cent confidence limits (0·2505) obtained by means of the Kolmogorov–Smirnov test.[10] Hence $D_{max}$ was significantly higher than expected.

$$\text{Calculation of 95 per cent level of significance } (0·05) = 1·36\sqrt{\frac{N1 + N2}{N1 \times N2}} = 1·36\sqrt{\frac{49 + 74}{49 \times 74}} = 0·2505.$$

the concentric circles decreased. Hence the presence of an excessive number of deaths from respiratory cancer round this area, when compared to the deaths from the 'control' disease, was indicated. The statistical significance of this excess, at the 95 per cent level of confidence, was demonstrated by means of the Kolmogorov–Smirnov test.

Of relevance to the finding that metallic air pollution was detected by the moss bags only within 500 m of the foundry was the observation that maximal significance for the clustering of respiratory cancer was attained at a radius corresponding to 600 m from the industrial area.

The Kolmogorov–Smirnov test has been used again to compare the radial distribution around the foundry area of mortality from respiratory cancer (1968–74) with the distributions of other cancers (1968–74) and also of respiratory cancer (1961–67). With both comparisons, the radial clustering of respiratory cancer (1968–74) was high. The statistical significance again exceeded the 95 per cent confidence limits.[15]

At the present stage it may suffice to state that the technique of radial clustering analysis requires only readily available mortality data, a map and simple mathematics and statistics. Appropriate concentric circles scratched on a Perspex sheet can be transferred readily from map to map, or from one industrial area to another; this avoids the task of re-drawing such circles frequently. Perhaps, therefore, this simple technique will prove a useful tool for general practitioners, occupational doctors and other 'humane physicians' who are interested in epidemiological studies of local patterns of disease in small communities, but who do not have access to computers.

## Detection of the zone of pericentral clustering

The clustering of deaths around other hypothetical sources of pollution in town V was examined.[15]

A series of about thirty concentric circles was drawn on the map, in a fashion similar to the method already described, but with smaller differences between the radial lengths. Their centre (or 'pericentre') was the position on the map of the obvious source of air pollution in town V, that is the foundry. The 'death residences' associated with respiratory cancer which lay outside each consecutive circle were counted again as described for radial clustering; these numbers were plotted graphically, but this time against the *square* of the radius ($r^2$) of the respective circles.*

*With the use of $r^2$ on one axis, a linear slope would illustrate random siting of residences, while a logarithmic curve would represent a zone of increasing concentration of such cases. Hence, if clustering were to be present around this foundry pericentre, the line derived from plotting $r^2$ against *number outside* necessarily would change its slope from linear to logarithmic as the pericentre ($r^2 = 1$) was approached.

Hypothetical sources of pollution in town V (*see Fig. 13.10*) were made, likewise, the map pericentres of similar series of concentric circles. Again the power of dispersion of respiratory cancer around each of these pericentres was represented graphically as the slope of a line. These sloping lines were compared with the line for the foundry's pericentre. The line with the greatest change in the number of deaths outside the consecutive circles, as $r^2$ approached unity, was identified as the site where pericentral clustering was maximal — the 'primary pericentre' (*Fig. 13.14*). That line was the one for the foundry.

The same procedure was used in the construction of similar lines for the pericentral distribution of male myocardial disease — the 'control' disease (*see above*) — both around the foundry and around the hypothetical sources of pollution (*Fig. 13.15*). With the use of the 'control' disease, the foundry was not identifiable as the pericentre associated with excess clustering, as had been the case with the distribution of respiratory cancer. Hence, it was unlikely that the observed clustering of respiratory cancer around the foundry was explicable in terms of an abnormal demographic distribution.

It was remarkable that the pericentral clustering of respiratory cancer around the foundry was present only when the value of *number outside* was greater than 14. When its value was less than 14, the

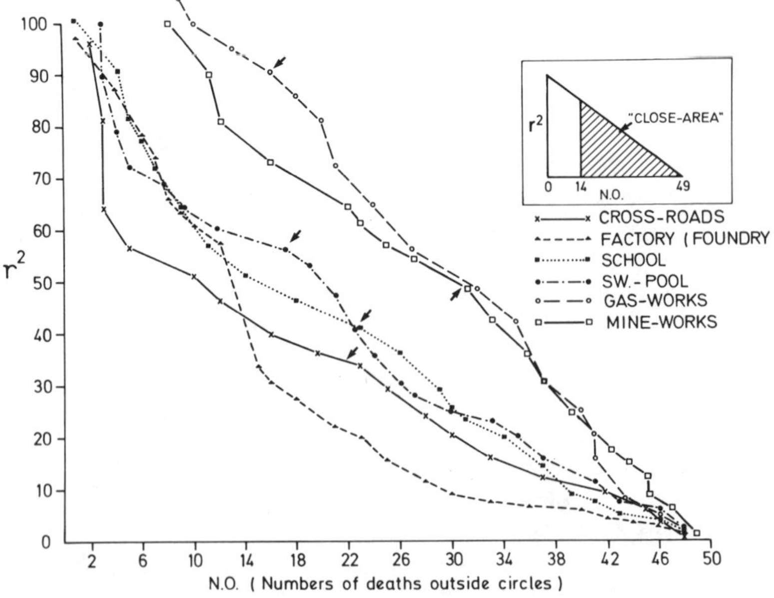


*Fig. 13.14.* Radial clustering (*n* = 49) around foundry and 'hypothetical' pericentres; diagram of 'close' and 'total' areas. (For explanation of arrows, *see* text.)

gradient became steep and linear. Hence it was deduced that there was no clustering of deaths from respiratory cancer at a distance from the foundry greater than where $r = 6$ cm approximately. This corresponded to a distance on the ground of about 630 m — the distance beyond which there was no evidence of pollution according to the moss bag survey.

## Quantification of differences in pericentral clustering

The differences in clustering of disease around the various pericentres were quantified as follows. It was demonstrated above that the smaller the areas below the lines (in particular the areas close to the pericentres), the greater was the clustering effect close to the pericentres of the respective lines. Hence the extent of the 'total' and the 'close' areas (*see Fig. 13.14*) beneath the various lines in *Figs. 13.14* and *13.15* would reflect the degree of clustering; the smaller the area, the greater the clustering. The areas beneath each line were therefore measured by planimetry (*Table 13.7*).

From these planimetric measurements, it was clear that the unequivocally minimal areas — both 'total area' and 'close area' — were observed beneath the lines representing pericentral clustering of respiratory cancer around the foundry. For the 'control' myocardial

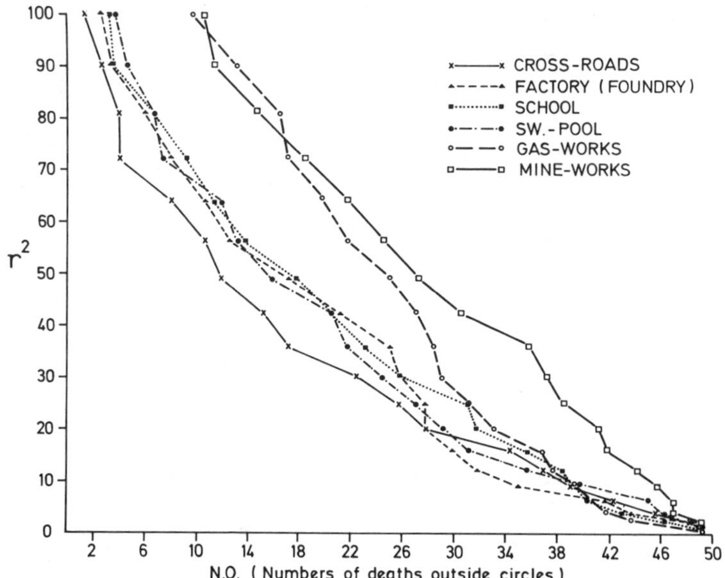


*Fig. 13.15.* Radial clustering around pericentres as in *Fig. 13.14* but for male heart disease ($n = 74$ adjusted to 49).

**Table 13.7.** Areas below lines around the probable and the hypothetical sources of pollution, derived by means of planimetry

| Disease areas (cm²) | Foundry | Cross-roads | School | Sw. pool | Gasworks |
|---|---|---|---|---|---|
| *Respiratory cancer* | | | | | |
| 'Close' areas (N.O.: 14–49) | 49 | 73 | 87 | 93 | 161 |
| 'Total' areas (N.O.: 1–49) | 145 | 156 | 189 | 195 | 299 |
| 'Close' area as percentage of 'total' area | 34 | 47 | 46 | 48 | 54 |
| *Myocardial disease* | | | | | |
| 'Close' areas | 76 | 69 | 90 | 85 | — |
| 'Total' areas | 186 | 164 | 198 | 185 | — |
| 'Close' area as percentage of 'total' area | 41 | 42 | 45 | 46 | — |

disease, in contrast, the foundry was not similarly exceptional. This observation was wholly consistent with the conclusions of the analysis of clustering by enumeration district (*see above*).

In *Fig. 13.14* the sloping lines representing pericentral clustering around the hypothetical sources of pollution (that is the non-foundry lines) showed transient changes in slope at intermediate distances along their lengths. Did these humps (marked in *Fig. 13.14* by arrows) indicate the geographical position in town V of the cluster of deaths from respiratory cancer in the area of housing 'at risk' from metallic air pollution? Could the precise location of that cluster be demonstrated by allowing the data to pinpoint it?

The distance from each hypothetical source of pollution to the hump on its line was measured. This measurement for each pericentre's line became the radius of a circle drawn around the corresponding pericentre (or hypothetical source of pollution) (*Fig. 13.16*).

Clearly, this procedure had identified the approximate position of the cluster of respiratory cancer within those enumeration districts which had also been implicated by more conventional means.

## THE PROBLEM OF INTERPRETATION

The fundamental findings were clear. In town V, general mortality had been high since 1961. Since 1967 the mortality from respiratory cancer had risen dramatically. Changes in the technology employed in the local steel foundry had preceded the sudden upsurge of respiratory cancer. The areas of metallic air-pollution and of clustering of respiratory cancer were shown to be similar.

How were these findings to be interpreted? They should be interpreted cautiously. For one swallow does not make a summer, nor does one epidemiological association necessarily imply causation.

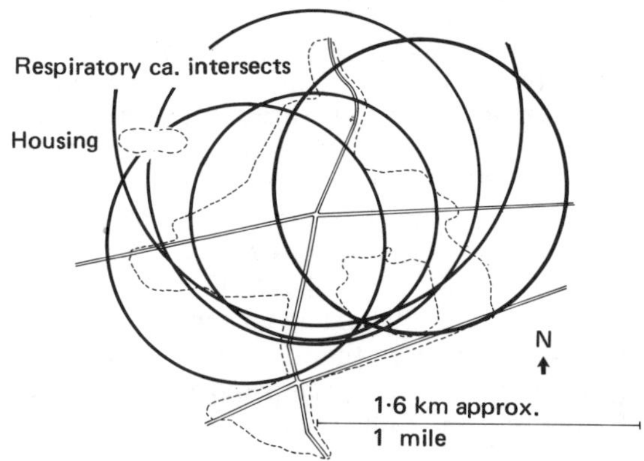


*Fig. 13.16.* Area of respiratory cancer clustering shown by intersecting arcs.

The results of similar work done elsewhere can be a valuable aid to interpretation. In this context, similar geographical associations between respiratory cancer and heavy industries (including metallurgical ones) have indeed been reported,[6,15-17] although some doubts remain because of methodological problems. Again, studies in occupational epidemiology have shown that 'hot-metal' workers have high rates of respiratory cancer.[6,18,19] In some cases, perhaps, these high rates were the result of changes in the steel-making process (Radford[18]). Of the specific metals trapped by the moss bags around the foundry in town V, nickel and iron compounds are linked causally with respiratory cancer in occupational fields.[6] Lastly, toxicologists have demonstrated convincingly the carcinogenic actions of metallic dusts in experimental animals.[20]

Yet, dubiety about the causal relationship in town V must remain until further analysis has been completed. For example, the role of occupation remains unsettled. As described earlier, there were no hints from the data on the death certificates that occupation was an important factor in town V as a whole; nor did foundry workers form the majority of the victims in the 'at-risk' area. It may be relevant to recall that, after liberation from a fume-stack into the general atmosphere, air pollutants can undergo physicochemical changes which alter their pathogenic properties. Hence, in the context of industrial air pollution, an *occupational* hazard (manifested purely on workers within the foundry) is not an essential bed-fellow of an *environmental* hazard (affecting residents in the neighbourhood). Nevertheless, more detailed analysis will be required to determine whether or not the cases of respiratory cancer had worked in the foundry during earlier years, before a convincing verdict on the occupational factor can be reached.

A more intensive investigation into the smoking habits of the deceased is also necessary. Here it will be important, however, not to allow prejudice to creep in.[6,21,22]

Finally, there is always the hope that local replication studies may be epidemiologically rewarding. Hence studies of other industrial communities are planned, and some are under way. It is of interest that preliminary findings of such studies are compatible with observations described in town V. However, again much more analysis will be required in order to delineate the 'reciprocal relationships between job, home and health' (the late Professor R. C. Browne's description of industrial health).

A small industrial community and an epidemic of a common disease with the cause as yet little more than a hunch:[23] many questions can certainly be generated by these circumstances. But whether or not answers will follow remains to be seen.

## Acknowledgements

I acknowledge gratefully the collaborative contribution to this research of the members of the Epidemiology Unit for Environmental Cancer: Joyce C. Langlands, Helen K. M. Tyrrell, Fiona A. Yule and M. Melody Lloyd. The typists were Mrs K. A. Anton and Miss J. C. Langlands. Research by the Epidemiology Unit for Environmental Cancer is supported financially by the Scottish Home and Health Department. Some of the material in this chapter first appeared in *Community Medicine* 1979; **1**:210–20.

REFERENCES

1. Jones RV. *Most Secret War*. London: Hodder & Stoughton, 1978.
2. Frank JP. *System einer vollständigen medizinischen Polizey*, 1779. Cited in: Underwood EA (ed.) *Science, Medicine & History*, vol. 2. London: Oxford University Press, 1953.
3. Knox EG. Epidemics of rare diseases, *Br. Med. Bull*. 1971; **27**:43.
4. Woodword RH and Goldsmith PL. *Cumulative Sum Techniques* (Monograph No. 3, Mathematical & Statistical Techniques for Industry). Edinburgh: Oliver & Boyd, 1964.
5. Lloyd OL and Barclay R. Hypothesis: a short latent period for respiratory cancer in a 'susceptible' population. *Community Med*. 1979; **1**:210.
6. Hueper WC. *Recent Results in Cancer Research*. Berlin: Springer-Verlag, 1966.
7. Figueroa WG, Raszkowski R and Weiss W. Lung cancer in chloromethyl methyl ether workers. *N. Engl. J. Med*. 1973; **288**:1096.
8. Armenian HK and Lilienfeld AM. The distribution of incubation periods of neoplastic diseases. *Am. J. Epidemiol*. 1974; **99**:92.
9. Report of a collaborative study on certain elements in air, soil, plants, animals and humans in the Swansea–Neath–Port Talbot area, together with a report on a moss bag study of atmospheric pollution across South Wales. Welsh Office Report, 1975.
10. Report of the Avon, Gloucestershire & Somerset Environmental Monitoring Committee on a Survey of Airborne Metals, 1977.
11. Lloyd OLl. Respiratory cancer clustering associated with localized industrial air pollution. *Lancet* 1978; **1**:318.
12. Leblanc F and DeFloover J. The relation between industrialization and the distribution and growth of epiphytic lichens and mosses in Montreal. *Can. J. Botany* 1970; **48**:1485.
13. Hogg AHA. Some applications of surface fieldwork. In: Hill D. and Jessen M (ed.) *The Iron Age and Its Hill-Forts*. Southampton Archaeological Society, 1971.
14. Siegel S. *Nonparametric Statistics for the Behavioural Sciences*. McGraw-Hill, Kogakusha, 1956.
15. Lloyd OL. MD Thesis, Edinburgh University, 1977.
16. Cecilioni VA. Lung cancer in a steel city. *Fluoride* 1972; **5**:172.
17. Lessard R, Reed D, Maheux B et al. Lung cancer in New Caledonia, a nickel smelting island. *J. Occup. Med*. 1978; **20**:815.
18. Radford EP, Milham S and Hirayama T. In: Saffiotti U and Wagoner JK (ed.) Occupational Carcinogenesis. *Ann. NY Acad. Sci*. 1976; **271**:228–37, 243–9, 269–72.
19. Office of Population Censuses & Surveys. Occupational Mortality 1970–1972. London: HMSO, 1978.

20. Saffiotti U, Montesano R and Sellakumar AR. Respiratory tract carcinogenesis induced in hamsters by different levels of benz-a-pyrene and ferric oxide. *J. Natl Cancer Inst.* 1972; **49**:1199.
21. Sterling TD. Does smoking kill workers or working kill smokers? *Int. J. Health Serv.* 1978; **8**:437.
22. Luh KT, Kuo SH, Lin CC et al. Primary lung cancer in Taiwan. *J. Formosan Med. Assoc.* 1974; **73**:129.
23. Alderson MR. Epidemiological studies of occupational carcinogens. *Arch. Toxicol.* 1980; Suppl. 3:3.

# 14. A LAWYER'S AND ENGINEER'S VIEW OF HAZARD MANAGEMENT

*Jonathan Plaut*

## THE MANAGERIAL PROCESS: ENVIRONMENTAL INPUT

The process of hazard management, hazard identification and control and risk assessment is critical for managers who are making or participating in business decisions with environmental input and impact ('environmental' includes pollution control, health, product safety and safety fields). It is easy, perhaps, to understand that the pollution control, health, product safety or safety professional (the 'environmental' professional) must be schooled in his or her particular field to understand the degree of hazard (including toxicity), the method of control and thus the population exposed, and the degree of exposure. It is more difficult to understand that many business decisions which appear to be non-environmental also require an input of hazard identification and assessment and a risk analysis of alternatives. This is often true of decisions to manufacture for a new use, to build or expand a plant, to acquire a business, or to market an existing product for a new use. Moreover it is also true of the everyday business and operating decisions of, say, the middle manager, or the plant superintendent or plant engineer with respect to such questions as the level of plant maintenance, or the effect of a cut in quality control or process monitoring, or the budget for night supervision, or allocation of resources for training.

The fact is that engineers and lawyers are by education and experience used to analysing facts and evaluating alternatives. They are trained to reach objective decisions based on analytical review. Because they are trained to do the essential job of the manager, that is to select the desirable alternative within the resources available, they should be equipped to deal with the risk management decision process. Ironically, there has been some distrust between lawyers and engineers dealing with environmental problems, even though their education and experience should lead them to understand each other's gifts and contribution to hazard identification and control and the risk management process.

## HAZARD IDENTIFICATION AND PERSPECTIVE

It is important to understand this system of hazard identification and routine analysis in setting priorities for resource application. Hazards may be divided into three categories:

The first is a *known hazard* for which standards exist and degrees of control can be measured, and against which policies and procedures may be issued. Measurement and 'auditing' of compliance in this defined category of hazard is possible.

The second are areas of *identified but usually unquantified hazard,* against which a professional in the field familiar with the operation can at least partially guard by the institution of controls. This might involve spill prevention measures, guarding against explosion potential, engineering to eliminate a potential health or safety concern, training operators as to quality control, or dealing with countless other situations involving incompletely quantified conditions which would arise except for control procedures.

The third deals with *the unknown* — that which has not been detected or specifically foreseen. Here neither the risk nor the control is known, and so overall systems of prevention as well as management attitude are pivotal to avoid or minimize the potential effects of such risks.

Business functions have a number of ways of dealing with these three basic levels of hazard. Environmental professionals in the disciplines of pollution control, safety and loss prevention, medical services, occupational health including industrial hygiene, product safety and toxicology are able to bring specific professional knowledge to the identification both of hazards and of possible hazards which would fall into one of the above three categories. The multidisciplinary approach and integrated effort of those professionals routinely considered as a part of the business function will make hazard identification and the control process more certain, and thus lower risk potential and improve the business process and its profitability.

### Pitfalls

In the conduct of its day-to-day business, in the light of environmental input, from the very outset there are some general pitfalls that have to be guarded against in the decision process or the operation will either suffer from preconceptions or try to ignore concerns to the detriment of hazard identification and risk management.

One such preconception is that *environmental costs are burdens to shun.* This is probably based upon negative experience which each business person seems to have somewhere in his background of environmental cost impacts. A corollary of this preconception is that one does not distinguish oneself by spending environmental capital on operating expense. Wisdom, however, dictates that the business entity, as well as the individual professionals within it, should adopt the philosophy that to *remain competitive* in the contemporary world, enthusiastic and practical attention must be paid to environmental concerns. It is true that unplanned cost *can* cause a general pressure on the environmental decision that can be the undoing of many a manager. The current wave of inflation which Western society is undergoing is frightening, and puts very great and often determinative pressure on what are often costly environmental decisions. Nevertheless, if environmental costs are not considered as a priority, the wrong business decision may be made in the longer run.

Secondly, there is the so-called *'goldfish bowl' effect*. There continues today a never-ceasing media interest, with incessant headlines on environmental stories which appear to take the facts out of context. Politicians are very responsive to scare stories — and there are good reasons for this — but this does place the hazard identification and risk decision process under pressure just at the time when it is hoped that prudent formulation of strategy will take place. Business must learn to live with this pressure and use the media because this pressure will persist.

Thirdly, there are an irritating number of *different pressures brought to bear on the environmental evaluation*. There is a preconception that says environmental pressures take the manager's options away. No business manager wants to make his decision based only on what he is told to do by professional health advocates, for example, or under the threat of the penalties of regulations. There are, after all, other priorities, such as capital resource allocation to research and energy. Environmental regulations come from many sources, including national and local agencies servicing different areas of the public interest, such as natural resource conservation, the rights of organized labour, international trade, the local community and so on. There is a great deal of overlap and sometimes confusion in the environmental decision process, and the manager dealing with what he might think is a simple decision related to handling and transportation of a waste material from his plant may be faced with a number of conflicting signals. However, a business must sort out the priorities in order to make the right decision in the long term.

Finally, the common general concern is that *controversy and even litigation seem almost to be the rule*. The judiciary process is drawn into decision-making far more often than would seem to be wise. Many decisions can be overwhelmed by the litigation threat. A lawyer's early advice in the balancing of pressures can be more valuable than the manager might otherwise think. Therefore lawyers must be consulted at an early stage if litigation is to be avoided.

These are some of the outside pressures placed on the manager in the hazard management process. However, they can be handled in a positive manner. If any of these pressures becomes too unreasonable or all-consuming, then the hazard management and control and risk decision process may become divorced from good management judgement. The lawyer, scientist and engineer are perhaps better equipped than most by education and experience to minimize these pressures.[1]

## CONTROL, PREVENTION AND RISK ANALYSIS

A short time ago we were tracing impurities in parts per thousand, and then parts per million. Now we are able to trace many impurities in parts per billion, parts per trillion or beyond. We do not always know what the significance of these trace impurities is, but our technical capacity has continued to increase dramatically, and with it our knowledge and our questions. In addition, we have markedly improved our testing procedures utilizing bacteria and animals and we are able to diagnose many more effects on these organisms or lower forms of life. What this all means to man is still not fully comprehended and is, therefore, in dispute. In the light of inadequate scientific knowledge, the task of identifying hazards and the risk they present to *man* is a great challenge. The 'goldfish bowl' syndrome can cause pressures which we are poorly equipped to handle at this time. When a business manager must decide either to abandon a product or process or to install a very costly mechanism to control only a marginally *suspected* hazard based on controversial animal evidence *or less,* he is confronted with the Hobson's choice of reacting in a manner which may be judged too slow in the light of good hindsight, or reacting too rapidly in a manner that may provide a disincentive to business. The environmental professional, often a scientist, health professional or engineer, schooled in hazard identification and control, must be involved early in this decision process if the business manager is to be helped.

The difficulty with this kind of a decision has been partly lack of resources and partly the number and complexity of the environmental concerns or apparent concerns. It is important to deal early with environmental concerns which present possibly serious hazards to the operation because of their characteristics, exposure potential or severity, for example. Control is the key. Only professional advice as to environmental, health and safety hazard identification *and control* can help the manager reach a decision about the potential concerns which merit attention with the resources available.

### Investigations

The hazard identification process should include investigation procedures which will measure causes of potential failure and costs.

Management cannot enhance reliability without such investigations: it should have a policy and formal programme for investigation of potential and actual failures. Such failure analysis will permit more accurate predictions about future failures. A strong policy of following up on any actual failures will reinforce and improve the control measures taken to eliminate or minimize failures.

Here is a simple example. Where machinery is used, patterns of strains in moving metal parts might be found. The application of strain gauges to monitor strain growth in such moving metal parts can be analysed according to the patterns. Where failure does occur, follow-up should be undertaken to understand the failure. The corrective measures should then also be subject to follow-up which will modify monitoring procedures and minimize future failures. This is simple and basic reliability engineering.

More elaborately, chemical hazard identification and control should be a subject of continuous study as to process reliability (including mechanical equipment), reliability of control measures, maintenance ability and human reliability. Through monitoring the reliability of four facets of the system — the process itself, the control measures, the maintainability of the system and the human input — many hazards can not only be identified but reasonably minimized.

On an even larger scale, a system of hazard identification and control made part of a routine may be used to deal with hazards such as spills, or emissions causing worker health hazards, or improper disposal. As in the case of monitoring metal fatigue, the key is to identify the area of concern, to construct a system to control the hazards, and then to monitor the occurrence of that problem in order to institute preventive measures where necessary. If the hazard identification system is to be effective over any period of time, investigation of incidents and follow-up is essential.


**The human element**

In all cases of hazard identification, the human element should not be ignored. This requires not merely training, but identification of the credentials needed to do the job. Too often people who are perhaps sufficiently trained but who do not have the credentials to carry out the job can be the cause of a hazardous incident, despite their training. That is a correctable error, and it is management's obligation to correct it by identifying what credentials are really required when dealing with a hazard after it is identified.

This system, both of hazard identification and of control analyses in a process of setting priorities for resource application, is an important one to understand. In the business function, only a certain level of resource is available. The question is to which concerns or potential concerns this resource will be applied. Total allocation of resources is 'zero-sum', that is what is spent in one place is taken away from another. Thus in the business management decision process, as also in government, priorities for the concerns should be set and the resources used where they can provide the most good. Where problems cannot be solved, or adequate control cannot be achieved, then the process or product must be abandoned, modified or limited to assure safety.

An example is given below of the environmental policy of one major business organization.[2] It should be noted that the policy specifically provides for the termination of a business activity where the resources available cannot provide an acceptably safe environment.

It is our policy to manufacture, handle and dispose of all substances safely and without creating unacceptable risks to human health or the environment. The Corporation will:

Establish and maintain programmes to assure that laws and regulations applicable to its products and operation are known and obeyed;

Adopt its own standards where laws or regulations may not be adequately protective, and adopt, where necessary, its own standards where laws do not exist; and

Stop manufacturing any product or carrying out any operation if the environmental costs are unacceptable.

To carry out this policy, the Corporation will:

1. Identify and control public health and environmental hazards stemming from its operation and products;

2. Conduct accident prevention and occupational health programmes to safeguard employees and the public from injuries or health hazards, to protect the Corporation's assets and continuity of operations, and to protect the environment;

3. Conduct and support research about the health and environmental effects of potentially suspect materials handled and sold by the Corporation and share promptly any significant findings with others, such as employees, suppliers, customers, government agencies and the scientific community; and

4. Work constructively with trade associations, government agencies and others to develop equitable and realistic laws, regulations and standards to protect public health and the environment.

Every employee is expected to adhere to the spirit as well as the letter of this policy. Managers have a special obligation to keep informed about environmental risks and standards and to advise higher management promptly of any adverse situation which comes to their attention.

### 'Acceptable risk'

The ultimate question raised, then, is what risks are considered acceptable?

One argument about what concerns should receive priority and how resource deployment should be maximized is the viewpoint expressed by a recent United States Environmental Protection Agency pamphlet.[3] The pamphlet raises the spectre of cancer and the potential destruction of environmental pollutants and suggests that regulation should not wait for 'scientific certainty'. Certainly there is the possibility that a lack of *control* could dictate such an attitude in a particular situation; but an unscientific approach can lead only to irrational dogma difficult to correct later, and wasteful application of resources, which can be put to better use in meeting environmental, health or safety concerns. The business manager should not respond to political stampeding; rather he should rely on rational analysis supported by input from environmental professionals.

Rational environmental, health and safety laws and regulations, and court decisions interpreting them, are based on a theory of reasonable risk, which is based on the institution and maintenance of controls being acceptable and unreasonable risk being unacceptable. Reasonable and unreasonable risk are, of course, relative concepts of risk–benefit and cost–benefit analysis.

*Unreasonable risk* can be analysed by:
>    Identifying hazards
>    Judging exposure and severity
>    Evaluating control measures available to deal with those
>        hazards and
>    Reviewing limitations and benefits.

Very few environmental, health and safety standards are zero-risk in concept, and those that are, are criticized greatly; for example, the so-called Delaney Amendment in the United States dealing with the impermissibility of animal carcinogens in food and food additives. Zero-risk, as generally understood, is the route to industry paralysis and societal misery, and the environmental professional, as well as the manager making the decision with the input of the professional, must think in terms of relative risk and benefit. One must also decide what the alternatives are and what the analysis is telling us about those alternatives. 'Reasonableness' to a business organization cannot usually be determined without an analysis of cost. Industry will need to determine whether the costs for adequate control can be justified. Safety must prevail, and costs will ultimately be passed on to the public. Government will have to decide at what level it requires control by legislative and regulatory fiat. Too much control may result in non-productive resource allocation, abandonment of innovation, economic stagnation and loss of jobs.

Dr Aaron Wildavsky, a University of California political scientist, framed the concern sufficiently, perhaps, when he said, 'How extraordinary. The richest, longest-lived, best protected, most resourceful civilization, with the highest degree of insight into its own technology, is on its way to becoming the most frightened . . . Chicken Little is alive and well in America'.[4]

While the approach of the reasonable to me as an engineer and lawyer is eminently the best approach, it brings with it the requirement for analysis and patience. The hazard must not be viewed in a vacuum, but in relation to the controls and weighing of alternatives, if a sense of the reasonable is to have meaning. Thus, reasonableness is not a concept of absolute security. Hazard control is a technical/ engineering/scientific approach, and evaluation of it should be independent of political factors. That requires not only industry and government doing its homework, but forbearance on the part of the public sector, including the media.

We should not underestimate the problem of dealing with a 'reasonable–unreasonable' risk decision process. To quote Professor Lester Thurow of the Massachusetts Institute of Technology: 'Basically we have created the world described in Robert Ardrey's *The Territorial Imperative* . . . We veto each other's alternatives, but none of us has the ability to create successful initiatives ourselves. Our political and economic structure simply isn't able to cope . . . The gains and losses are not allocated to the same individuals or groups . . . Each group wants government to use its power to protect it and to force others to do what [it perceives] is in the general interest.'[5]

The reasonable requires constant explanation, education and forbearance with respect to one's own *perceived* interest. The government and organized labour must stand up with industry and the scientific community on this point over the coming critical years.


## INTERNATIONAL IMPLICATIONS

Decisions involving hazard identification and risk assessment in each business and in each country are difficult simply from the domestic viewpoint. Current and future international relationships will further complicate the process. Adoption by a United States Environmental Protection Agency (EPA) or by a European Economic Community (EEC) of a labelling policy, for example, has far-reaching implications. Domestic regulatory decisions bear directly upon not only domestic health and safety, but international trade as well. The EEC becomes a prime party of interest when the EPA makes labelling decisions, and industry on the two continents is involved. Likewise, analysis as to testing, including good laboratory practices, requires interaction between industries of different nations and their governments if the manager is to be able to make a sensible decision about the research or product development process needed to market abroad. It is not hard to understand a similar interest on the part of the Japanese, or for that matter Third World countries, which often reason that standards imposed by the developed countries will function as trade barriers to their developing economies. It is thus important for the manager dealing with hazard identification and risk management to identify comprehensively the international issues affecting his decision.

What will be the implications in the future to the business manager of the utility operation in one country (say the United States) as a result of the acid rain controversy and the environmental pressures in a neighbour (Canada)? Can the business manager in England or Switzerland add into his health and environmental fate testing for a new product the requirements of the good laboratory practices which will be adopted under the Toxic Substances Control Act for the product to be marketed in the United States? Has the plant manager in a central European country taken into account the effect of his effluent discharges on the Mediterranean in a new plant addition which he is about to construct? Are there new health concerns being voiced in the United States, Japan or Sweden as to workers' contact with particular allegedly mutagenic materials which should be understood and evaluated for process materials made in an English plant and exported to the United States, Japan or Sweden? Is there an international union involved which intensifies the potential that domestic concerns will be discussed in another country? Can confidentiality of the identity of a new product produced in France be protected against the requirements of disclosure under Occupational Safety and Health Administration regulations in the United States if the product is to be marketed there? And what of the different conclusions reached by esteemed scientific panels in different nations as to unreasonableness of proceeding in the marketplace based on animal test data, such as in the case of cyclamates?

National boundary lines do not delineate environmental, health or safety concerns. The explosion at Flixborough, England, had ramifications in terms of safety risk analysis all over the world. The good manager understands that hazard identification, control and risk analysis takes into consideration these concerns across national boundaries. Environmental impact assessment of new plants and industries is a subject under review and discussion in all parts of the world, and is seen, for example, as a sound development tool by the United Nations Environment Program.[6]


## ORGANIZATION APPROACH

The basic elements of hazard identification and control have been described. Again, the key in risk assessment is an understanding of methods and opportunities for control, and an evaluation of alternatives, including costs. The choice must be made within applicable laws and regulations. The specific design and emphasis of a risk assessment system should be tailored to the needs and characteristics — such as organization size, complexity, nature of product line and technical resources — of the organization.[7] However, in all cases a clear sense of purpose, resolve and authority of the decision-makers should be understood within the organization.

Hazard identification and control may be undertaken by an individual or by a multi-disciplinary approach including professional pollution control, health, product safety and safety input. Environmental professionals should have the degree of independence necessary to ensure their integrity in the assessment process by the business manager. But whatever the organization structure, it is essential that the hazard identification and control and risk determination and avoidance procedure be communicated throughout the organization as a tool *favourably* looked upon by management. To avoid abuses, it is also essential that management make it clear in policies and procedures that it will, in all aspects of its operations and manufacture, handle and dispose of all substances safely, and without creating unacceptable risks to human health or the environment, and within all laws and regulations applicable to its products and operations. A corollary of this management policy is that management will not continue operations in which the costs are unacceptable to bring the risks within the laws and regulations *and* to a level acceptable to the management.

Management should not rely fully on the fact that its operations will conform with company policies and procedures, no matter how independent its professionals, how well organized its people, nor how thoroughgoing the training and indoctrination. It may wish to institute a system of surveillance to sample on a routine basis whether in fact the operations are complying with the laws, regulations, company policy and procedure. A regular and routine programme of plant inspections and reviews of operations, in addition to those by government instrumentalities, will tell the management whether its operations are within the parameters of the risk assessments that it has chosen in the past. Together with rigorous follow-up on all deficiencies found, the surveillance programme will confirm to the employees of the company that the management means to carry out its environmental policies and programmes. Perhaps most important, this surveillance process will enable management to assure itself that it is not creating liabilities or future areas of costly concern by its present practices.

## SCIENTIFIC REPRESSION

Finally, as a lawyer and engineer I would like to voice a personal concern that I have raised in the past. I think that engineers, scientists, lawyers, bureaucrats and individual citizens should be worried about the effect on the development of science and technology of an *over-reacting* government regulatory process.

I do not believe the Lysenko case in the Soviet Union during the Stalin era should ever be too far from our minds. Lysenko attacked Mendelian genetics and his views received official support. This is a case where a pseudo-scientific bureaucrat prescribed by fiat the 'accepted' reaction to scientific questions involving substantial doubt to the overwhelming detriment of his country's agriculture, and thus to its people and government, for decades to come.

One lesson to be learned from the Lysenko case is that the legislator or government bureaucrat may become so enamoured with his laws, rules and regulations as to freeze or tend to freeze science from further development, contradiction or open flow of contrary attitudes. Reasonable risk assessment allows for the consideration of *changing* concepts in science.

The power of government is very great and one can come face to face with dictatorial power that assumes the worst to be scientific fact and then regulates on that assumption. As the *New York Times* put it, 'The emphasis on absolute safety in existing law can undermine sensible regulation. The mantle of scientific leadership will pass from industry and . . . indeed from western society, if that hardening of attitudes as to scientific assumptions is not checked.'[8]

Bureaucracy must leave room for science to express, debate and in time prove or disprove doubtful hypotheses. We will have to ask the questions and do the patient basic research which will tell us whether, for example, massive doses for animals in specific cases really are adequate for predicting results in man. There are many unknowns, and regulation must allow for future work and changing ideas if science is not to be frozen by present dogma, and society, unwittingly, is not to be the loser.

There is a growing awakening to this concern of science being controlled by an overzealous official dogma of the regulators. Truths do turn into dogma if they are applied as truths after they are legitimately disputed. Unfortunately, there is the desire on the part of many — and not just those in government — to further their personal career with the conversion of theories into absolutes in the environmental field by the establishment of dogma. Many a research grant supports dogma by its very existence. Thousands of animal feeding projects to establish carcinogenicity in test animals may not answer the question of relevance to man, but they do reinforce the accepted dogma of applicability.

Business organizations will sometimes ask their managers to give them a risk-free formula if they are to continue or to begin in a business under attack because of environmental concerns. The business manager, supported by the professionals within the organization, will have to carry the load of convincing the management that the risk is reasonable, and that zero-risk is unattainable and unrealistic for them as a goal. The manager will need to convince his superiors that risk assessment not only results in good environmental management, but justifies the application of continued resources. This requires risk analysis, skill and patience, using the talents of scientists, engineers and lawyers, and it will also require developing a level of trust with the management.

REFERENCES
1. The potential effect of these and other pressures on a regulatory agency are also discussed in the following article: Plaut J. The implications of TSCA for industry and society. *The Royal Society of Medicine International Congress and Symposium Series* 1980; **17**:55–63.
2. Environmental Policy of Allied Chemical Corporation, signed by Allied Chemical's Chairman, Edward L. Hennessy jun. on 17 March 1980.
3. *Pollution's 'Invisible' Victims: Why Environmental Regulations Cannot Wait for Scientific Certainty.* OPA 119/0. Washington, DC; United States Environmental Protection Agency, 1980.
4. Wildavsky A. *American Scientist* 1976; **67**:32.
5. Thurow LC. *The Zero-Sum Society.* New York: Basic Books Inc., 1980.
6. *Industry and Environment.* UN Environment Programme, 1980; **1**:1.
7. Wallum HJ and Plaut J. *The Dispute Management Manual.* New York: Center for Public Resources, 1981.
8. Editorial. *New York Times.* 11 March 1979.

# 15. THE CHANGING WORLD OF WORK
*Liam Gorman*

It would be hard to exaggerate the importance of work in our lives. Many of us spend a sizeable proportion of our lives at work, and, of course, a considerably higher proportion of our waking time. For many of us, work is the source of most of our income. In return for the sale of our skills or effort we receive a greater or lesser income. In these days most work from which income is derived is performed in organizations, that is, social systems that have been specially set up to achieve some purpose, usually the production of goods or services. However, much work is done for which people do not get paid, for instance, housework and voluntary work. The subject of this chapter will be 'paid work' and the changes which are taking place or may take place in the coming decade in the way in which such work is carried out, or the changes which may take place in organizations and which will significantly affect how work is carried out.

As well as our income, work confers other significant benefits. Through our work we become connected with a wider society and community. One of the distressing aspects of unemployment apart from the obvious economic privation is that the unemployed person feels cut adrift from the community and his self-esteem is badly dented because he is unable to contribute to the wellbeing of society and play a clearly defined role in the community. These psychological and social disabilities accompanying unemployment often give rise to or are accompanied by physical symptoms, such as high blood pressure, diabetes, peptic ulcers, headaches or heart attacks.[1]

One reason, perhaps, why people put up with the boredom, humiliations and/or dangers of their working lives is that a job has to be very bad before it is worse than no job at all.

For many of us our work is a major source of our identity. This point might be made simply by the fact that we usually tell people what job we do when we try to give others insight into ourselves. Equally, when seeking information on others in order 'to place them', we usually begin by finding out what their occupation is.

Our health and wellbeing, as has already been noted, are affected by whether or not we are employed, but, equally, may be affected by the kind of work we do (*see*, for example, the work of Vertin[2]). This study showed that foremen had an ulcer rate seven times that of the first level employees studied. Other writers have shown that the way in which we spend our leisure can be affected by the kind of work we do.[3] In general, these studies show that lack of involvement in work leads often to a lack-lustre use of leisure. However, not only are the wellbeing and life opportunities of the employed person affected by the type of work he* does but the opportunities of his family are clearly related to that work through the differences in income society deems various kinds of work to deserve.

In summary, our work is an extremely important element of many of our lives. So, let us turn now to the main theme of this chapter, that is, the way in which the world of work is changing.

## CHANGING LEADERSHIP PATTERNS

A major factor affecting the climate in which work is performed is the changing norms concerning how authority or leadership should be exercised. Until the recent past two major bases for the power of authority figures were generally acceptable to followers. These were, first, the power to threaten and, if necessary, to carry through the threat to deprive those in subordinate positions of something they valued, for instance, money, a promotion, desirable assignments or even a job. The second acceptable base of authority was position power, that is, subordinates were disposed to obey those who had a higher hierarchical position or position of status simply because they held those positions. However, with regard to the latter basis for getting compliance, there has been a marked decline in deference — few people in Western Europe under 30 years of age call anybody 'sir' any more. The institutions such as the home, the school and, in some countries, the church, in which previously the young served an apprenticeship in deference have changed their norms regarding obedience so that young people no longer come to work organizations disposed to concede automatically that wisdom resides in those in authority. Equally, the capacity to threaten punishment and, if necessary, to punish, has been curtailed by legislation, by collectivism and by norms that regard this way of maintaining social control as not very acceptable. To exercise authority now, those in responsible positions must look for new bases for their power. Now leaders must earn their power; power is conferred not by position — or only in small measure by position — it is granted to leaders by followers.

Authority figures must now earn the loyalty of their followers, in the main through the personal qualities and competences they manifest in their own work lives. For instance, to get their followers' commitment they must be administratively and technically competent, they must

*When a particular gender is used in the text it is intended to cover both the masculine and feminine.

have personal qualities that command respect, they must be able to project a vision of a desirable future to which they can bring those they lead, they must rely more on reward and positive control processes and they must share information more liberally than in the past.

## ASPIRATIONS FOR MEANINGFUL WORK

A related change in the aspirations or characteristics of many employees which has critical implications for work in organizations is the desire for jobs which satisfy distinctly human needs and not merely provide financial recompense to meet material needs. The rising educational levels, the change in the ratio of skilled, professional and semi-professional employees to unskilled employees in the workforce and the better information available through the various media are some of the factors that account for these raised aspirations. So far trade unions have not given a strong lead in the demand for the redesign of jobs and the restructuring of work systems. However, as more white collar, graduate and managerial employees become unionized as is happening in several European countries, at least some trade unions may be expected to include the nature of work in the issues on which they seek to negotiate.

Concern about the nature of the work they will have to perform is particularly strong among young people. I recently visited twenty secondary schools and talked to over three hundred pupils in their final year at these schools about their perceptions of particular careers. The extent to which this group emphasized the importance for them of jobs in which there was some scope for initiative, opportunities to take responsibility, to exercise some discretion and to develop skills and competences was striking. Some writers claim that money and security become relatively more important than intrinsic job factors to people as they become older, marry and take on the responsibilities of a family. While this may be so, there does seem to be an overall relentless quest by many to escape the routine, repetitive, 'soul-destroying' jobs which made men and women appendages of machines. This has been a feature of the design of production systems so far in this century but it will not do for the 1980s.

Many of the work systems which ignored man's needs for achievement, for problem solving, even for social contact, and for responsible action are, no doubt, continuing to operate, some would say successfully. But at what cost in human misery, in social conflict, in waste of materials, grossly suboptimal performance and sabotage? Because of the many dysfunctional features of these production systems some employers have taken initiatives to create work more compatible with the needs of man and the growing aspirations of our young people for a humanized work environment.[4] For instance, Bang and Olufsen Ltd, the Danish electronics group, dismantled their assembly lines and provided opportunities for production operators to carry out a significant number of steps in a long production process and then take responsibility for the quality of the goods they produced. In this way operators experienced greater variation in their tasks, longer work cycles, more influence on the method of work and speed of working and as a result had greater satisfaction in their jobs while the company had the benefit of greater productivity.

As well as efforts by some employers, several governments have set up agencies to promote the development of humanized work systems. Thus in France, legislation was passed in the early 1970s setting up the Agence Nationale pour l'Amélioration des Conditions du Travail (National Agency for the Improvement of Working Conditions) which strives in various ways to encourage the restructuring of work systems so that they will meet more distinctly human needs. Nevertheless, efforts so far to tackle the problem of meeting the mounting aspirations of the young and to alleviate the problems caused by the frustration of those working dehumanized work systems have not been terribly impressive. The size of the problem scares many employers. They are stuck with the technology, in the design of which there was scant attention given to the needs of the people who would operate it. Recessions and stiff competition put the emphasis on survival. The trade unions which, particularly in, say, Britain and Ireland, are strongly craft-based are hesitant because work restructuring would make necessary the elimination of demarcations, thus threatening the existence of many of these unions.

Despite the extent to which employers, trade unions and governments ignore the work design problem, it is likely because of the widespread aspirations especially among young people for a radical approach to this problem that this will be one of the issues concerning work which will preoccupy us in the remainder of this century. Too many employees want to live to work and not just work to live for the problem to go away.


## THE CONTRACTUAL ORGANIZATION

The desire of individual employees for more freedom and flexibility, added to the problems organizations have in the management of people, will lead to the emergence of what some writers have referred to as the contractual organization as opposed to the employment organization, that is the conventional organization as we know it at present.[5] The contractual organization will pay for *output* in the form of fees; at present employees tend to be paid wages or salaries for time spent in a particular organization, that is for *input*. The idea is that more and more organizations will attempt to give their employees the status that professionals, artists and some craftsmen have: for a given service or finished product fees will be paid. The employees will thus experience the independence, self-control, responsibility and flexibility of the self-employed and the organization will avoid the costs of supervision and bureaucracy and the restrictions of legislation. Such systems are already common in the construction industry but many hard-pressed companies will come to understandings with their more independent-minded employees to develop the idea vastly in the coming decades. The spread of management by objectives, the growth of autonomous work groups at shop floor level (that is, work groups which undertake responsibility for the completion to a satisfactory quality of complete products) and the spread of 'flexitime', which gives employees discretion in determining when they will work, are signs of the emphasis switching to output as distinct from input. This switch may make easier the emergence of the contractual organization in the future. Large organizations will also use franchising arrangements to keep their employee numbers from rising. Cottage industries but with more powerful technological bases will be encouraged by large organizations to assemble their products and sell them into the large organization. A related development is the growth of the 'black' or 'free' economy (the label you give it depends on your outlook). Despite the disapproval of government and trade unions but not of employers, it is likely that more and more work will be carried out in the black (or hidden) economy as more individuals try to escape what they would regard as a penal tax system and the dependence, restrictions and 'bureaucracy' of the large organization. In this regard, society is faced with a difficult problem: how to provide for the weak and yet not stifle the initiative of those able and willing to work.


## DEMAND FOR PARTICIPATION

A phenomenon related to the movement for leadership by consent and for intrinsically satisfying work is the increasing desire by employees to influence decisions in their organizations, particularly those decisions that affect them directly. One way in which the desire for participation is manifest is in the increasing desire of employees to have a right of veto over organizational and technological changes which affect them. It is not my view that employees wish to impede technological change but they do wish to exercise a veto if they feel that appropriate consultation about changes has not taken place with them and if they feel that major decisions about their lives are being taken without their views being canvassed. Another manifestation of increasing participation by the workforce will be the added emphasis on disclosure of financial and other information which will be a changed feature of organizational life in the 1980s.

The European Economic Commission has elaborate proposals for the type of disclosure that will be required from companies of a reasonable size. Structural solutions to the demand for participation such as the appointment or election of worker directors or various types of consultative committees will be another path that will be pursued in some countries sometimes with the support of legislation at national level. Such solutions are likely to accomplish little if the work restructuring process and participation at the shop floor level brought about by the new ways of exercising authority which have been alluded to already are not pursued.

In this connection, *see* the study of Brannen et al. of worker directors in the British Steel Corporation.[6]

## SOCIAL DIVISIONS AT WORK

It seems likely in the 1980s that some of the gross social divisions in our organizations and their accompanying privileges for some and restrictions for others will be less acceptable. Some examples: a survey in Britain showed that 98 per cent of operatives were required to clock in whereas 6 per cent of managers had to book or clock in (usually book in!), similar percentages of each group lost pay if late for work.[7] While pay differentials will probably continue to be acceptable if they reflect differential effort and performance levels faithfully it is likely that the thousand and one small distinctions — for example, parking spaces — and not so small — for example, payment during illness — will become less and less acceptable and that a movement towards all employees being made salaried employees will gain momentum. Employers, while seeing some obvious difficulties in such a change, for instance, possibly raised absence levels and other cost increases, will also see benefits.

The changed psychological contract will open up the possibility of employees committing themselves more to their organization, thus making possible more flexibility in labour practices so vital for organizations that will be coping with highly competitive and changing environments. Employees, in my view, will increasingly want to be cherished equally by their organizations, otherwise they will draw the obvious conclusion and the conflicts, absenteeism, apathy and sabotage which are features of much organizational life will be exacerbated. In my view a step in the eighties towards renegotiating psychological contracts for all employees which would eliminate the apartheid between salaried and unsalaried employees that is such a prominent feature of many of our organizations would be as important in improving the quality of life at work as the restructuring and redesigning of work and jobs to satisfy human needs.

## ASPIRATIONS FOR DEVELOPMENT

Those coming into the workforce in the 1970s, and there is little reason to think this trend will not continue in the 1980s, have been characterized by a greater desire to explore their potential abilities and capacities and to some extent through their work to realize their potential. They are the 'me' generation. They do not wish, or much less so than previous generations, to live their lives vicariously through the experience of others or future generations. On the other hand, organizations have increasingly complex, turbulent, unpredictable and changing environments to cope with. As a result employees are required to be prepared to learn new ways if they are to help their organizations to cope with these changes. Individuals, therefore, will have to be prepared to acquire new skills, retrain and change careers if they are to retain a useful role in organizations in a demanding environment. Organizations will be required to provide more extensive opportunities for development, not just through the expansion of training and educational facilities but through the provision of jobs which do not lead individuals into blind alleys and leave them with obsolescent skills. Job mobility levels will rise, particularly among specialists and skilled groups as people seek the job that optimally uses their talents or provides the right balance between work and other aspects of their lives. Career planning will become of more importance not just to those who are concentrating on advancement within a particular enterprise but for those (the majority) who define success in other ways but who can still provide a useful service for the organization when they understand the relative balance of personal needs, family and work issues in their lives. The balance of emphasis in manpower planning, which in the past was exclusively concerned with the needs of the organization, will change. The needs of the individual for self-realization will have to receive attention as well as those of the organization for performance and relevant skills. To create wider career options, development opportunities and to make greater individuality in treatment of employees possible, innovations such as the following will become more common: multiple career ladders, that is career ladders of equal standing for specialists' careers as for those with line responsibilities; sabbaticals at intervals to generate a sense of renewal; secondments to voluntary organizations or from large to small organizations or state to private enterprises; more part time and contract work; individualized payment packages allowing employees to have different balances in the benefits they receive; and so on. Legislation making those in employment entitled to some time off work each year to be spent on their own development may well become widespread.

## CHANGING WORK ASPIRATIONS OF WOMEN

A striking feature of employment patterns in Western European and North American countries since the end of World War II is the increasing proportion of women in the workforce. Not only has the female labour force grown in numbers but it has undergone some interesting structural changes, for instance, the age composition has altered so that the re-entry to work of women in their middle years is now a common pattern. Despite the large numbers of women in paid employment, however, few until relatively recently saw themselves as pursuing a career or doing work which would have to be given equal priority with their husband's work if questions of, say, relocation or care of children were to interfere with either of their careers or work. A high proportion of women are employed part-time; the great majority say they have financial reasons for working to secure specific extras or a somewhat raised general standard of living.[8] The average earnings of women tend to be about 60 per cent those of men. They are concentrated in low-paying jobs in which men are usually not found and in which there is often no on-the-job training, a factor in making career progression possible for men.

A growing feature of the world of work in the 1970s and a trend gathering momentum is what has been called the dual-career family in which both partners are strongly committed to their work or careers. Some of the strains such families are subject to have been documented by the Rapoports;[9] for example, the excess of responsibilities and demands from home and work, pressure on wives from being in jobs where women as yet are not accepted, criticism of wives for allowing others to care for their children and criticisms of both husbands and wives for breaches of sex role stereotypes.

While so far the dual-career family is a middle-class phenomenon, it seems likely that more women from all classes will break with the conventional view of women's role and give work outside the home a higher priority than they have in the past. This changed role for women, a change made more likely because of failure to include housewives' services in the national income, thus lowering its social status, will make available many new potential abilities to organizations but will bring many transitional stresses and strains not least to the wives and their husbands who forge these dual-careers. Organizational strains will arise because as more women take their careers seriously, more will get advancement and men with conventional views may have difficulty in accepting roles subordinate to these women. The proportion of women who are managers at present in European countries is something between 5 and 10 per cent. In the coming decade, through the new aspirations of women themselves and legislative pressures, such proportions may well be doubled or trebled and more and more jobs previously men's preserve will be entered by women. A woman's place will no longer be at the typewriter or the work bench. It is noteworthy that 32 per cent of engineers in the Soviet Union are women.[10] But these changes will not take place without considerable readjustment on the part of many in our organizations. Pursuit of work careers by women will also have major implications for the roles of husbands and wives within the home and many long-held stereotypes of these roles will have to be radically altered if women are not to be badly overloaded with responsibilities arising from their work at home and their work in the market.

## IMPACT OF TECHNOLOGICAL DEVELOPMENTS

While much of our material prosperity has been based on technological advances, technology has also brought grim, routinized existences to many employees. But the newer technological advances may not prove so ominous for the quality of work life in organizations. Certainly there will be major dislocations. The 'chip' technology will greatly extend man's capacity for information processing and when we consider how much of our work is of an information processing nature it is possible to see the ramifications of chip technology for organizational life. Women workers who form such a large proportion of clerical employees will particularly experience disruption since the office will be one of the principal foci for change in this techno-logical revolution. But few will go untouched. Certainly the new technology will bring some new employment; in the main traditional areas of employment, however, there will be disruption, though the unemployment caused will not be as great as some commentators have suggested. Employment will be created in new or expanded industries, some of which will be based either centrally or peripherally on the new technology but other employment will be created in newly developed service industries. Industries that will become increasingly important will be the educational and leisure industries, which will expand greatly to cope with the self-actualization age we are entering and the greater leisure opportunities which the new technology will make possible.

Unlike other technological advances, white collar and professional employees will be greatly affected by the microprocessor development and the deskilling which earlier technology brought about for the shopfloor employee will now affect a different type of employee. At the same time many boring and repetitive processes will be taken over by the microprocessor and the community will have the advantage of increased productivity, uniform quality products and the elimination of dead-end jobs.

Widespread applications of microprocessor technology to clerical employment will come early in this decade while the applications to manufacturing processes will come more slowly. It has long been possible to apply computers to control industrial processes but expense and bulk have been deterrents. With the new technology, those obstacles no longer exist. Many mechanically based industries will have to apply the microprocessor technology if they are not to slip down to dangerously low levels of performance. In the longer term this technology opens up possibilities for greater leisure and the products of the technology ought to enhance the enjoyment of this leisure while aiding whoever, husband or wife, is responsible for doing the housework. One final comment on this technology is that it will make possible the development of the contractual arrangements alluded to earlier by making it easier to carry out work previously done in organizations in the home. As a consequence, it may be easier for women to participate in the world of work and for men to take an expanded role in the running of the household.

## IMPACT OF CHANGES IN ORGANIZATIONAL FORMS

A feature of organizational life in the 1960s and 70s was the increasing recognition that organizations were highly interdependent with their environments and that for the organization to prosper a substantial amount of time must be invested in managing the interface between the organization and its environment. Environments became more demanding in the 1960s and 70s through, for instance, the emergence of environmental and consumer pressure groups, through various forms of legislative restrictions and through increased competitiveness with the loosening of trade restrictions.

To cope with these environmental demands, some senior people and not so senior people in organizations have to spend a substantial proportion of their time (for senior people about 90 per cent of their time) managing aspects of the environment so that these aspects do not destabilize the organization. Many specialists had to be employed or their numbers greatly increased to assist in managing the organization–environment boundary, for instance, public relations people to liaise with the media; research people to ensure that the implications for the organization of scientific discoveries being made in universities and research institutes were recognized and if necessary acted on; lawyers to interpret the implications of new legislation; negotiators and liaison people of various types to deal with regulatory agencies and pressure groups and so on. But problems do not always come in a form that personnel with one professional background can cope with them, so task forces, often of a multi-disciplinary nature, have to be formed to deal with problems. When these problems have been handled, these task forces are disbanded and new ones formed to deal with new issues that arise.

In short then, the growing complexity of the environments of organizations and the increase in the rate of change have meant that the proportion of scientific, professional and managerial personnel in workforces has been increased so that a new group, demanding in the quality of work they seek and the type of relationships they want, are added to our organizations. For this group the foci of their loyalties are often not the organizations they work for but professional associations to which they belong. Furthermore, their own career advancement is of paramount importance to them.

Organizational innovations such as the continuous formation and disbandment of task forces have weakened some traditionally accepted 'principles' of organizations, for example, that each employee should have one boss. One could be a member of several task forces with different leaders and a leader of one task force might be a follower in another. Distinctions between line and staff and other notions about span of control and unity of command case to have the same relevance that they had in our once-simpler organizations, particularly those with few products and operating in few markets. These new, more flexible organizations can suit some but put powerful pressure on others. Individuals have to be prepared to enter new groups, adjust quickly and achieve results, then disband as working relations have been formed and proceed to enter an entirely different group. Leaders gain position by competent performance. Conflict resolution and negotiation skills are important in ensuring that those with a particular professional background accept the validity of the viewpoint of those from very different academic disciplines and backgrounds. These are a few of the implications for work in organizations with increasingly complex environments. These organizations are forced to deal with this complexity by the creation within of the requisite 'matching variety' through the employment of 'knowledge' workers, the creation of new departments and the replacement of the simple bureaucratic structures of the past with more flexible organizational structures.


## CONCLUSION

Earlier in this chapter the changed nature of leadership required in the world of work was discussed. The increasing demand for less routinized work, more participation and development and less social division of employees was outlined. The possibility of the contractual type of organization as against the employment organization emerging was mentioned. The impact of technology, of greater complexity of environments and of the changed aspirations of women in the world of work was referred to. In this concluding section we return to the theme that this chapter started with, the importance of work in our lives, or more precisely, the question of whether or not work is becoming less important to people and whether other areas of their lives, say their leisure or family activities, are of more importance to them than their work activities. Some writers have claimed to show that the workers they studied had an 'instrumental' view of work — seeking a high economic return from work but having their home and family, not work, as the central interest of their lives.[11] Undoubtedly there are individual differences among workers, and indeed managers, in their wants. Some are keen to have meaningful jobs and responsibility at work while others are prepared, if the wage is right, to submit to very unsatisfactory work situations. There is some evidence of work becoming less central in people's lives even among managerial groups. For instance, in a recent survey of 3000 managers by the American Management Association, again, the family seemed to be a more central life interest for a majority of this group than their work. Nevertheless, there is still a widespread desire among people at all levels, including on the shopfloor, for more meaningful work and more control over their own work situations.[12]

There have been civilizations, for example ancient Greece, where work was devalued, and we have more recently had societies in which 'gentlemen' did not work. It does not seem, however, as if any such crisis faces us. To mention just one piece of evidence, when people are asked whether or not they would continue to work even if they were so rich as to make it unnecessary four out of five of even unskilled people agree that they would.[13] The reasons the skilled and professional workers give for continuing to work are different from those given by the unskilled. Nevertheless, the answers suggest there is no mass movement away from the desire to work or a decline in the 'Protestant ethic'. In my view there is, certainly, a more critical workforce in organizations that are not prepared to be loyal to inefficient, uncaring, untrustworthy and exploitative systems, but explanations for our organizational problems in terms of workers' philosophies are suspect. There are sufficient shortcomings in the system — witness the sorry tale of deaths and injuries at work as only one stark shortcoming — to explain why many workers should be alienated or, at least, disillusioned.

Management must be looked to, however hard-pressed with other issues, to create with the cooperation of trade unions the meaningful work and the sense of community in organizations which, even if not a panacea, could eliminate some of the dysfunctional features of work life and which, where tried systematically, has been shown to have widely satisfactory productivity effects for the firms and work satisfaction outcomes for employees.[14] Certainly, such advances will not be made without costs in terms of effort and finance. Some will say we cannot afford such costs and diversion of efforts, but if the outline given here of changed attitudes, of new aspirations and changing environments approximates the truth, can we afford not to make the effort?

REFERENCES
1. Kasl SV and Cobb S. Blood pressure changes in men undergoing job loss. *Psychosom. Med.* 1970; **32**:19–38.
2. Vertin PG. Bedryfsgeneeskundige Aspecten van het Ulcus Pepticum. (Occupational Health Aspects of Peptic Ulcer.) Thesis, University of Groningen, Netherlands, 1954.
3. Friedmann G. *The Anatomy of Work.* London: Heinemann, 1961.
4. Cooper CL. Humanizing the workplace in Europe: an overview of six countries. *Personnel Journal* 1980; **59**:488–91.
5. Handy C. Through the organizational looking glass. *Harvard Business Review* 1980; **58**:115–21.
6. Brannen P, Batstone E, Fatchett D et al. *The Worker Directors.* London: Hutchinson, 1976.
7. Field F. *Unequal Britain.* London: Arrow Books, 1974.
8. Klein V. *Britain's Married Women Workers.* London: Routledge, 1965.
9. Rapoport R and Rapoport R. *Dual-Career Families Re-examined.* London: Robertson, 1976.
10. Dodge NT. *Women in the Soviet Economy: Their Role in Economic, Scientific and Technical Development.* Baltimore: Johns Hopkins Press, 1966.
11. Goldthorpe JH, Lockwood D, Bechofer F et al. *The Affluent Worker: Industrial Attitudes and Behaviour.* Cambridge: Cambridge University Press, 1968.
12. Tannenbaum AS. *Control in Organizations.* New York: McGraw-Hill, 1969.
13. Whelan CT. *Employment Conditions and Job Satisfaction: The Distribution, Perception and Evaluation of Job Rewards.* Dublin: Economic and Social Research Institute, 1980.
14. Walton RE. From Hawthorne to Topeka and Kalmar. In: Cass EL and Zimmer FG (ed.) *Man and Work in Society.* New York: Van Nostrand–Reinhold, 1975.

# 16. WORKING ABROAD
*A. Downie*

The motivation of the individual who chooses to work abroad is most often financial, though in certain instances there may be a desire to escape from a difficult life situation, or a genuine vocational reason for wishing to work in an underdeveloped country. The expatriate is usually a skilled man, willing to provide his expertise in the country of his choice, though unlike the emigrant his roots remain in his country of origin to which he will eventually return. Formerly large trading companies, the Colonial Service, the armed forces and certain agricultural plantations were major sources of employment for those working abroad. Today this role has largely been taken over by the construction industry and the multinational corporations, especially those involved in petrochemicals.

The employer of expatriate labour has usually the opportunity to make considerable profits but against this must expect high overhead expenses. To attract skilled labour he has to offer not only a high salary but also a variety of fringe benefits which attempt to equate with home country conditions. What must be provided obviously depends on local standards of living and the size of the organization, but since many of these companies operate in underdeveloped countries it may represent a sizeable expense. The provision of housing, food, health care, recreational facilities and adequate home leave may be costly, but can be of equal importance to a high salary in attracting and keeping good personnel. To the company an employee who quits before the end of his specified contract represents an unwanted expense as replacements are always costly.

## SELECTION OF THE EMPLOYEE

Selection is obviously important to the employer, many of whom make use of professional recruiting agencies since they do not have offices in the country of hiring. It is vital that such agents are fully aware of the employer's requirements and the environmental conditions and it is an advantage if they themselves have visited the country.

The employee must obviously be qualified in his trade or profession and ideally should have had a number of years' experience. There has already been a degree of self-selection by the time the candidate has applied for the job, in that he is prepared to leave his home environment to work in a foreign country. His work history and especially previous overseas employment is significant. Frequent changes of employer, particularly in mid-contract, for someone having worked abroad should indicate a high-risk employee. Before being offered employment, the candidate should be given an accurate account of the living and working conditions and a concise job description. Any attempt to gloss over problems is unwise and misunderstanding over job requirements may later lead to considerable resentment and poor work performance.

Many expatriate posts require that the employee should leave his family in the home country, but in those where family status is offered, it is important to consider the wife and children. A wife's ability to adapt to living conditions or the educational facilities provided for children may often be the deciding factor in whether or not the employee fulfils his contract.

## PRE-EMPLOYMENT MEDICAL EXAMINATION

This ritual examination tends to be carried out with little variation in content no matter whether the employee will work in a cold, temperate or hot climate. It is an attempt to protect the employer from hiring an unfit employee and at the same time to protect the candidate from becoming ill while working abroad. As a rule the employer has more faith in the efficacy of the examination than the doctor performing it, since it has little value in predicting subsequent health and less value in the prediction of subsequent absence from work attributed to ill health.[1]

There are of course alternatives, the most common being the pre-placement examination where, with the knowledge of the job content and the environmental conditions, one attempts to assess the fitness to do the particular job. The standards obviously would vary greatly between selecting divers and clerical staff but there is a genuine attempt to protect the employee from being assigned to a job that might be harmful. Other reasons given for this type of examination are to provide a baseline for further examination and to identify and correct remedial physical defects by referring on for treatment.[2,3] Some firms have had satisfactory experience by replacing the examination with a health questionnaire either for all applicants[4] or for the younger professional group.[5] If the decision is made that a pre-assignment health appraisal be carried out it is recommended that the health status, including assessment of the emotional status, be evaluated.[6] The latter aspect is the most difficult of all functions to perform when a person is seen only once, though it may be of paramount importance in relation to the stresses encountered in working abroad. The family practitioner with his long term knowledge of the individual is better equipped to perform an emotional assessment and in many respects is the ideal person to perform a pre-employment medical. Alternative attempts at personality screening, such as the Minnesota Multiphasic Personality Inventory (MMPI),[7] can be carried out. While originally intended as an objective aid in routine psychiatric diagnosis and as a method for the determination of the severity of the condition, this 566-question test can, with skilful interpretation, suggest personality difficulties and may be helpful in job screening.

The Occupational Medical Practice Committee of the AOMA[6] suggests that physician recommendations to management regarding the assignment of an employee to a job be based on (a) medical history, (b) occupational history, (c) assessment of the organ systems likely to be affected by the assignment and (d) evaluation of the description and demands of the job to which assignment is being considered.

The examining physician must have a clearly defined company policy regarding the acceptance or rejection of candidates as the legal situation in relation to job discrimination varies from country to country. A previous history of psychiatric illness or former use of alcohol or drugs, a past history of myocardial infarction in someone without angina and with a normal exercise tolerance test or a former history of irritant dermatitis are all examples of conditions which might or might not cause failure of the candidate, depending on the attitude of the company. A comprehensive self-administered health questionnaire followed by more direct questioning on relevant issues will point to the more significant chronic and recurrent health problems. Additional information may be obtained from the family practitioner or specialists if this seems to be appropriate.

How seldom one sees an occupational history accurately recorded in a pre-employment examination! With the present tendency for companies to be sued by employees and former employees for occupational illness it is vital that a precise working history be obtained. This should include the time spent in the job with an outline of the job itself and the likely exposures. On the basis of this history, specialized examinations may be carried out to give baseline values at the time of joining the company.

As for the examination itself, there is a tendency to do a blanket screening from head to toe, to throw in tests of visual acuity, audiometry, haematological and biochemical profiles, chest X-ray, occasionally back X-rays, ECG and spirometry in the hope that any significant defects will be caught in the net. It is unnecessary here to go through the full details of the clinical examination but a few points are worthy of mention. Ideally this procedure should be somewhat variable depending on the history obtained from the applicant, the requirements and potential exposures of the job and the environmental conditions of the country.

Almost without exception the candidate has a chest X-ray, though the prevalence of tuberculosis in many areas is sufficiently low to make such examinations unproductive. In addition, the candidate wishing to work abroad often has several of these done for different companies. The Energy Technology Committee of the AOMA[8] suggests that rather than subject an individual to repeated chest X-rays a system of borrowing films be used and that the normal, healthy, unexposed adult with a negative tuberculin test who is not at increased risk for other lung diseases probably only needs a chest X-ray when specific justifications appear under the age of 40. One might wish to vary this in those with increased risk of chronic lung disease or cancer, such as cigarette smokers and anyone exposed to pulmonary irritants or carcinogens. Pulmonary function tests are relatively simple and safe procedures that give more relevant information than a single chest X-ray. A few screening examinations have included routine lumbar spine X-rays in an attempt to exclude employees with potential back problems. This is, however, regarded as poor practice[8] as the procedure should be solely for the evaluation of existing disease. The examination of the skin is a simple but often overlooked part of a pre-placement medical. It is accepted that skin disease is the most common of all occupational illnesses and that within this group the greatest number of cases result from the exposure to irritant chemicals. Climatic conditions, such as extremes of temperature, wind chill or ultraviolet exposure, in susceptible individuals may aggravate the effects of irritant chemicals. There are a number of chronic skin conditions which, depending on the circumstances of employment, may present a reasonable case for exclusion in those working abroad. Shmunes[9] suggested a list of 'patients with recalcitrant and severe cases of atopic eczema, psoriasis, intertrigenous inflammatory eruptions, chronic exfoliative dermatitis, lichen planus, acne conglobata and chronic miliaria . . . discoid and systemic lupus erythematosus, polymorphous light eruption, light sensitive porphyria and solar urticaria'. While by no means all-embracing, it provides a useful guide to selection. In examinations where biochemical and haematological profiles are standard, a combination of increase in MCV, serum gamma glutamyl transpeptidase ($\gamma$GTP) and uric acid might make one suspicious of someone with a potential alcohol problem. It is not reasonable to diagnose the condition on these markers alone[10] but they can be valuable where there are other pointers towards alcoholism.

The decision on which conditions cause a candidate to fail his medical examination is often a complex one and may in part depend on the standard of medical care available in the country where he will be working. It also may prove easier for a senior executive being transferred from a parent company to be accepted with significant health problems than it would be for a new candidate joining the affiliate directly. The chronic health problems which cause difficulties in home country employment are likely to be aggravated when there is the additional factor of extremes of temperature. For those working in cold climates, peripheral vascular disease, pre-existing cardiac disease, mental disturbance and TB are normally considered the major contraindications to employment.[11,12] The Arctic environment is a particularly harsh one where man's ability to adapt depends totally on insulation from the effects of cold. Perhaps Robert W. Service best sums up the pre-employment requirements when he says, 'This is the law of the Yukon and ever she makes it plain, Send not your foolish and feeble, send me your strong and your sane'.

In hot countries man has the ability to acclimatize after an initial period, but the older age group, the obese and those with chronic skin disease display more than their fair share of problems.

If a medical examination of the employee is required, it is only reasonable that a similar screening process be used for his wife and family. The environmental stresses also affect families and if the local medical facilities are inadequate, repatriation costs can be high. It is common practice for firms to exclude relatives with chronic or recurrent conditions which require regular specialist treatment. Once again a clearly stated company policy is required by the examining doctor. Most firms do not pay the cost of dental treatment for their employees and those who have embarked on extended courses of orthodontic treatment for their children are unlikely to find such facilities in developing countries. Few companies are prepared to pay repatriation costs for what is mainly a cosmetic dental treatment.

## PROBLEMS RELATED TO LIVING AND WORKING ABROAD

### Language

One of the major difficulties may be with language and the disinclination of those whose mother tongue is English to learn a foreign language. In certain multinational corporations the inability to speak the language of the country is not perceived as a handicap since the common business language is English and there may be a sufficient number of one's colleagues living close enough to be regular social contacts. Any effort made by the employee is well worth while as an attempt to speak the language, even badly, usually invokes a sympathetic reaction from the local inhabitants, be they shopkeepers, neighbours or working colleagues.

### Culture

Cultural differences are often another difficulty to be overcome by the newcomer and it is easy to criticize those whose basic values appear far removed from our own. Attitudes to life and death, morality, crime and punishment, the role of the male and female in society, hygiene and the acceptance of a religious influence on the laws of the state may have to be viewed from the standpoint that they evolved for a different lifestyle from that which may now exist. It is worth remembering that in some of the oil-rich countries of the Middle East until a generation ago life had changed little over the previous 300 years. These differences as a rule are known to the employee before he goes to a country but often he fails to realize their impact on the way he is expected to live.

## Climate

Extremes of climate have their own problems. In Arctic regions good indoctrination and a well-organized work regimen should keep incidents of hypothermia to a minimum, but of course there is always the danger of an accident and especially an immersion incident. The wind chill factor in addition to the low ambient temperature has to be borne in mind. A 20 mph wind at a temperature of $-40°C$ will give the same effect as a temperature of $-70°C$ in still air.[12] In hot climates up to 15 days may be required for acclimatization. This process tends to be more difficult in the obese — who have a low ratio of surface area to body weight — in the elderly and those with degenerative cardiovascular disease.[13] Unlike those working in cold areas, there is an actual physiological response, with the body gaining the ability to increase the sweat output and at the same time decrease its salt content. During this period of acclimatization the individual has to increase his fluid and salt intake, wear loose lightweight clothing and restrict the amount of physical hard work he performs. In addition there may be difficulty with concentration and clear thinking and there is much to be said for carrying out such tasks early in the day.[14]

## Health

Health problems obviously vary with the levels of endemic disease in the country. The most common of all complaints is of course diarrhoea. The causes of this condition are numerous and may include dietary or alcoholic excess, exotic foods, altitude or fatigue. It may be viral in origin but one should remember invasive *Escherichia coli,* shigella, salmonella, *Giardia lamblia* and amoebae as possible causes. If the condition is severe the single most important feature in treatment is fluid and electrolyte replacement. Intestinal parasites may include trichuriasis, ascariasis, tapeworm and schistosomiasis. Prevention of these conditions demands scrupulous attention to food and drinking water. With regard to locally grown fruits and vegetables, peeling or scrubbing with a chlorine containing disinfectant such as 'Clorox' is recommended. If the local drinking water is of doubtful purity bottled water can usually be purchased. Flies and cockroaches may be abundant and care must be taken to cover food, clear up crumbs and dispose adequately of garbage. Local meat should be well cooked and never eaten rare.

Malaria remains a problem in many areas and one cannot stress too highly the importance of prophylactic anti-malarial drugs and their continuing use for the appropriate time after leaving the malarial zone.

Hepatitis, brucellosis and leptospirosis are other less common conditions which may occur in certain areas.

Wolfe[15] recommends the following screening procedures for returnees from exotic places: stool examination for ova and parasites, stool cultures, complete blood count (eosinophilia in the absence of other causes being an indication to perform further parasitological investigation) and liver function tests. There may be further indications to perform TB screening or serological screening for schistosomiasis and filariasis.

Occupational health problems may arise due to different work practices, for example sandblasting is fairly common outside the UK and protective equipment is often less than adequate. The climatic conditions may make the wearing of such equipment much more difficult and therefore there is the tendency for the individual not to use it. Whatever environmental standards may be in use are usually set for temperate countries and do not allow for factors such as an increased respiratory rate and a modification of the physiological function of detoxifying and excretory organs due to the effects of heat. The regulations concerning the exposures of the workforce vary greatly from country to country and in developing countries there may be a willingness to tolerate less than adequate conditions in order to attain higher standards of living. Where legislation does exist there is seldom adequate enforcement. Many hazardous jobs, such as insulation or removal of old asbestos, sandblasting, chemical tank cleaning and disposal of toxic chemical waste, are subcontracted to smaller firms. This is relatively inexpensive to a large organization who as a rule insist that it is the contractor's responsibility to take adequate health protective measures. The contractor often has a large turnover of relatively unskilled labour and does not have the facilities for health and hygiene monitoring. Provided that the work is completed on schedule and there are not too many episodes of acute exposure, little is done to ensure safe working practice. Certain members of the workforce may be at increased risk with chemical exposures because of anaemias, pre-existing liver disease, glucose-6-phosphate dehydrogenase deficiency or sickle cell disease. Fasting during the month of

Ramadan may make it extremely difficult to perform much physical work in hot summer weather. The new supervisor going out to a tropical country must be aware of the occurrence of the above and he would also do well to appreciate the unfamiliarity with tools of the first generation industrial worker. This probably has much to do with a high accident rate among workers in the tropics and he may find that workers require much more personal supervision than he might give in his home country.

## The family status employee

In a multinational corporation a transfer to an overseas affiliate may be a sign of success within the organization or it may be given to an individual as an opportunity to prove his managerial capability. Employees offered family status are often among the more senior within the company since the expense in relocating a whole family is usually considerable. In addition to the salary incentive to go abroad it is valuable if the employee can be given some idea of how such a move will affect his long term career planning. If all the family members are included in the decision-making process it is more likely that such a move will be successful. There will probably be some anxiety on the part of the wife and children who have the need to establish a new identity within a completely different community. The husband has the challenge of his job to occupy his attention but the wife may have to give up her own career to go to an area where work may not be available for her. The younger the child the easier it will be for him to adapt and the extrovert sporting type of child is usually quickly accepted. For some teenagers the change may be traumatic and lead to a severe educational setback. Schooling is always a concern and it may be necessary for the child to attend a boarding school in the home country if the local education facilities are inadequate. There is always liable to be some uncertainty as to living conditions in a foreign country and it has been suggested[16] that an informed contact in a new area can do much to allay fears and advise on what the family might bring from home.

Housing will obviously vary depending on the size of the company. It may be provided furnished or the individual may be allowed to ship his own household goods. In developing countries there may be problems related to the maintenance of basic services such as water, drainage and power which can render even the most luxurious accommodation inadequate. Familiar items of food may prove difficult to purchase and facilities for recreation are likely to be different from the home situation. The wife who has to send her children to boarding school, perhaps give up a job and also for the first time have cheap household help may be in a particularly vulnerable position. Her husband is possibly fully occupied adjusting to his new job and sheer boredom can make her liable to depressive illness or alcohol-related problems. It may be difficult for her to go anywhere and she has nobody among her family to whom she can turn for support. A family which is unstable and has problems at home is unlikely to do well in a foreign country as the additional stresses are likely to intensify its problems.

There is likely to be some concern over the provision of health care facilities especially in a developing country. The screening process makes it unlikely that there will be much chronic disease in the expatriate population and there are no geriatric problems. The major worry is likely to be the ability of the medical facilities to handle illness in young children. Companies, depending on their size, make varying provisions for health care of their employees. The larger ones may attempt to provide the major specialist facilities available in the home country, some rely totally on the local medical facilities and others provide a good standard of general practitioner who has the authority to ship out cases requiring specialist treatment. With present day air transportation the latter facilities usually prove adequate for most emergency situations.

## The single status employee

The employee who leaves his family in the home country has much more disruption of his life than one who brings them with him. As a rule he is a skilled tradesman who is offered a salary far in excess of that which he could earn at home. To the company the cost of hiring such an employee is much less than the expense of providing family accommodation. In spite of this there is a tendency for firms to expect employees to share bedrooms. This is a totally unacceptable practice and is to be deplored as the individual is entitled to his privacy even if the room provided is a small one. Housing is an area where even some of the larger multinational corporations do not come up to standard. As a rule, where food is provided the quality of catering is reasonable and in many areas the problem may be one of overfeeding. Some form of recreational facilities are usually available but often the employee prefers to work long hours for overtime rates. Boredom is common and the employee may smoke or drink more than he would at home. Frequently the individual merely exists between his home leaves. The ratio of work to leave periods varies from company to company but ideally the employee should go home a minimum of three times per year. Such a life can be disrupting and family breakdown is not uncommon. The average employee may stick such a life for a few years and often has a goal such as paying off a mortgage which makes it possible for him to continue.

## The job

The employee himself must be prepared for some adaptation to the local working practice. His job may be in part supervisory but he will probably have a requirement to train his potential successor. It is a common mistake for someone to come from the USA or UK and expect his local employees to perform in the manner he is used to at home. Adjustments have to be made in order to get the best out of the employees in the prevailing conditions. It has already been stated that it is vital to have an accurate job description before going to work abroad. There is an unfortunate tendency in a few companies whose profit margins are massive to recruit the most highly qualified employees and place them in jobs far below their capability. This lack of challenge and apparent lack of importance to the organization when combined with the high salary leads to what Garland[17] described as 'Fraser's disease'. Work satisfaction is important to the mental health of the individual and thereby to the wellbeing of the company. Ideally each employee should feel he has really earned his salary. There are, of course, at the other end of the scale, some who are promoted beyond their ability and their stress comes from being unable to match up to the requirements of the job. This situation combined with a Type A behaviour pattern[18] of high achievement, competitiveness, devotion to work, time urgency and a preoccupation with deadlines usually spells disaster.

In spite of the many problems, it has been my experience that the employee working abroad has less time off work than his counterpart in the UK. A comparison of 'non-effective rates' (average number of employees/1000 absent daily) published by Exxon[19] for its various affiliates would tend to support this observation.

## Alcohol

Finally, I should like to make brief mention of a problem which is frequently ignored but is certainly no less prevalent in those working abroad. Alcohol-related disabilities are perhaps the least well diagnosed of all medical conditions. As a rule the relative cost of alcohol related to salary earned is extremely low and drinking is a socially accepted norm, often more so than in the home country. These factors combine to make the depressed lonely housewife and the bored single status employee particularly vulnerable to excess consumption of alcohol.

REFERENCES
  1. Gardner AW and Taylor PJ. *Health at Work*. London: Associated Business Programmes Ltd, 1975.
  2. Zenz C. *Occupational Medicine*. Chicago: Year Book, 1975.
  3. Shepard WP. *The Physician in Industry*. New York: McGraw-Hill, 1961.

4. Williamson SM. Eighteen years' experience without pre-employment examinations. *JOM* 1971; **13**:465-7.
5. Schneider RF and McDonagh TJ. Experience data on university recruits hired without pre-employment examinations. *JOM* 1971; **13**:363-70.
6. Karrh BW, Bennett BH, Briefer C et al. Scope of occupational health programs and occupational medical practice (committee report). *JOM* 1979; **21**:497-9.
7. Hathaway SR and McKinley JC. *Minnesota Multiphasic Personality Inventory Manual*. New York: The Psychological Corporation, 1967.
8. Lincoln TA, Kelly FJ, Lushbaugh CC et al. Guidelines for use of routine X-ray examinations in occupational medicine (committee report). *JOM* 1979; **21**:500-2.
9. Shmunes E. The importance of pre-employment examination in the prevention and control of occupational skin disease. *JOM* 1980; **22**:407-9.
10. Whitehead TP, Clarke CA and Whitefield AGW. Biochemical and haematological markings of alcohol intake. *Lancet* 1978; **1**:978-81.
11. Crippen EF. Industrial health problems in a frigid environment. *JOM* 1971; **13**:492-5.
12. Clothier JG. Medical aspects of work in Arctic areas. *Practitioner* 1974; **213**:805-11.
13. Minard D. Physiology of heat stress. In: *The Industrial Environment — its Evaluation and Control*. Washington: US Government Printing Office, 1973.
14. Wyon DP, Anderson IB and Lundqvist GR. The effects of moderate heat stress on mental performance. *Scand. J. Work. Environ. Health* 1979; **5**:352-61.
15. Wolfe MS. Management of the returnee from exotic places. *JOM* 1979; **21**:691-5.
16. Olive LE, Kelsey JE, Visser MJ et al. Moving as perceived by executives and their families. *JOM* 1976; **18**:546-50.
17. Garland T. Fraser's disease — an occupational discontent. *Trans. Soc. Occup. Med.* 1966; **16**:83-4.
18. Friedman M and Rosenman RH. *Type A Behaviour and your Heart*. London: Wildwood House, 1974.
19. Review of Medical Statistics, 1979. *Medical Bulletin* 1980; **40**:67-107.

# 17. EDUCATION AND OCCUPATIONAL MEDICINE

*Dennis Malcolm*

## INTRODUCTION

This chapter is based on what I believe my needs would be if I were training now to deal with the range of problems I have encountered during the past 30 years in this field, and might be likely to face in the foreseeable future.

Our status will depend on our achievements in the field in which we work and how our colleagues on an international scene rate our contributions in our specialty and in the field of general medicine. We will not gain status by awarding ourselves degrees. The quality of our specialty will depend to a large extent on the quality of the teachers and the calibre of the students they are able to attract.

When the Society of Occupational Medicine in the UK was invited by the Royal College of Physicians to take part in discussions on higher medical training and specialist registration, occupational medicine was one of fifteen specialties. The basic qualification for the other fourteen was Member of the Royal College of Physicians (MRCP) UK, followed by suitable clinical training and experience as registrar and senior registrar in the chosen specialty.

The difficulty for occupational medicine was that specialization had developed on a less formal basis. The Diploma of Industrial Health (DIH) was introduced in 1945 based first on a 1-year academic course, but subsequently on a 3-month intensive course or a 2-year part-time course. In 1970 the Master of Science (MSc) was introduced to raise the level of academic training for full-time specialists. With specialist registration and the formation of a Faculty of Occupational Medicine it would seem appropriate or even essential, that future consultants in the specialty have clinical training to MRCP UK or equivalent level, and then proceed to the MSc or equivalent course. The DIH is to be replaced by an examination for Associateship of the Faculty of Occupational Medicine (AFOM).

The changes will result in a three-tier system. The first level will be appropriate to part-time doctors in occupational medicine whose main specialty will be general practice or some other specialist branch of medicine. The second tier will be those who have an adequate basic training and who proceed to Membership of the Faculty of Occupational Medicine (MFOM) or MSc or its equivalent, and are practising full or part time as company medical advisers, employment medical advisers or lecturers in academic departments.

The third tier will be a consultant grade and will comprise senior occupational physicians who have distinguished themselves in the field of occupational medicine by their research or published work and who hold senior posts in industry, research or the academic field. It is probable that specialist accreditation will normally be confined to the second and third tiers.

Similar changes in the education of future specialists are taking place within the European Economic Community (EEC) and in the United States, as well as in a number of other countries. European and American education in this field will continue to attract a number of overseas students.

## TRAINING

### Undergraduate training

With so many demands for time there is not going to be opportunity to teach the undergraduate a great deal about the specialty of occupational medicine. I believe that a most important aim in all medical training is to bestow a comprehension of the effects of health on work and the effects of work on health. At a recent postgraduate course on chest diseases it was clear that all the physicians at the Brompton Hospital were thoroughly familiar with this approach as part of their normal daily clinical practice. This attitude of mind should become general in all our teaching schools. If occupational health services become universal in the Health Service, then a better understanding of the place of occupation in relation to health will spread.

The occupational physician responsible for undergraduate training should seek to ensure at an early stage that a succinct occupational history is included in the medical case notes. Such a history is important for all patients, not only those who may have occupational diseases. This information will establish the patient's social and work background and will help to indicate the type of individual who is seeking medical help — an important part of getting to know the patient. Is he a conscientious individual who enjoys his work, or does he regard work as something to be endured or even avoided? What is discovered about personality and attitudes in this way will also be relevant to the management of the patients and their cooperation in treatment and rehabilitation. Finally, one can also learn in this way a great deal about the therapeutic and rehabilitative effects of suitable work. The effectiveness of such teaching will depend very largely on the qualities of the physician as a teacher.

Biochemistry, including microbiology, taught to a standard to understand the basic principles of pharmacology and toxicology is today an essential part of undergraduate training. Medical statistics are now taught to a standard so that the average graduate should understand their use in the literature. Epidemiology is taught in community medicine, but there are only two specialist departments of epidemiology in the UK.

As an extension to the effects of health on work, the need to include in treatment appropriate rehabilitation, both physical and mental, which will enable the individual to lead a full, normal life again, is vitally important. Too often hospital or even general practice treatment stops at the stage of 'curing the disease'.

An important part of medical education is for the doctor to know and understand what people do when they are well. A few carefully planned factory visits, preferably including a coal mine, will be a great advantage to any doctor in his understanding of his patients.

Rightly or wrongly, most doctors provide certificates indicating fitness or otherwise for following a normal occupation. Such opinions would be much better founded if the physical and mental demands of occupations were better understood.

The presence of a specialist in occupational medicine on ward rounds or case study discussions could provide a useful extra dimension to medical practice and make our future doctors aware of occupational aspects of health.

## Postgraduate training for part-time doctors in industry

While some authorities consider that all occupational medicine in future will become organized as a field of full-time specialists, I believe that a great deal of the work in the UK, and probably elsewhere, will continue to be covered by part-time doctors. Postgraduate training is essential for all these doctors. At present many have not had any formal training. If EEC proposals are accepted, all doctors will have to have adequate specialist training.

The need for training such doctors will probably be met by continuation of part-time day release courses over 2 years for the equivalent of the DIH or AFOM.

In addition to their basic academic training, such doctors often need to become specialists in their own fields. Many have done this with distinction by reading, meeting their full-time colleagues and attending suitable specialist seminars. Continuous short term refresher courses are often run in the UK by the Society of Occupational Medicine or by the local postgraduate medical federation which is responsible for postgraduate medical education in the regions. It is very important to try to organize such courses to meet the needs of the part-time doctors working in the area.

Another very useful type of training is the periodic in-company medical meeting, where special problems of the industry are discussed. Doctors in charge of medical services of groups should ensure that their part-time doctors receive reprints of important literature as it is not possible for part-timers to cover a wide specialist field.

Similar patterns of postgraduate training will need to be fitted into the appropriate organization of medical teaching in other countries. A number of useful international conferences on special subjects are already organized which are very helpful to both part-time and full-time specialists.

### Specialist training full time

The basic clinical and academic training of specialists up to MFOM standard or its equivalent has been discussed in the introduction.

There will be a need as in other specialties for post-academic experience under the guidance of doctors experienced in the practice of the specialty. In the UK it is proposed to set up registrar posts which are similar to US boards. One of the difficulties so far has been to find sufficient training posts of suitable calibre to satisfy the need.

A flexible approach is needed and suitable appointments could be offered by large companies, by the Employment Medical Advisory Service or by a 2-year field research project based on a university department. Further posts could be arranged by several smaller companies combined with academic departments.

Whether such experience would be best obtained before or after the academic specialist course can be debated. There is no doubt that some practical experience in industry is of great help to anyone taking an academic course.

One of the great benefits of my own early factory training was the chance to work as an operative in each of the works departments for my first 3 months in industry. This is strongly recommended to all doctors intending to practise medicine in relation to occupation. Practical work like this gives one an insight into some of the difficulties of normal production work which no reading or listening could teach so effectively. For instance, why should working on a furnace during a hot summer be very cold at 2 a.m.? But it was.

## SPECIALIST POSTGRADUATE CLINICAL TRAINING IN OCCUPATIONAL MEDICINE

### The part-time specialist

Part-time specialists may already have some occupational training and experience, but most will be general practitioners. They will bring with them the clinical skills they have. In a short course of about 3 months there will not be much opportunity to improve their clinical skills. Attendance at outpatient clinics, especially those devoted to occupational diseases, is the best help that can be given at the time. There is, however, within the EEC and the USA an increasing interest in providing postgraduate training and clinical teaching.

Short refresher courses in specialist subjects should have an adequate clinical input. In the UK such courses are usually based on national clinical institutes.

## The full-time specialist

Assuming that the future specialist will have clinical training and experience up to the equivalent of the MRCP (UK) or American specialist boards, clinical skills should be of a high order. The need for further training in occupational diseases will not be great. The well-trained clinician with a good knowledge of toxicology will be able to recognize the effects of occupation on health.

The majority of occupational diseases which have occurred over the past 10 years have been chest diseases and skin diseases. For this reason, clinical experience of industrial chest disease and monitoring of workers in the asbestos and coal industries is essential. It should be recognized, however, that preventive measures already in being will reduce the present rate of incidence of disease in these industries rapidly over the next 10 years, with the possible exception of skin diseases.

## Further training for specialists

For the full-time consultant, meeting the need for professional contact must be largely the responsibility of the individual doctor in relation to his particular job and the facilities available. Symposia on special subjects are held in many countries both by academic and industrial organizations.

The courses run by the British Postgraduate Medical Federation are today, generally speaking, excellent. In the early stages one was often asked what doctors in industry wanted to know about. This understanding has been considerably helped by the end-of-course discussion on content and presentation, instituted by Dr Gauvain. During the discussion on course content one heard the usual request to be brought up to date on some basic aspects of lung function testing and other fundamental training. I believe that basic reading and up-dating is the responsibility of individual doctors and not of high level postgraduate courses. Recent experience shows an excellent understanding of the place of occupational medicine.

Such courses should be run by experts in the field of the subject being considered. Their aim should be to review recent important advances and developments in the specialist fields and deal particularly, but not exclusively, with those aspects which might have some bearing on occupational health practice. The standard should be high enough to interest those of consultant status who may or may not have special knowledge in the field.


## SPECIALIST TRAINING IN OTHER COUNTRIES

### Proposed EEC training in occupational health

The EEC in a draft Charter[1] stated that every medical student should be introduced to occupational health by teaching and examination. The extent to which this is carried out at present varies from virtually no specific lectures to twenty or more hours of formal lectures.

The recommendations include the interaction between health and work, including the effects of work on common illnesses. Information should also be given on the relevant statutory provisions and national bodies responsible for occupational health.

For doctors who decide to practise in the field of occupational health, adequate postgraduate clinical training is regarded as essential, followed by 2 years' further training in occupational health. This would consist of a 1-year theoretical course in occupational health and an additional year under the guidance of a qualified occupational physician. Qualified physicians are obliged to undergo regular further training.

Future UK ideas on specialist training would include all the essential elements of the EEC proposals. The detail in the EEC proposals tends to put more emphasis on work methods, safety, hours of work and work layout.

There appears to be a lack of awareness in the detailed subjects of the distinction between the advisory role of the doctor in industry and the role and responsibilities of line management. There are areas such as working hours which are settled by industrial bargaining not based on scientific evidence. The doctor must avoid being drawn into bargaining situations where he will find it very difficult to retain an impartial attitude. This does not prevent him from giving scientific advice.

Under EEC proposals, only doctors trained in occupational medicine will be able to practise in this field.


### Finland

Finland devotes about 20 hours of lectures to occupational health at undergraduate level. Specialist training comprises 1 year of postgraduate clinical training, 3 years in a special medical subject and

6 months as a plant physician. Much of the training effort goes into part-time specialist subject courses for part-time plant physicians. There is no specialist grading in occupational medicine for part-time plant physicians.

## United States

When the Occupational Safety and Health Act was passed in 1971 an Occupational Health Accreditation Commission was set up.[2] Gauvain[3] describes a decline in the membership of the American Academy of Occupational Medicine and the American Occupational Medicine Association. She also describes action being taken to review training programmes. It is clear that the National Institute for Occupational Safety and Health (NIOSH) wishes to reverse this trend.

## USSR

In the USSR countries specialist training in preventive medicine starts in the third undergraduate year, but all students receive at least 100 hours of formal training in occupational medicine.

## Other countries

The present position in a number of countries is well described by Gauvain.[3] In the US there is a very good corps of experienced full-time doctors in industry, as well as a great deal of good research. Like the UK, however, it can be difficult for small companies to find doctors with any occupational training or experience who will advise on a part-time basis. In Canada, Australia, New Zealand and South Africa standards are very good, as are also those of Singapore and surrounding countries.

Conditions in Africa vary from country to country but generally in the care of lead workers in the battery industry, they are not too far from European standards. A number of African and other overseas students have of course studied occupational medicine in Europe.

## SUBJECT MATTER

The subjects discussed are not exhaustive. They include the majority of disciplines which can be regarded as essential for a rounded training in occupational medicine. The level which should be achieved should be adequate for doctors entering industries to study these special subjects in greater depth, for instance, radiation medicine, which in this chapter would be included as part of toxicology.

The subsequent order of headings indicates my view of the subjects' relative importance, but this does not imply that those at the end of the list are unimportant.

Occupational medicine is the study of a variety of scientific and other disciplines aimed at providing the specialist with an adequate base from which to practise in this field, so as to be able to make accurate diagnoses and good judgements which can lead to effective and acceptable action in preventing harm to health or to deal with the results of ill health in the best way to help both the individual and the organization which the doctor is advising.

## TOXICOLOGY

Toxicology as considered in this chapter is the study of the effects of physical, chemical and other materials or environmental conditions on the health and wellbeing of persons liable to be affected by contact with or absorption of such materials. The study includes immediate and chronic effects as well as long term sequelae including carcinogenesis, teratogenesis and effects on the reproduction of the species.

In the UK the Health and Safety at Work Act (1974) requires, under s. 6 (4), suppliers to ensure so far as is reasonably practical that substances supplied are without risk to health when properly used, to carry out any necessary research to comply with the first requirement and to give adequate information regarding the conditions of safe use.

One has only to look at random in Sax[4] to find that the toxicology of many substances used in industry is unknown. Prolonged use without acute or chronic ill health may be some comfort, but how much is known about possible long term effects, such as the excess incidence of cerebrovascular accidents found among lead workers.[5] The information supplied by manufacturers varies from comprehensive and excellent to bland, unsupported statements, such as 'this material is safe if properly used'.

In addition to the above requirements, the setting up of health and safety committees and the training of representatives have made it necessary for management to be adequately informed and advised about the potential toxicity of all materials used at work. This is becoming one of the occupational physician's most important and exacting advisory roles.

At present it is not possible to comply fully with the requirements of the Act. Similar legislative requirements are being introduced within the EEC, and the Occupational Safety and Health Administration (OSHA) in the United States have laid down some very strict criteria. The view that most cancers are due to environmental causes and figures quoted that occupational exposure may be responsible for anything from 1 to 40 per cent of all cancers is likely to cause considerable concern to employees throughout industry. The need to maintain a proper sense of proportion based on sound information is clear.

## Undergraduate training

It is doubtful whether most medical schools will find time in the syllabus for studying industrial toxicology. Basic training in bio-chemistry, including microbiology, sufficient to understand the basic principles of pharmacology and the toxic effects of some of the prin-cipal drugs and the safety testing required, is as far as most undergraduate teaching is likely to go. If the undergraduate is also aware of some effects of work on health and has some general knowledge of metal toxicology and dust diseases, together with some general knowledge of causes of cancer and awareness of teratogenic effects of drugs and chemicals, the newly qualified doctor should have a good grounding to study toxicology at a more advanced level for postgraduate specialization in occupational medicine.

## Part-time doctors in industry

For the part-time doctor in industry, wide, detailed training in toxicology is not feasible. For an Associate of the Faculty of Occupa-tional Medicine however, an understanding of the basic principles of toxicology is needed. Some of the major areas of toxicology such as metal toxicology, industrial dust diseases, organic chemicals, solvents and potential carcinogens, can be covered in short courses, but it is not necessary to try to cover a wide field. For those whose respon-sibilities cover less usual problems, help is always available from full-time colleagues and university departments. Surrey University runs an information service. Short weekend courses, such as the seminars on asbestosis run by Turner Bros. Asbestos at Rochdale in 1970–71 or the meeting on occupational exposure to cadmium organized by the Cad-mium Association in London in March 1980, form ideal training for the part-time doctor with special interests in such fields.

## Postgraduate training

Toxicology, like medicine, is a multidisciplinary subject and is growing rapidly. It will not be possible during a 1–2-year course to train postgraduate doctors as professional toxicologists, but it is necessary that they should thoroughly understand the basic principles under-lying the criteria and methods for evaluating the toxicity of chemicals.[6] Those who have a special interest and whose basic training includes 'A' level chemistry, biochemistry or pharmacology should receive an ade-quate basic training to enable them to specialize as professional tox-icologists. Teaching within university departments for MFOM, MSc or equivalent qualifications should aim to achieve such standards. There is also a responsibility to prevent harm to the community outside the factory. The ecological aspects of potentially toxic materials and waste have become of considerable importance. Industry now has a legal as well as a moral obligation in the UK and many other countries to prevent harm to persons living in the environment. Doctors in industry will have a responsibility to advise management on how and to what standards they can meet such obligations. There is a great deal of over-reaction to many supposed environmental problems. The basic principles of evaluating the risks posed by industries using toxic materials or disposing of toxic wastes need to be understood by the specialist in occupational health.

Those who are going to work in industries with toxic problems should become experts in their own fields. Postgraduate teaching or discussion on standard-setting and legislation should equip the specialist to carry out relevant research and enable him to give sound advice on the setting of future standards. Much of the work presented or reviewed for present standard-setting does not provide reliable data to answer the questions.

## STATISTICS

### Basic knowledge at undergraduate level

Today this subject is taught in schools and at undergraduate level, so doctors coming into the field of occupational medicine should be better equipped than in the past.

### Postgraduate teaching

Statistics seem to be unpopular with many doctors training in occupational medicine. Perhaps the professional statisticians have not always realized the difficulties of their less numerate students. Good teaching is necessary to achieve adequate comprehension by the less numerate students, and it is important for those organizing courses in occupational medicine to ensure that the standard of teaching is comprehensible to the students. It is suggested that a working knowledge of the following is required for the full-time specialist in occupational medicine:

mean, modes, medians, variation, probability and binomial distribution, normal distribution, tests of statistical hypotheses, types of error, correlation and regression, confidence limits, non-parametric statistics, analysis of variance.

Practical exercises in the use of statistical methods and testing of significance of results can be very helpful to the student, as well as indicating which tests of significance are appropriate to particular problems.

The part-time doctor in industry should be aware of the problems where statistical knowledge is important and know where to get professional advice.

One of the primary reasons for a better understanding of statistics is the importance of properly designed epidemiological studies in establishing the safety or otherwise of control standards for potentially toxic substances, especially over long periods of exposure. This subject is discussed in Chapter 12. Where an environmental or biological estimation has a large variation, the mean of a working group has often been used. If the standard deviation and confidence limits are also known, it is possible to show whether control is adequate or not.

Probability and the binomial distribution are the basis of most epidemiological studies which show whether a studied group differs significantly from a control group.

In standard-setting, the degree to which a measurement such as blood lead correlates, if at all, with lead in air, is important. Too often correlations have been calculated without first establishing the accuracy and precision of the two measurements. Only by adequate comprehension of such problems and the value of a proper statistical approach shall we make realistic progress in these fields.


## EPIDEMIOLOGY

Epidemiology is closely linked to statistics through the study of environmental conditions and the measurement of any effect on health or biological systems of exposed populations. It is basic to the study of populations.

While adequate toxicological testing may reduce the possibility of environmental exposure from having effects on health in the future, this is merely a hope and is a long way off. The principal method of relating exposure to behavioural or environmental effects on health and mortality will remain epidemiological. A great deal of useful information is stored in medical and other industrial records and remains unused.


### Undergraduate training

Undergraduate studies introduce this subject as an important aspect of community medicine and where it is adequately taught as an important tool in preventive medicine, this approach should be adequate.


### Part-time doctors in industry

For the part-time doctor in industry, the basic principles of epidemiology must be included in the examination for Associateship of the Faculty of Occupational Medicine.

**Specialist training**

The specialist's training should introduce the design of data collection for epidemiological studies, the possibility of record keeping in line with the needs of epidemiological studies and also of relative costs of different methods of recording valuable information.

This is becoming increasingly important today when industry is becoming more cost conscious and collection and storage of data are increasingly expensive. A good guiding principle is 'don't collect any data that you are unlikely to use'.

Record linkage and cooperation between doctors and others in related industries is particularly important. As an example, the present knowledge of the relationship between cadmium exposure and prostatic cancer is inadequate because most industries using cadmium are relatively small. To increase the size of the population being studied, a number of companies are pooling their data for a much larger mortality study sponsored by the International Lead Zinc Research Organization.

For the specialist in occupational medicine this subject is an essential part of the MSc or equivalent specialist academic course. It should be taught by those specializing in epidemiology to a standard where the trained specialist can plan and carry out his own epidemiological studies and also know how to plan his record keeping to give the information required in a complete and easily extractable form. The most important part of such studies is to have a clear idea of the aims of the study as this will often determine how, and in what form, the data should be collected.

Some procedures carried out in occupational health services are of doubtful value. These activities include pre-employment medical examinations, periodic senior executive health checks, influenza vaccination and so on.

The best way to define answers to such questions more clearly is to try to quantify the benefits or lack of them, by suitable epidemiological studies. Such topics form useful material for discussion groups in teaching epidemiology.

## LEGISLATION

**Undergraduate level**

There is little place for the teaching of legislation in relation to occupational health. Undergraduates should be aware of duties to report prescribed diseases and legislation on rehabilitation and disabled persons. The EEC proposals extend the requirements to occupational health legislation.

**Part-time doctors**

Here the needs will vary with the type of industry. A 1- or 2-day course could best be devoted to the basic principles of appropriate legislation and a guide to where and how to consult information in more detail. Associate Members of the Faulty will have to become familiar with particular legislation related to their practice although all doctors taking the proposed future examination will need to be familiar with the Health and Safety at Work etc. Act, the Factories Act, Disabled Persons Legislation and such similar legislation which may exist in their own countries.

## The accredited specialist

Legislation is changing and differing regulations apply to different industries. There is a continuous process of harmonizing occupational and environmental legislation within the EEC. The United States did not have Federal legislation on occupational health until 1971 when the Occupational Safety and Health Act was passed, and at the same time the National Institute of Occupational Safety and Health was set up as the scientific and research organization. In the past, particularly in the Commonwealth, the pattern of UK occupational legislation had a considerable influence. Eastern Europe has a different approach to occupational health and limit values have tended to be much stricter, although such information as is available on the control of occupational diseases suggests that the standards have been set as targets to be achieved some time in the future.[7]

All these trends have led to new approaches to industrial and environmental legislation. The basic principles and aims behind these various forms of legislation should be studied as well as how effective the existing or proposed legislation is, or is likely to be in achieving its aims. Naturally, this will also depend on the methods and effectiveness of enforcement.

The Health and Safety at Work etc. Act and earlier legislation have adopted a pragmatic approach to setting standards and making regulations. Where the word 'practicable' is used it has a stronger force than the term 'reasonably practicable'. Here there is a need to weigh what is required against the risk involved and also the state of technical control available at the time. Since both technical knowledge as well as what society accepts as a reasonable risk is changing with time, what is practicable or reasonably practicable also changes with time. This approach, together with high standards of inspection and enforcement, has resulted in very high standards being achieved in Britain.

Recent lead regulations in the United States have set standards which are not technically feasible in many industries, certainly within the next decade or so. It remains to be seen whether such rules will bring the law into disrepute, particularly as in the opinion of many specialists with practical experience in the field, such standards are unnecessary to safeguard the health and wellbeing of lead workers.

The teaching of legislation to an MSc or equivalent course in occupational medicine needs to be broad and based on an analysis of aims and how well these are achieved. A brief introduction on the development of health and industrial legislation in Europe and elsewhere should lead up to trends in future development. The aim should be to give the specialist a working knowledge of how the law is developing and functioning rather than any expertise on any detailed aspects. I found the reading of Factories Acts difficult and extremely boring. The detailed study of regulations applying to a particular industry should be part of post-specialization job training.

Undoubtedly, the Environmental Health Criteria of the World Health Organization (WHO) will have some effect on regulations where the subject has been covered in this series. However, the concept of total safety, which is the basic aim of these criteria, and its feasibility need to be carefully studied.

## OCCUPATIONAL PSYCHOLOGY

### Undergraduate teaching

Undergraduate teaching will not normally be concerned with a specialized occupational approach. However, basic psychological training is necessary to understand the study of psychiatry. In view of the high incidence of psychological illness in modern society, improved understanding in this field is required.

### Specialist training for part-time doctors

Basic principles should include methods of psychological testing in an occupational setting and the interpretation of results. Modern concepts of stress and adaptation, group behaviour and an introduction to social psychology should also be included, together with an introduction to behavioural assessment in relation to effects of toxic agents.

### Training for accredited specialists

Psychological testing is becoming increasingly used in selection and job placement in industry. Too often such testing and interpretation is carried out by individuals who have no formal training in this field. Often the doctor may be the only person in an organization with any formal scientific training in psychology. However, apart from pre-employment assessment and fitness for service overseas, the doctor should not become involved in selection procedures.

Specialists in occupational medicine need to understand the scope and limits of scientific knowledge in the areas of psychological methods related to selection, motivation and organization, in order to advise management on the scope and limits of applied psychology.

The nature of stress and how to adjust or avoid unnecessary stress are important. The study of alcohol- and drug-dependence including diagnosis and treatment should be included, although in my experience the problems are not so widespread as some authorities seem to believe.

It is possible that much more study in the field of work satisfaction and human relationships could be undertaken with better basic training in these subjects.

It is certain that the behavioural effects of many toxic substances will become of increasing importance over the next decade or two. The methodology is still in the early stages of development and with the rapid developments in the electronics field, behavioural measurements are likely to become much more sophisticated. At present, our understanding of the results of what we believe we can already measure is poor. For example, does lead at low levels of exposure really make people more aggressive or interfere with flicker fusion?

In the first case it is known that increased irritability was commonly found among people who had early symptoms of lead poisoning. Work on the behavioural effects of lead[8] has been contradictory and subject to a good deal of methodological criticism. If lead within the present acceptable limits does alter the threshold at which flicker fusion takes place, do we really understand what this means to the wellbeing of the individual? At present if one questions the meaning of such fine measurements, one tends to get reactions of psychological stress in the researcher rather than scientific enlightenment.

We need to know the reliability of estimates of a small fall in IQ level, supposed to be due to increased lead exposure, if more adequate evidence is produced that such changes really occur. The evidence is at present contradictory.

It would seem probable also that exposure to many narcotic solvents may be causing measurable behavioural problems. We need to know better how such effects can be measured more reliably and if any significant changes are found, what do they mean to the health and wellbeing of the patient? How can we relate any such effects to those considered socially acceptable, such as one or two pints of beer?

For postgraduate training in occupational health the basic principles of psychological assessment and measurement should be studied. The effects of heredity, society and work in individual and group mental health should be included, together with some understanding of the limitations of knowledge and methods of investigation in these fields. In addition, the presently rapidly developing field of behavioural effects of toxic substances and harmful environments on individuals or groups will become increasingly important. This field will need a joint approach to teaching by psychologists and toxicologists.

## REHABILITATION

At present the interest of the medical profession in rehabilitation is very variable. Effective rehabilitation is of vital importance to the patient. As at present taught and organized, treatment is often too little and too late.

### Undergraduate training

Rehabilitation is for every patient and not just for the disabled. Its aim should be to reduce the incidence of disability to a minimum and to maximize the capacity of those who are inevitably going to have some disablement, either physical or psychological. It follows that the need for adequate advice and treatment should be an integrated part of all undergraduate medical training. The attitudes of the whole profession can probably be improved if they are made aware of the effects of health on work. Teaching in this field is not likely to improve until many more members of the profession are convinced of the need. Perhaps the root of the problem is that the majority of patients undertake their own rehabilitation more or less successfully.

### Postgraduate training

For the specialists and the part-time doctors in occupational medicine, postgraduate teaching should include legislative requirements, the present organization for rehabilitation under the Employment Medical Advisory Service (EMAS), the responsibilities of Disablement Resettlement Officers (DROs) and some understanding of what can be achieved by doctors in industry in cooperation with general practitioners and hospitals as well as with DROs and the EMAS.

There are some excellent examples of good cooperation and rehabilitation practices based on cooperation between doctors in industry, hospitals and general practitioners. The armed forces also show what can be achieved.

Postgraduate teaching should also include adequate information on special services for re-training, fitting artificial limbs and re-training of the blind and others with special handicaps, together with the social support available.

Visits to appropriate institutions are an essential part of postgraduate specialist training. Part of the registrar period for trainee consultants could well be spent in a rehabilitation centre.

It is suggested that one of the major gaps in the field of rehabilitation includes a lack of realization by some general practitioners and consultants of the benefits of adequate communication and simple good advice. For instance, I have no doubts that advice to the post-hernia operation patient to take it easy for 6–12 weeks is interpreted by many patients as to remain seated most of the day. Advice to walk daily and to aim to increase the distance to 5 miles per day before returning to work, as well as graduated lifting, ensures that many men are fit for their normal work when they return, and only those engaged in heavy lifting may require some protection for a few weeks. On the other hand, some of the older people, especially those with poor abdominal musculature, may have some permanent limitations.

The therapeutic benefits of an early return to work could be better studied and documented.

## ERGONOMICS

### Undergraduate training

At undergraduate level the basic scientific education in anatomy, physics, physiology and psychology is appropriate to the later study of ergonomics, but a formal study in this field would not be appropriate except perhaps to understand how industrial injuries occur, including those due to lifting patients.

### Postgraduate training

This should include a formal as well as a practical study of the basic principles. For the accredited specialist, the subject should be studied in greater depth.

This is an area where existing knowledge has generally been inadequately applied. It is also a field where a great deal of esoteric work has been carried out which is not normally applicable to everyday use. If industry wishes to learn how to use ergonomics effectively, let it take a good look at modern dental ergonomics. The systems designed are not only much better for the dentist, but also for the patient. High speed air drills with diamond bits and water cooling work faster and with much less discomfort. The dentist can work sitting down, which helps him to reduce fatigue and so perform better work later in the day. Modern lighting improves his vision and is not uncomfortable for the patient. Above all, his productivity and quality of work has improved. It is probable that the use of ergonomic theories and knowledge will not find great scope in many industries until management and engineers have also studied the contribution this body of knowledge can make to industrial production. Some answers may be found by studying the Japanese car and other industries. We need to understand the role ergonomics can play in everyday work and to answer some of the simpler questions, recognizing that the answers may not be easy. How can we reduce lifting of excessive weights throughout industry, and what is an excessive weight for the average man or woman and what are the effects of age? Some useful research is being carried out at the University of Surrey with the help of a pressure-sensitive radioactive pill on the effect of lifting methods on intra-abdominal pressure and the relationship to back injury. Naturally, complex questions will need to be referred to a specialist in the field of ergonomics, but the occupational physician with his training in anatomy and physiology and his knowledge of people should be able to formulate many of the right answers in his normal visits round the works, and to help the engineers to come up with some effective solutions in the field of work efficiency and lifting. Often the best help in defining the problem comes from the man doing the job.

## SAFETY

### Undergraduate training

Now that the Health and Safety at Work etc. Act in the UK applies to medical schools and hospitals the undergraduate should become aware of a positive approach to this subject. No one who has worked in a casualty department can fail to see the need for accident prevention.

### Postgraduate training

It is probable that very few accidents are directly related to physical health.  The clearest statistics probably relate the effects of epilepsy and coronary heart disease to road accidents. Considering the number of epileptics who are actually driving, the evidence is that they have relatively few accidents. On the other hand, the effects of alcohol on driving accidents would appear to be very important.

The effects of alcohol and emotional factors would seem to be important areas for research into man's relationship with risk and accidents at work. So far, little study has been made of the effects of alcohol in causing accidents at work. This is an area where epidemiology studies by doctors in large industries could contribute to our knowledge.

While there are many theoretical ideas about emotional states, fatigue, shift work, ergonomics, boredom and so on, many of the possible factors involved in industrial accidents are difficult to quantify, even if one can collect objective evidence. Probably our understanding in this area will progress slowly because of these difficulties. When contrasted with travel and accidents in the home, much of industry has a very good safety record.

Probably the most important single factor affecting safety in any industrial undertaking is the attitude of employees towards safe working practices. Such attitudes are usually generated by the genuine interest taken in safety by top management, which leads us to the study of human relations and how these can be affected by management policies.

This area is one where opinion is paramount and scientific knowledge is inadequate. Guidance on the possibility of scientific advance in this field needs to be considered and the full-time specialist should be able to make a valid contribution as part of a team interested in the field of safety.

At present the availability of suitable teachers would seem to be limited.


## FIRST AID

### Undergraduate teaching

In the past first aid has not been taught in medical schools but its teaching in the early stages of medical education could bring home to students the importance of simple life-saving procedures which they might be called upon to use at any time in their careers.


### Specialist training full and part time

An important responsibility of the doctor in industry is to ensure that first aid provisions are of a high standard and this often involves teaching. A number of industries run competitions and the enthusiasm of the teams helps to achieve a high standard of practical efficiency.

All doctors in industry should be able to teach first aid in an interesting and practical way. An excellent practical approach is described in *New Essential First Aid*[9] and *New Advanced First Aid*.[10]

There is also need to explain first aid and treatment for international travellers, which must include time zone changes, gastrointestinal effects and an adequate range of inoculation, and how to prevent malaria, bilharzia and other tropical diseases to which the traveller may be prone.


## ENVIRONMENTAL ASPECTS OF WORK

### Undergraduate teaching

At undergraduate level training will probably be limited to the physiological and psychological adjustment to adverse and extreme environmental conditions and the cause of heat stroke and effects of cold.


### Specialist training

The part-time doctor should understand the basic principles of environmental comfort, effects of working under high atmospheric pressure and similar abnormal climates. Good guidance on the practical application of basic principles should be given. Teaching should be done by experienced doctors and hygienists who have expertise in the environmental field.

The full-time specialist should be trained to a standard where he can undertake the measurements required to analyse the basic nature of environmental comfort and give sound scientific advice on how to correct or deal with health or comfort problems arising from the environment. European doctors have tended to concentrate on work physiology and this could be included in teaching in other countries.

Modern advances in environmental control involve safety and comfort in flying, space travel, working at high barometric pressures, adverse temperatures and humidity. Environmental control also includes work layout, colour, noise, windowless buildings, time zone changes and shift work adaptation.

Teaching should be shared by experts in the field including architects and engineers. Future cooperation could be improved by this joint approach.

## THE ORGANIZATION OF INDUSTRY

### Undergraduate training

This subject would not be appropriate to undergraduate training although it could be helpful to have an outline of the aims, structure and functioning of the National Health Service. This would be taught as an essential part of community medicine.

### Postgraduate training

The way in which an industry is organized is equally important to the part-time doctor and the full-time specialist.

The doctor who is medical adviser to industry will be able to advise more effectively if he understands how that industry is organized. An outline of the various types of industrial organization and how they are supposed to function is an essential part of postgraduate specialist training. Consideration should also be given to where the doctor fits into an organization as well as some guidance on preparing reports or recommendations for management and how to get cooperation and support from other non-medical specialists such as engineers, hygienists, architects and personnel and production departments. For specialist training the business school of the university concerned should be able to provide the best teaching.

The organization and legitimate aims of trade unions should also be included. In many industries the importance of adequate communication with trade unions can be vital to the success of protecting and monitoring the health of employees. The advantages of an open and honest approach should be stressed in creating trust. The doctor must learn how to remain 'neutral' in what is often an adversarial situation between management and unions.

A further advantage of understanding the organization of industry is that much of the stress on individuals arises out of inter-personal relationships and either wrong organization or, more often, situations in which the practice does not match the theoretical organizational model. The company doctor may well be in a unique position to help sort out problems arising in individuals as a result of excess stress at work.

Stress problems arising among specific occupational groups employed in small industries could be studied by departments of occupational health or group health services. Such studies could be an excellent training following the specialist academic course.

## TRAINING FOR ASSOCIATED PROFESSIONS

Discussion of training in associated professions has not been included in the chapter as they have been adequately covered elsewhere.[3,11] What is important for the occupational physician is to have an adequate understanding of the roles of colleagues working in similar fields, particularly those of occupational hygiene, nursing and safety. In many cases the doctor or nurse may be working on their own and so may have to give elementary advice on hygiene. In such circumstances it is important to know where to go for professional advice and the ways in which such advice can be helpful.

## CONCLUSION

Occupational medicine is an area of operation where the doctor with good clinical training can make a considerable contribution to the health and effectiveness of organized work.

The subject matter suggested is mostly outside the scope of normal graduate training but has proved invaluable to those who have attended for diplomas or specialist accreditation in helping them to make contributions in a wide variety of occupations, research and teaching.

With the present world wide interest in improved health and safety at work, many more doctors adequately trained in this specialty will be needed. The need for improved standards of training have been largely recognized. This chapter has presented a personal view of how appropriate standards might be achieved.

REFERENCES
1. Standing Committee of Doctors of the EEC. *Revised Draft Charter on Occupational Health* (Document CP80/153E). Dublin Meeting, 1980.

2. Eckardt RE. Occupational Health/Safety Programmes Accreditation Commission. *J. Occup. Med.* 1976; **18:**822–4.
3. Gauvain S. Training in occupational health and safety. In: Ward Gardner A. *Current Approaches to Occupational Medicine.* Bristol: Wright, 1979.
4. Sax NI. *Dangerous Properties of Industrial Materials.* New York: Reinhold, 1968.
5. Dingwall-Fordyce I and Lane RE. A follow-up study of lead workers. *Br. J. Ind. Med.* 1963; **20:**313–15.
6. Environmental Health Criteria 6, Parts I & II. *Principles and Methods for Evaluating the Toxicology of Chemicals.* Geneva: World Health Organization, 1978.
7. Grandjean P. *Standards Setting.* Copenhagen: Tryteknik, A/S, 1977.
8. Repko JD and Corun RC. Critical review and evaluation of the neurological and behavioural sequelae of inorganic lead absorption. Cincinnati, Ohio: NIOSH Research Report, HEW ITR-74-26, 1976.
9. Ward Gardner A and Roylance PJ. *New Essential First Aid.* Bristol: Wright, 1979.
10. Ward Gardner A and Roylance PJ. *New Advanced First Aid.* Bristol: Wright, 1979.
11. Schilling RSF. *Occupational Health Practice.* London: Butterworths, 1973.

# 18. HEALTH EDUCATION AT WORK
*James McEwen*

## INTRODUCTION

The fact that there is no single acceptable definition of health education is indicative of the uncertainty that is still attached to this field. The definition adopted by the World Health Organization[1] provides a useful basis:

> The focus of health education is on people and action. In general, its aims are to persuade people to adopt and sustain healthful life practices, to use judiciously and wisely the health services available to them, and to take their own decisions, both individually and collectively, to improve their health status and environment.

In this chapter health education is used in the broadest sense possible incorporating a variety of approaches. It aims to encourage health-promoting behaviour and activities and to discourage harmful ones; to provide information which will enable individuals to participate meaningfully in choices relating to health; to promote appropriate and effective self-care; to devise measures leading to improvement in the quality of life; to encourage individuals to assume greater responsibility for their own and others' health; to enable patients to make the best use of existing (including alternative) health services; to stimulate the participation of professionals and non-professionals in continuing care; to enable individuals and groups both to participate in the planning and administration of health services and to have a significant voice in the development of policies that in any way affect health.

Health education is a continuous process which begins in early infancy and continues throughout life. Formal teaching of health education occurs mainly in school but this is a small part of a much wider process involving family, friends, peer groups, professionals and the media. Recently there has been interest in offering a health education component through existing adult and further education sessions. As his occupation consumes a significant part of an individual's life, the health implications associated with it and the influences of those working in the fields of health and safety must make an important contribution to an individual's attitude to health.

## THE OCCUPATIONAL SETTING

It is necessary to relate a discussion of health education in the occupational setting to a wider context, since a subject like this transcends divisions in health services. In Britain, health education is clearly a responsibility of the National Health Service but is not the prerogative of any single part of that service. The role of education in enabling individuals to maintain a healthy way of life was one of the key assumptions in the Royal Commission on the National Health Service.[2] Accordingly, health education is not solely the responsibility of health education officers and their staff, but should relate to virtually all National Health Service staff who have contact with patients. However, this is not always achieved and there is ambiguity between a general and specialized commitment. Clearly, health education officers are responsible for advising, supporting and encouraging those who in their normal work have an opportunity to provide health education. Such support should be available to all health professionals, whether or not they are working with the National Health Service, as the responsibility to provide health education is to the public as a whole, not sectors of it.

The Health and Safety Commission[3] have the following objectives and responsibilities:

to identify health hazards;
to advise on environmental control;
to advise and inform workers and employers of risks; and
to advise on medical aspects of employment.

This clearly implies an educational responsibility although the scope involved, or how it is to be achieved, is not clear. With the small number of staff involved in the Employment Medical Advisory Service it would be unlikely that they could be directly involved in a wide scale educational process within industry and it would seem that their prime function must be to provide expert advice to other key people who can disseminate it: unions, management and occupational health staff.

Unions, who are responsible for the welfare of their members, have indicated their involvement in health and safety matters by the development of training programmes for safety representatives, the publication of booklets and pamphlets, and at the central level by participation in policy making.

Local authorities have a responsibility for environmental health which includes a health education component. Local authorities also have certain legal responsibilities imposed on them by the Health and Safety at Work Act for aspects of the working environment.

With regard to occupational health services, since there is no legal requirement to have one, there is obviously no legal commitment to any particular function. The nature of an occupational health service is determined by management and the professional staff. The official ILO Recommendation[4] provides a guideline:

1. Protecting the workers against any health hazard which may arise out of their work and the conditions in which it is carried on.

2. Contributing towards the workers' physical and mental adjustment, in particular by the adaptation of the work to the workers and their assignment to jobs for which they are suited.

3. Contributing to the establishment and maintenance of the highest possible degree of physical and mental wellbeing of the workers.

This suggests a wide-ranging health educational approach.

Where there are no occupational health services there must by law (if the establishment is over a certain size) be the appropriate number of first aiders. In neither the current nor the Proposed First Aid at Work Regulations[5] is there a suggestion that the first aider's role should include health education.

The responsibility for health and safety as laid down by the Health and Safety at Work Act 1974 is now well known. It appears that legally there is a definite responsibility on employers to include an educational component as a major aspect of their health and safety provisions to employees and to the public. The exact scope is not mentioned nor is the method to be employed. In general it can be said that everybody involved in work has some responsibility for health, although the nature of the responsibility will vary.[6] In industry it is generally accepted that responsibility is associated with status and a person's position in the structure of the organization. It is expected that individuals as well as the organization are liable to allegations of negligence if they do not exert the responsibility that is commensurate with their position — ignorance or failure to prepare themselves for responsibility is no defence. An educational policy to prepare people to act responsibly in matters of health and safety within an organization is essential. Legislation provides the framework on which an educational approach can be developed.

In other countries the service provision varies but the results are similar: few health professionals are trained in health education and even fewer are found in industry. Some European countries do not have defined categories of health education, health education being included within preventive medicine, while in industry the emphasis is placed on routine medical examination. In the United States there are many different specialist health educators, often with a nursing background. Examples of the health education programmes that result from these varied service situations will be given later.

Rather than seeing the workplace as a captive population of high-risk individuals who can be subjected to ill-informed and ineffective health education, the challenge is to devise something that meets the expressed needs of groups within the occupational setting using all the skills of a wide-ranging team.

## WHY HEALTH EDUCATION?

In almost every country, rich or poor, developed or developing, there is a realization that existing health services are failing to meet the health needs of the population. New initiatives are required if the commitment of the World Health Assembly to 'health for all by the year 2000' is to be achieved.[7]

While the differences between countries in the pattern of disease and the provision of services provide the opportunity for critical evaluation, it is salutary to be reminded of the differences *within* a country. In Britain this has been highlighted by the recent report *Inequalities in Health*.[8] Why, in a country like Britain with a national health service, has the inequality in health between socioeconomic groups, as measured by mortality and morbidity, increased since World War II? The arguments and the conclusions of the report are too lengthy and complex to discuss here, but the adverse influence of the working environment, the failure to use preventive health services and ineffective health education are considered to be contributory factors.

The problems in different countries vary, but in attempting to find solutions, some similarities are found. The recognition of an ecological approach to health which acknowledges a multifactorial contribution to ill health necessitates a new approach to intervention. The control of schistosomiasis depends on an understanding of individual and group behaviour; on the effects of nationally planned agricultural irrigation schemes which have extended the waterways providing new breeding grounds for the snail; and on detailed knowledge of the parasite. The common psychosocial health problems of the Western world may relate to altered family structures, inadequate community support, personal behaviour and an inimical working environment.

Both in prevention and care there is a complex interplay of individual and collective behaviour and environmental influences which may span the world of work and non-work. Solutions based on traditional medical care, which in many countries is limited in availability, expensive or unacceptable, at the best tend to be palliative. For many problems a change in individual behaviour will be required if ill health is to be prevented and a collective approach is necessary if unhealthy environments are to be remedied. Where disease and disability exist an approach bringing together a variety of services and the active involvement of the individual patient and his family and friends may be needed to ensure full coping, adaptation and rehabilitation.

At the international level through agencies such as WHO and ILO, and at the national level through government and voluntary agencies concerned with health, the importance of encouraging positive prevention of illness and participation in health care is becoming recognized. Can health education and those who make use of it meet some of the challenges of the changing health problems that are presented by society today?

## WHAT IS HEALTH EDUCATION?

In the introduction to this chapter the scope of health education was illustrated, but it is useful to group these various activities. Draper and his colleagues[9] consider that

there are at least three types of health education. The first and most common is education about the body and how to look after it. The provision of information and advice on human biology and hygiene is vital for each new generation. The second is about health services — information about available services and the 'sensible' use of health care resources. But the third, about the wider environment within which health choices are made, is relatively neglected. It is concerned with education about national, regional and local policies, which are too often devised and implemented without taking account of their consequences for health.

While few would disagree with the traditional preventive medical approach consisting of information and encouragement to change individual behaviour, some professionals are concerned about encouraging self-care and informed decision-making. Many professionals and members of management are alarmed at the more radical approach to health education which encourages or demands a role for individuals in planning and social policy.

Draper considers that this radical approach of health education is in the true tradition of public health, but that sadly this approach has recently been moribund. This type of health education is sensitive to the dynamic relationship of people with their environments and must involve social change. Responsibility for public health is not met simply by letting people 'look after themselves' or 'adopt healthy lifestyles'. Such social changes tend to be opposed by the essentially conservative forces in society — financial interest of individuals, industrial or business concerns and the professional groupings of health workers.

It is impossible here to do more than refer to such debates, but one controversy merits brief mention. Is health education 'altering peoples' attitudes' or should the aim be to encourage people to 'make decisions' for themselves or is there some relationship between the two that is yet to be investigated? Is the approach to be based on propaganda or education? These important questions posed recently by Sutherland[10] are a matter of constant argument and debate for all concerned with health and education. Tuckett,[11] in a paper entitled 'Choices for health education', notes that the strategies adopted by health educators are related to their own views of the aims of health education — 'One sort of health education will attempt to convey what standards are appropriate, the other will attempt to convey various standards and consider the implications of each'.

While there are ambiguities about the aims and the strategies and these are associated with the complexities and uncertainties of the process of ill health, it is generally recognized that the basis of any intervention must be existing knowledge, no matter how limited it may be. In the middle of the last century a great debate, supported by such notables as Thackrah and Noble,[12,13] raged as to the relative influence of 'the factory system' and 'the great town' on the 'great mass of disease constantly witnessed in the districts where manufactures prevail'. The practice of epidemiology was able to provide an understanding of disease processes which led to effective intervention, both in towns and factories, which in turn produced a marked reduction in communicable diseases and industrial poisonings.

Today, the expanded skills of epidemiology, sociology, psychology, education, ergonomics, communication, management and administration all contribute to knowledge and provide possible interventions. More research is required, including reviews of relevant literature, analysis of existing records, study of sickness absence and reported illness and accidents, accurate analysis of employee attitudes and perceptions of health needs, quality of life and use of health services. The importance of developing well-designed and usable recording systems and the establishment of broad-based epidemiological studies in all occupational health services cannot be overstated.

In the context of Dubos' definition of health, 'A *modus vivendi* enabling imperfect men to achieve a rewarding and not too painful existence while they cope with an imperfect world',[14] Newens[15] notes:

> There is no easy dichotomy between individual and political solutions to health problems. Clearly opportunities exist to influence the choices available to people in coping with and adapting to the world around them. Increasingly there is also recognition of the importance of social and political institutions and policies for the health of individuals. Traditionally, health education has tried to influence 'imperfect men'. It could also help people contribute more to the re-shaping of their imperfect world.

These debates, indicative of a new and rather uncertain discipline, do point to a discipline that is determined to clarify its theoretical basis and to pursue actively new initiatives and relevant technologies. For those in special situations in occupational health services, it is necessary to develop appropriate aims and objectives, the staff must gain the necessary skills and cope with the political and ethical problems. All views concerning health and illness in society have individual and collective implications and are closely related to concepts of responsibility. If health professionals fail to be active in health education, then they may either be dragged unwittingly into it by a more concerned public or excluded because of their lack of concern and their ineffectiveness.

## THE SCOPE OF HEALTH EDUCATION AT WORK

Unfortunately it is not possible to describe accurately the extent of health education at work. The only national study of occupational health services,[16,17] carried out by EMAS, was based on reported activities and used broad grouping of activities. Health education ranked fifteenth (out of eighteen activities) and was noted to be performed regularly in 8·5 per cent of all firms, or if related only to firms with occupational health services occurred on a regular basis in only 31·9 per cent. In the context of the survey, health education was rated as a non-occupational activity and additional categories of preventive education which included training first aiders and participation in the prevention of accidents and occupational disease ranked ninth and fourth respectively.

As there is no defined account of health education in the workplace, it may be most useful to look at a few widely differing examples and then to see whether these can be integrated into an overall schema, which may be useful in examining current activities and future plans.

### Broad-based health education programmes

Health education programmes have been developed extensively in the United States and this may reflect the high cost to industry of illness and the difference between Britain and the US in national health care systems. Many companies now organize broad-based programmes during working hours. These programmes include general prevention, patient care and counselling.[18-21] The occupational health nurse usually has a key role. The content may be based on the knowledge of past problems presented to the health service or on some form of questionnaire distributed to the workforce to elucidate what they consider to be the most important issues. Films, videotapes, panel discussions, talks by nurses, lectures from eminent guest speakers, visits from members of voluntary organizations and role play have all been used. Active participation by all concerned in the planning of the courses is encouraged and courses are publicized through notices or work newspapers. Discussion during the sessions is encouraged. The aim is to make the sessions relevant and interesting.

Some sessions, such as a film on low back pain, clearly have a specific occupational link, while other topics, such as skin care, common diseases, sexual and family problems, alcohol abuse, self-examination for cancer, coping with stress and problems with children, indicate a much wider concern. Courses on pre-retirement for older age groups have a long tradition in many countries.

## National or community preventive medicine

Several European and Scandinavian countries regard the workplace as one target for a wide-ranging and integrated educational programme. This is usually based on a strong national preventive medicine approach[22] and anti-smoking or the related problem of heart disease has often served as the focus. These programmes are usually centred on an extensive mass media campaign, using radio, television and the press and are backed by government action, which may include increased tax on tobacco, a ban on advertising cigarettes and legal enforcement of no-smoking areas. A series of comprehensive and related programmes are aimed at different audiences — children, teachers, health workers, pregnant women and defined adult groups.[23] By integrating all these activities it is hoped to achieve maximum impact. It has been suggested that it is possible to aim to abolish smoking in one generation. Specific programmes designed for workers are part of the whole national project.

## Exercise and positive health

Sports activities have a long and honourable history in many companies, although some have tended to emphasize the competitive elements of team sports, rather than the personal benefits of participation. More recently, there has been a renewal of interest in the benefits of both exercise and relaxation for those with sedentary, demanding and boring jobs, or employment deemed to be stressful in any way. While this has often been designed for the executive, many other groups of workers have similar problems and programmes are now being made more widely available. Some evidence suggests that this does not improve 'health' although 'wellbeing' may improve.[24]

## Patient education

Few studies have examined patient education as it is a routine and significant aspect of normal occupational health. Many studies[25] have noted the failures of communication, the misunderstandings and the lack of knowledge of patients about important aspects associated with their personal health problems — the necessary and unnecessary limitations, diet, exercise, medication, side effects and so on.

While most of this education, support and advice takes place on a one-to-one basis on either a casual or regular review, there is the possibility of group sessions for selected groups with a common concern. Such sessions can be included within a broad-based health education programme related to screening or monitoring programmes.

## Education on hazards at work

The new interest and active involvement by unions in health and safety is clearly seen in the recent 'spate' of reports and publications on a wide range of safety issues including: *Not One Minute Longer: the 2,4,5-T Dossier* from the National Union of Agricultural Workers;[26] *The Prevention of Occupational Cancer* and *Vinyl Chloride* by the Association of Scientific, Technical and Managerial Staffs[27,28] and *Shiftwork* by the General and Municipal Workers' Union.[29] These are designed to provide union safety representatives and shop stewards with comprehensive information on particular hazards. Publications such as *Health and Safety at Work* (MacLaren Publishers), *The Hazards Bulletin* (The British Society for Social Responsibility in Science) and *The Health and Safety Information Bulletin* in the Industrial Relations Review and Report, provide a review of current topics for a wider readership.

Earlier publications tended to be more general, such as *The Hazards of Work*[30] and *Health and Safety at Work* (TUC).[31] A current series, *Health and Safety in the Workplace,*[32] is designed to provide concise, practical guides about hazardous factors in the working environment of interest to trade union safety representatives.

It is likely that safety representatives will become increasingly well informed as the TUC and individual unions expand their education and training functions.[33] The current debate on occupational cancer may serve to indicate future trends.

In the foreword to *The Prevention of Occupational Cancer* published by ASTMS[27] it is stated that:

> The document attempts to make all aspects of the debate accessible to our members without oversimplifying any of the arguments. There is too much emphasis on curing cancer and too little emphasis on prevention. The basic assumption of this document is that cancer is a preventable disease. To prevent it we must take steps to control carcinogens.

Accepting that there are genuine divergences of scientific opinion, it is clearly stated that:

> This document should provide safety representatives and union officers with the information to argue the case with employers [and also to revise] a number of issues which must be dealt with at national and international level.

Although some people may feel that points are overstated or arguments oversimplified, it is tragic that the Chemical Industries Association response[34] is only to carp over the interpretation and the accuracy of the data reported and to claim that it is a 'brazen use of health issues for political ends' and an example of 'confrontation writing'. It does not make a single constructive comment on 'prevention' or accept that the aims of the document are reasonable. Perhaps if relevant health professionals had produced a well-argued document with a positive approach to prevention and participation several years ago, such 'confrontation writing' or the use of 'political means for health ends' might not have been necessary. The sort of approach which is outlined in the next example might have been appropriate.

An example of a participatory approach is found in an American study[35] which 'was designed to inform and educate rank and file and low level union leadership' and encourage workers to inform their fellows about the nature of the problems at work.

The components of the programme included:

1. Training workers to identify specific occupational hazards and to use legal resources in having them corrected.

2. Developing contact between workers and occupational health specialists.

3. Stimulating understanding of and interest in job health by helping workers write leaflets and hazard information sheets.

4. Setting up accident and health hazard reporting systems.

5. Facilitating links among unions or worker groups facing similar problems.

## First aid as a focus

The renewed interest in first aid is seen by some as a way of encouraging a more positive approach to health as well as providing training in emergency care. A few enhanced first aid programmes in industry include the prevention of dermatitis, the problems of stress and simple ergonomics and accident prevention.[36]

Another approach[37] has been to encourage large numbers of staff in a factory to undergo first aid training: this was subsequently extended to teaching first aid to a community. While it will be necessary to follow such an approach for a long period of time, there is a suggestion from this study (known as *First Aid Community Training*) that there has been a reduction in accidents at work, which may indicate a more positive attitude to prevention and safety.

## Screening and monitoring

Three types of activity can be considered under this heading and all offer the opportunity for and indeed probably require an educational component. While the debate on screening cannot be discussed in detail here, it is necessary to recognize the limitations of any activity, to recognize the high risk groups and to be aware of the expectations of the public who may have too great a trust in the procedure.[38]

The long-established monitoring for occupational disease is clearly linked with environmental monitoring, education and training of operatives, provision and use of protective equipment and the development of safe attitudes to work, and is well known to all who practise occupational health.

While screening for general health has been common in selected groups such as executives and those involved in overseas travel, more recently demands have been made for this screening to be extended to include cervical cancer, breast cancer and hypertension, either as individual programmes or in some combined general health screening.

The third type of activity is the review of special categories of patients such as diabetics, hypertensives and those on regular medication. This usually includes both review of the disease condition and the working environment of the individual.

In all these categories, the medical review must of necessity be intermittent and prevention of illness or daily control of the health problem must rest with the individual and depend upon his understanding of the relevant issues. Because of the continuing contact between patient and occupational health staff there are special opportunities for a good relationship and repeated education and discussion.

## The care of pregnant women at work

As a result of the concern over perinatal mortality and morbidity, and the Spastics Society Campaign, 'Save a Baby', questions are being asked about the attitudes of working women to antenatal care, accessibility of antenatal clinics and the potential of the place of work as a means of access to women at high risk.

Two British companies employing large numbers of women have investigated the problems of pregnant women at the workplace and have developed similar practical programmes to meet the perceived needs.[39] In one company it was found that: there was an appalling lack of knowledge about the importance of antenatal care; there was a lack of knowledge about the timetable of pregnancy and the terms used by those providing the medical supervision; attendance at clinic was not helped by the knowledge that a loss of earnings would be involved; and there was lack of a personal approach by medical staff in the clinics and the feeling of being a statistic rather than a person.

The second company, following discussions with management, unions, occupational health staff, the Spastics Society and the health authority (this included the district dietician, sister in charge of antenatal clinic, nursing officers, community medicine staff and health education officers), produced a policy document with the following aim:

by offering a programme of formal health education to all female employees of childbearing age, and stressing the importance of early attendance at antenatal clinics for pregnant employees, the levels of perinatal morbidity and mortality may be reduced.

The objectives are:

To increase the awareness of the importance of antenatal care for female staff.

To increase rubella vaccination uptake prior to pregnancy.

To encourage earlier antenatal attendance and to lessen defaulting at antenatal clinics.

To improve nutrition during pregnancy.

To increase infant birth weight.

To reduce smoking during pregnancy.

To reduce excessive alcohol intake during pregnancy.

To improve prescribed drug compliance during pregnancy.

To decrease non-prescribed, non-approved drug taking during pregnancy.

To increase knowledge of rights, for example maternity grants, benefits and allowances.

To support and advise occupational health staff.

A detailed educational programme was devised and a number of incentives were offered to pregnant employees.


## OPPORTUNITIES IN THE OCCUPATIONAL SETTING

These few examples illustrate some recent and varied developments which may be found in workplaces. Indeed the diverse nature of the workplace and the variation in service provisions are likely to result in wide-ranging approaches to health education. It appears that most activities can be covered by three broad (and not necessarily mutually exclusive) groupings: general health education; patient education; and specific occupational health education. These groupings tend to reflect a number of artificial barriers which seem to have little theoretical justification or practical utility.

For historical reasons occupational health practitioners are separated from their colleagues in the National Health Service, not only in their employment, but by their career structure, their training and their responsibilities, and it is likely that this will continue in the near future. As already mentioned, it is becoming increasingly difficult to separate health problems into work and non-work categories and few individuals at work, whether healthy or sick, appreciate the division. A second barrier with a historical cause is between prevention and cure. The professionals are not only separated in practice but in status. The third barrier is the separation between professional care and self-care. While sometimes it arises in response to professional failure or unacceptable professional behaviour, it is due sometimes to the recognition that non-professional skills and resources have an important, albeit different, contribution to make. The self-care movement is not limited to the health field.

The occupational setting, particularly where there is an occupational health service, has several advantages which would help to overcome some of these barriers and give rise to the possibility of new initiatives in promoting health and safety which do not exist elsewhere.[40] In this setting there is a well-defined population with recognized subgroups, where individuals and groups already know each other. It is possible either to use existing groups or to re-group according to identity of need.

It has been traditional for those professionals committed to health education to define the problems as they see them and then devise appropriate methods to put over the desired message. The occupational situation provides an easy route to a better educational strategy — of 'starting where people are' — to discovering needs and demands as perceived by the consumers, and what individuals or groups define as the priority issues. Some of the innovative examples outlined above started in this manner.

Health education has often failed because it has been an isolated activity, consisting of a few unrelated lectures given by outsiders. The occupational setting provides the possibility of continuity, active participation and an opportunity to show the implementation of the message. There is little point in having a nutrition programme if the works canteen is unhygienic, provides unbalanced, unattractive meals with no choice and has advertisements for sweets and cigarettes. In addition to the suitable selection of canteen staff, their training and surveillance, a health policy may encourage choice of meals (such as salads), no-smoking areas, dental health education, slimming competitions and discussion of alcohol-related problems.

Just as it is essential for all health-related matters to be approached in an integrated manner in an organization, so it is necessary for links to be established with the changes in health care and attitudes in the community. The links with primary care, where prevention and health education are also being emphasized,[41] need to be strengthened. This is particularly true of developing countries, and in a review of primary care schemes in several countries, Newell[42] is able to say, 'In no example presented here is there a separation of the promotional, preventive and curative health activities at the primary health care level'.

Another message from developing countries is that the doctor need not be the first person of contact in a community health service. Although this has long been accepted in occupational health it is only now being discussed in the context of primary care in the developed countries.[43]

If it is gradually becoming possible to break down some of these barriers, would it not be possible to co-ordinate some of the health education functions in community and workplace for the benefit of those who need information, help and advice? One example of an integrated community approach has been the American heart disease prevention programme[44] which involved the mass media and local groups encompassing industry, sport facilities, health services, municipal authorities and voluntary organizations. In Britain on a more limited scale, some local groups (including industry, offices and clubs) have taken advantage of the 'climate of opinion' generated by the positive health campaigns of the Health Education Council or the Scottish Health Education Unit to provide additional support at the local level for those who wish to participate in exercise programmes, stop smoking, reduce obesity etc.

The programme 'Health Choices' and the associated book *The Good Health Guide* produced by the Open University[45] as part of its non-degree courses could provide an occupational health service with the opportunity to make use at local level of excellent material from a national source. *The Good Health Guide* is clearly directed at people in their own situation. It is 'about the ways in which you and your family can live a healthy and satisfying life. It is *not* concerned with illness and cures: it is about the choices you can make about the way you live, and the way you feel about yourself'. It includes sections on work and health, stress and emotion and a healthy community.

## EVALUATION – DOES HEALTH EDUCATION WORK?

An examination of the history of health education and a review of some of its present activities show the changing philosophies and some of the resulting confusion. It is not surprising that those who are involved in health education as well as the sceptics ask, what has been achieved in the past and, perhaps more cogently, what is likely to be achieved by the present endeavours? Two monographs[46,47] which mainly review published studies have attempted to answer these questions. While Tones[48] notes that 'there is a dearth of well-documented success in health education', it is accepted that there are some excellent studies which have shown beneficial changes in knowledge, attitudes and behaviour, and in which the importance of group methods, participation, repetition and continuity were clearly demonstrated. The information contained in these two volumes on theory, practice and methods would be most valuable for anyone seeking to undertake health education.

As Green[49] points out

It is premature to expect most health promotion programmes, at their current state of development, to have measurable health outcomes for evaluation. Evidence linking some recommended behaviour to specific medical or health outcomes is still somewhat tenuous. The probable benefits of many practices are not established sufficiently to ensure medical outcomes within the short periods and with the small samples available for most evaluations of health preventive programmes or methods.

The uncertainty of the message is usually the result of our lack of knowledge of disease causation. How certain does one have to be before action? Ischaemic heart disease is perhaps the best example of the dilemma. The necessity to do no harm is obviously crucial and it is interesting to note that Burkitt's[50] views on bran — it may do you some good and it will certainly do you no harm — appear to have been influential in changing dietary patterns.

## DISCUSSION

Some people may feel that the ideas and approaches outlined in this chapter are too radical or depend on a participatory approach that is unlikely to be available at present; others may consider that what has been described is common practice.

There is no single model of health education that can be applied to the occupational setting. The examples in this chapter illustrate the value of starting where people are, determining needs, making full use of existing services and trying to develop a relevant strategy. Health education in schools shows clearly what can be achieved in a relatively short time. Occupational health education is in a similar stage of development to school health education of 20 years ago, where programmes, if they existed at all, included a few lectures on hygiene, physical fitness, dental health and a tangential and embarrassed look at sex. The Schools Council Project in England and Wales[51] aims

to help the children make considered choices on decisions related to their health behaviour by increasing knowledge and clarifying the beliefs and values which they hold. Health education should not only concern itself with the provision of information but should also involve children in the process of making choices or decisions.

If this and other current developments in school health education are implemented, then a new generation of people will be entering the world of work with differing attitudes to health, participation and health service provisions.

In a Department of Education and Science Working Paper[52] health education is described as part of 'the education of the individual', relevant to a consideration of 'education for family life and living in society' and part of the education 'for the world of work'. Here is an opportunity for those in industry and commerce to participate in the educational process at schools and colleges before people arrive at their first working environment. Indeed it has been suggested that occupational health staff have a responsibility to expand their concern beyond the factory to participate in the education of the public on topics such as the meaning and determination of safe levels and the problems of environmental contamination.[53]

It is necessary to examine the overall relationship between health and work and the differing aims and objectives of health professionals, industrialists and educators. To some[54] 'the fight for the realization of health is very much at the centre of the conflict between capital and labour which takes place at the workplace and heightens in moments of crisis'. Others see the workplace as one where people can cooperate to promote health, prevent illness and seek to improve the quality of life. Health education will never be effective if those who advocate it avoid the controversial issues: it has long been recognized that health is a political issue.

The Health Education Council was responsible for setting up a workshop which brought together occupational health nurses, doctors, teachers of occupational health and health education, health education officers and representatives of the TUC, to consider 'health education in the workplace'.[55] Controversy, scepticism, enthusiasm and apathy were all evident in the discussions but many forward-looking recommendations resulted from it and it was felt that there was a need for action and a need to publicize the problems and the opportunities.

Discussions continue between interested groups and individuals but it is essential that new initiatives take place at the local level making use of appropriate resources and strengthening existing links. Priorities at both a local and a national level must be established. A top priority must be to seek to reduce inequalities in health, and the workplace may open the way to some of those most at risk.

This will inevitably have implications for occupational health staff. New skills will be required to determine needs and to devise appropriate educational strategies. Greater flexibility and new working relationships with other professionals and non-professionals will have to be established. There will be requirements for change in professional education and continuing education.

The opportunities are enormous, but for any who seek to participate in new initiatives, there is a responsibility to include a well-designed evaluative component in any endeavour and a commitment to communicate the results to their colleagues.

REFERENCES

1. World Health Organization. *Planning and Evaluation of Health Education Services.* Technical report series No. 409. Geneva: WHO, 1969.
2. Royal Commission on the National Health Service. *Report* (Merrison). London: HMSO, 1979.
3. Department of Employment. *Employment Medical Advisory Service: a Report of the Work of the Service for 1973 and 1974.* London: HMSO, 1975.
4. International Labour Office. *Recommendation Concerning Occupational Health Services in Places of Employment.* Recommendation 112. Geneva: ILO, 1959.
5. Health and Safety Commission. *First Aid at Work.* Consultative document. London: HSC, 1979.
6. Farmer D. Personal responsibilities. *Health and Safety at Work.* 1978; **1**:34–5.
7. Mahler H. Health for all by the year 2000. *WHO Chron.* 1975; **29**:457–61.
8. Department of Health and Social Services. *Inequalities in Health.* Report of a research working group (Black). London: DHSS, 1980.

9. Draper P, Griffiths J, Dennis J et al. Three types of health education. *Br. Med. J.* 1980; **281**:494–5.

10. Sutherland I. History and background. In: Sutherland I (ed.) *Health Education: Perspectives and Choices.* London: Allen & Unwin, 1979.

11. Tuckett D. Choices for health education: a sociological view. In: Sutherland I (ed.) *Health Education: Perspectives and Choices.* London: Allen & Unwin, 1979.

12. Thackrah CT. The effects of the principal arts, trades and professions and of civic states and habits of living on health, longevity, with suggestions for the removal of many of the agents which produce disease and shorten the duration of life. In: Meiklejohn A *The Life, Work and Times of Charles Turner Thackrah, Surgeon and Apothecary of Leeds, 1795–1833.* Edinburgh: E & S Livingstone, 1953.

13. Noble D. *Facts and Observations Relative to the Influence of Manufactures upon Health and Life.* Shannon: Irish University Press, 1971 (Facsimile reproduction, first edition 1843).

14. Dubos R. *Man, Medicine and Environment.* Harmondsworth: Pelican, 1968.

15. Newens M. Education and participation in health. In: Sutherland I. (ed.) *Health Education: Perspectives and Choices.* London: Allen & Unwin, 1979.

16. Health and Safety Commission. *Occupational Health Services: the Way Ahead.* London: HMSO, 1977.

17. Phillips M and McEwen J. *Private Occupational Health Services in Britain. The EMAS Survey 1976.* Report produced for the Health and Safety Executive. Nottingham: Department of Community Health, 1979.

18. Thomas MR. A vital link in an employee health program. *Int. J. Occup. Health Saf.* 1980; **49**:52–5.

19. Felton JS. Contemporary health education at a heavy industry navy worksite. *J. Occup. Med.* 1976; **18**:681–4.

20. Dedmon RE. Employees as health educators. *Int. J. Occup. Health Saf.* 1980; **49**:18–24.

21. Strasser AL. Health education in industry: the key to a vigorous workforce. *Int. J. Occup. Health Saf.* 1980; **49**:25–6.

22. Hellberg H. Government attitudes to health education: a crucial factor in effective action. *Int. J. Health Educ.* 1980; **23**:76–81.

23. Lorawsky DN and Karataeva NB. Health education of young people in the USSR. *Int. J. Health Educ.* 1980; **23**:162–6.

24. Corroll V. Employee fitness programmes: an expanding concept. *Int. J. Health Educ.* 1980; **23**:35–41.

25. Haynes RB, Taylor DW and Sackett DL (ed.). *Compliance in Health Care.* Baltimore: Johns Hopkins University Press, 1979.

26. National Union of Agricultural Workers. *Not One Minute Longer: the 2,4,5-T Dossier.* London: NUAAW, 1980.

27. Association of Scientific, Technical and Managerial Staffs. *The Prevention of Occupational Cancer: an ASTMS Policy Document.* London: ASTMS, 1980.

28. Association of Scientific, Technical and Managerial Staffs. *Vinyl Chloride.* London: ASTMS, 1979.

29. General and Municipal Workers' Union. *Shiftwork: a GMWU Discussion Document.* Esher: GMWU, 1980.

30. Kinnersly P. *The Hazards of Work: How to Fight Them.* Workers handbook No. 1. London: Pluto Press, 1973.

31. Trades Union Congress. *Health and Safety at Work.* London: TUC, 1975.

32. Le Sevre A and Doyle M. *Health and Safety in the Workplace.* (Four volume series.) Walton-on-Thames: Thomas Nelson & Son, 1980–81.

33. Bibbings RE. Health and safety at work: the trade union view. *J. Soc. Occup. Med.* 1980; **30**:90–7.

34. Howard JK. *The Prevention of Occupational Cancer: an ASTMAS Document — a Review*. Background paper. Chemical Industries Association, 1980.
35. Wegman DH, Boden L and Levenstein C. Health hazard surveillance by industrial workers. *Am. J. Publ. Health* 1975; **65**:26–30.
36. Wheeler C. An ounce of prevention. *Int. J. Occup. Health Saf.* 1980; **49**:58–68.
37. McKenna SP and Hale AR. The effect of emergency first aid training on the incidence of accidents in factories. *J. Occup. Accid.* **3**:101–14.
38. D'Souza MF. The value of screening and health surveillance to employment. *J. R. Coll. Phys. Lond.* 1978; **12**:230–9.
39. Health Education Council. *The Care of Pregnant Women at Work*. (Leaflet to Area Health Authorities.) London: HEC, 1980.
40. World Health Organization. *Health Aspects of Wellbeing in Working Places*. Report of a WHO working group Euro reports and studies. Copenhagen: WHO, 1980.
41. Bennett FJ. Primary health care and developing countries. *Soc. Sci. Med.* 1979; **13A**:505–14.
42. Newell KW. *Health by the People*. Geneva: WHO, 1975: 194.
43. Saint-Yves IFM. Need general practitioners be patients' first contact with health service? *Lancet* 1980; **2**:578–80.
44. Maccoby NM and Farquhar JW. Communication for health: unselling heart disease. *J. Commun.* 1975; **25**:114–26.
45. Open University. *The Good Health Guide*. London: Harper & Row, 1980.
46. Gatherer A, Parfit J, Porter E et al. *Is Health Education Effective?* London: Health Education Council, 1979.
47. Bell J and Billington DR. *Annotated Bibliography of Health Education Research Completed in Britain from 1948 to 1978*. Edinburgh: Scottish Health Education Unit, 1979.
48. Tones BK. *Effectiveness and Efficiency in Health Education*. Edinburgh: Scottish Health Education Unit, 1978.
49. Green LW. How to evaluate health promotion. *Hospitals* 1979; **53**:106–8.
50. Burkitt DP and Trowell HC. *Refined Carbohydrate Foods and Disease. Some Implications of Dietary Fibre*. London: Academic Press, 1975.
51. Williams T. *Think Well. Introduction and Policy Booklet*. Schools Council Project. Health Education 5–13. Clacton on Sea, 1978.
52. Department of Education and Sciences. *Health Education in the School Curriculum*. Working paper. London: DES, 1977.
53. Johnson MN. Occupational physicians can help to educate the public about 'acceptable' exposure levels. *Int. J. Occup. Health Saf.* 1978; **47**:41–7.
54. Navarro V. Work ideology and science. The case of medicine. *Soc. Sci. Med.* 1980; **14C**:191–205.
55. Randell J and McEwen J. *Health Education in the Workplace. Report of a Workshop*. London: Health Education Council, 1980.

# Index